ENVIRONMENTAL VALUE TRANSFER: ISSUES AND METHODS

THE ECONOMICS OF NON-MARKET GOODS AND RESOURCES

VOLUME 9

Series Editor: Dr. Ian J. Bateman

Dr. Ian J. Bateman is Professor of Environmental Economics at the School of Environmental Sciences, University of East Anglia (UEA) and directs the research theme Innovation in Decision Support (Tools and Methods) within the Programme on Environmental Decision Making (PEDM) at the Centre for Social and Economic Research on the Global Environment (CSERGE), UEA. The PEDM is funded by the UK Economic and Social Research Council. Professor Bateman is also a member of the Centre for the Economic and Behavioural Analysis of Risk and Decision (CEBARD) at UEA and Executive Editor of Environmental and Resource Economics, an international journal published in cooperation with the European Association of Environmental and Resource Economists. (EAERE).

Aims and Scope

The volumes which comprise *The Economics of Non-Market Goods and Resources* series have been specially commissioned to bring a new perspective to the greatest economic challenge facing society in the 21st Century; the successful incorporation of non-market goods within economic decision making. Only by addressing the complexity of the underlying issues raised by such a task can society hope to redirect global economies onto paths of sustainable development. To this end the series combines and contrasts perspectives from environmental, ecological and resource economics and contains a variety of volumes which will appeal to students, researchers, and decision makers at a range of expertise levels. The series will initially address two themes, the first examining the ways in which economists assess the value of non-market goods, the second looking at approaches to the sustainable use and management of such goods. These will be supplemented with further texts examining the fundamental theoretical and applied problems raised by public good decision making.

For further information about the series and how to order, please visit our Website www.springer.com

Environmental Value Transfer: Issues and Methods

Edited by

Ståle Navrud
Associate Professor
Department of Economics and Resource Management
Norwegian University of Life Sciences

and

Richard Ready
Associate Professor
Department of Agricultural Economics and Rural Sociology
Pennsylvania State University

A C.I.P. Catalogue record for this book is available from the Library of Congress.

ISBN-10 1-4020-5405-X (e-book)
ISBN-13 978-1-4020-5405-1 (e-book)

Published by Springer,
P.O. Box 17, 3300 AA Dordrecht, The Netherlands.

www.springer.com

Printed on acid-free paper

Cover photograph of frozen apples taken by Ståle Navrud

TABLE OF CONTENTS

S. NAVRUD AND R. READY

PREFACE

The transfer of environmental values in time and space has increased rapidly with the widespread use of cost benefit analysis in project evaluation and regulatory assessments over the last three decades. Over the last 15 years, other policy uses like environmental costing, greening of systems of national accounts and natural resource damage assessments after oil spills and other pollution accidents have also contributed to the increased demand for environmental values. However, most early transfers were conducted in an uncritical manner, often lacking sound theoretical, statistical and empirical basis, and did not question the validity and reliability of the transferred values.

What appears to be the first environmental value transfer exercise estimated damages, and illustrates the point that what is generally termed benefit transfer, should rather be termed value transfer in order to capture both reductions and increments in environmental quality and natural resources. This first attempt to transfer environmental values seems to be the calculation of lost recreational value from the Hell's Canyon hydroelectric project more than 30 years ago, as described by John V. Krutilla and Anthony C. Fisher in their book (Chapters 5 and 6): *The Economics of Natural Environments Studies in the Valuation of Commodity and Amenity Resources.* (John Hopkins Press, Baltimore, 1975). The first large-scale user of value transfer was the USDA Forest Service. In preparation for the 1980 Resource Planning Assessment (RPA) the Forest Service launched a large-scale effort to collect data on the economic values associated with recreational use of forest lands, in order to balance these against timber production and other uses. This research effort and policy use was expanded in the later RPAs, conducted every fifth year.

The set of papers in the 1992 Special Issue of *Water Resources Research* on benefit transfer organized by David Brookshire, and the 1992 Workshop on Benefit Transfer of the Association of Environmental and Resource Economists (AERE) in Snowbird, Utah were instrumental in questioning the prevailing practice of value transfer and putting uncertainty of value transfer on the agenda. Since then there has been a steady growth in the literature on testing validity of benefit transfer, the development of transfer methods and statistical techniques, and applications of these to both use and non-use values of different environmental goods. This book attempts to present the state-of the-art in value transfer after more than 10 years of more directed research into the applicability and validity of different value transfer techniques, and concludes by identifying issues and challenges for value transfer as policy applications of environmental values are expanding.

The idea of this book first emerged after Olvar Bergland and Ståle Navrud organized a workshop on benefit transfer in Lillehammer, October 14–16 1999. The workshop was the seventh in a series of nine workshops organized under the Concerted Action *Environmental Valuation in Europe (EVE)*, funded by DG Research of the European Commission. The papers from the workshop got the book project started, but more than half the chapters were invited after we became part of this book series on *The Economics of Non-Market Goods and Resources.*

We would like to thank all contributors to the book for their keen interest and enthusiasm in contributing to this book, and the book series editor Ian Bateman for his patience and encouragement in finalizing the book. Thank you also to the *Journal of Environmental Management* and editor A. Gill for permission to reprint the paper in Chapter 7, to *Land Economics* and editor Dan Bromley for permission to reprint the paper in Chapter 8, and to the people at Springer for their positive attitude and support.

Drøbak / State College Ståle Navrud Richard Ready

A.C. FISHER

FOREWORD

It is very kind of the editors to attribute to John Krutilla and me the first exercise in value transfer, in our 1975 volume on the economics of natural environments. In fact the material appeared even earlier, in our 1972 article, with Charles Cicchetti, in the American Economic Review. As they note, we were attempting to estimate the recreational value that would be lost as a consequence of putting a dam in the Hells Canyon reach of the Snake River, part of the Columbia River system in the Pacific Northwest of the U.S.

By modern standards, the attempt was of course very unsophisticated. First, we worked only with unit values, rather than a value function, much less a meta-analysis of studies of value functions. Second, we did not exactly transfer the values. But this was in my judgment a strength of the approach, and one I think still has some merit, as I shall explain.

The recreational activities we were looking at were primarily hunting and fishing of various kinds. We first developed estimates of the number of visitor days in each type of activity that would be lost due to the project, and then imputed values to the visitor days. Although some early results on estimation of the imputed value of these activities were available, we focused instead on amounts actually paid in settings where this occurred, where rights to game and fish are privately vested. For example, grouse shooting in the United Kingdom was known to go for $750–1200 per week, and the privilege of taking the red stag in central Europe for $5,000. Similarly in the case of fishing, we found that the cost per rod day on the better artificial ponds in the United Kingdom was about $9, and as much as $200–500 for Atlantic salmon fishing.

Of course the characteristics of the recreation experience, and almost certainly of the recreationist, were different in these instances from their counterparts in our study. What we did, however, was to impute very much lower values to the big game, bird hunting, and fishing days in the Hells Canyon area: $25, $10, and $5 respectively, and then argue that these were very conservative estimates, in that they were very much lower than the prices paid for comparable if not exactly equivalent activities elsewhere. Further – and most importantly – with these very conservative estimates, the resulting value of lost hunting and fishing days, when added to the other costs of the project, resulted in an overall negative net present value for the project. Better estimates could not reverse the ranking of the project/no project alternatives.

The procedure was undoubtedly crude, and much more sophisticated and more defensible methods are now available, as this excellent volume, a thoughtful exposition and critical analysis of the now large literature on benefits or value transfer makes clear. But where what is required is a ranking of alternatives, such as different projects – or no project – and research resources are limited, our strategy may yet have merit.

Athony C. Fisher is professor, Department of Agricultural and Resource Economics, Universiity of California, Berkeley.

S. NAVRUD AND R. READY

REVIEW OF METHODS FOR VALUE TRANSFER

1. WHY VALUE TRANSFER?

Increased use of Cost-benefit analysis (CBA) in the environment, transport and energy sectors have increased the demand from decision makers for information on the economic value of environmental goods. Policy uses of environmental values include: CBA of projects and policies, environmental costing to calculate the social optimal level of pollution (and optimal size of e.g. green taxes), green accounting at the national, community and firm level; and calculating compensation payments after a pollution accident, for example Natural Resource Damage Assessment (NRDA).

Due to limited time and resources when decisions have to be made, new environmental valuation studies often cannot be performed, and decision makers try to transfer economic estimates from previous studies (often termed *study sites*) of similar changes in environmental quality to value the environmental change at the *policy site*. This procedure is most often termed *benefit transfer*, but could also be transfer of *damage* estimates. Thus, a more general term would be value transfer.

Value transfer is now common in applied analysis, and the practitioner can refer to excellent guides that cover the practical steps and key aspects of conducting a value transfer (see especially Desvousges, Johnson and Banzhaf 1998 or Rosenberger and Loomis 2003). However, there remain important methodological questions about value transfer, and over the last five years there has been an increasing body of research advancing transfer methods and exploring the limits of for what types of environmental goods and in what situations environmental value transfer is valid and reliable. The purpose of this book is to assemble the latest research results and take stock of the state of the art in environmental value transfer, introduce new methods to advance the methodology, and to further the research agenda in this area.

Three important issues surrounding the practice of value transfer will be recurring themes in this book. First, how should value transfer be done? The case studies included in this book demonstrate several different value transfer methods. Second, how much additional uncertainty is introduced by value transfer, relative to the uncertainty that is inherent in all non-market value estimates? Here, we are interested in knowing both whether value transfer is valid in a statistical or theoretical sense, as well as in measuring the potential error that value transfer can introduce. Third, what level of additional uncertainty (or transfer error) is acceptable in a policy analysis? Here, no clear guidelines exist, but the level of precision in environmental values that is required in a policy analysis likely will differ from case to case.

S. Navrud and R. Ready (eds.), Environmental Value Transfer: Issues and Methods, 1–10.

2. VALUE TRANSFER METHODS

There are two main approaches to benefit transfer:

1. Unit Value Transfer
   i) Simple Unit Transfer
   ii) Unit Transfer with income adjustments
2. Function Transfer
   i) Benefit Function from one study
   ii) Meta analysis

Several classification of value transfer methods are used, for example unit value transfers is sometimes termed benefit value transfer or just value transfer, and all types of function transfer (including meta analysis) can be seen referred to as value function transfer. The main difference between the two main approaches is, however, always that the first one is based on a single point estimate or value range from a summary of studies, while the second approach transfers an estimated model describing how benefit measures (from one or more studies) change with the characteristics of the study population or the resource being valued.

2.1. Unit Value Transfer

Simple **unit transfer** is the easiest approach to transferring benefit estimates from one site to another. This approach assumes that the marginal value to an average individual at the study site from the environmental good is the same as that which will be enjoyed by the average individual at the policy site. Thus, we can directly transfer the mean willingness-to-pay (WTP) estimate from the study site to the policy site.

For the past few decades such a procedure has often been used in the United States to estimate the recreational benefits associated with multipurpose reservoir developments and forest management. The selection of these unit values could be based on estimates from only one or a few valuation studies considered to be close to the policy site, or based on mean values from literature reviews of existing values. Walsh, Johnson and McKean (1992, table 1) presents a summary of unit values of days spent in various recreational activities, obtained from 287 contingent valuation (CV) and travel cost (TC) studies.

The obvious problem with this transfer of unit values for recreational activities is that individuals at the policy site may not value recreational activities the same as the average individual at the study sites. There are two principal reasons for this difference. First, people at the policy site might be different from individuals at the study sites in terms of income, education, religion, ethnic group or other socio-economic characteristics that affect their demand for recreation. Second, even if individuals' preferences for recreation at the policy and study sites were the same, their recreational opportunities might not be.

Unit values for non-use values of for example ecosystems from CV studies might be even more difficult to transfer than recreational (use) values for at least two reasons. First, the unit of transfer is more difficult to define. While a common choice of unit for use values are consumer surplus (CS) per activity day, there is greater variability in reporting non-use values from CV surveys, both in terms of WTP for whom, and for what time period. WTP is reported both per household or per individual, and as a one-time payment, annually for a limited time period, annually for an indefinite time, or even monthly payments. Second, the WTP is reported for one or more specified discrete changes in environmental quality, and not on a marginal basis. Unit value transfer of non-use values is feasible only if the initial level of environmental quality at the policy site and the magnitude and direction of the change are similar to that which was valued at the study site.

For health impacts the question of which units to transfer seems somewhat simpler. With regards to mortality the standard unit is the Value of a Statistical Life (VSL), though values for years of life lost (YOLL) have also been proposed and used. For morbidity, it is more complicated, since the health impacts of environmental quality are so varied, and can be measured in different ways. For light respiratory symptoms like coughing, headaches and itchy eyes, the concept of symptom days (defined as a specified symptom experienced one day by one individual) is often used. When these symptoms limit activities, they may result in restricted activity days and/or work-loss days. Values for more serious illnesses are reported in terms of value per case (for example per hospital admission or per case of chronic bronchitis). However, the description of these different symptoms and illnesses varies, for example in terms of severity. A better alternative for acute illnesses would therefore be to construct values for episodes of illness defined as type of symptoms, duration and severity (described in terms of restrictions in activity levels, whether one would have to go to the hospital etc. see chapter 5).

On the issue of units to transfer, one should also keep in mind that often the valuation is one step in a chain of analytical models, where physical and ecological models predict impacts, which are then valued. A familiar example is the damage function approach (see Figure 1), Here we try to find values for the endpoints of dose-response and exposure-response functions for environmental and health impacts, respectively, due to changes in for example emission of air pollutants. Thus, a linkage has to be developed between the units the endpoints are expressed in, and the unit of the economic estimates. This has been done successfully for changes in visibility range (Smith and Osborne 1996), but is more difficult as complexity of changes in environmental quality and natural resources increase.

In addition to difficulties in defining the units in which values should be measured and transferred, and assuring that these units represent the same good at both study and policy sites, we must also consider whether the impacted individuals are different at the different sites. One of the most obvious ways in which the affected populations could differ between study site and policy site is in their incomes, particularly when transfers are between countries. A common, simple extension of unit transfer to such situations is to adjust unit values for differences in mean income between the study site and the policy site. Two practical issues when adjusting for differences in income between countries are: (i) determining the income elasticity

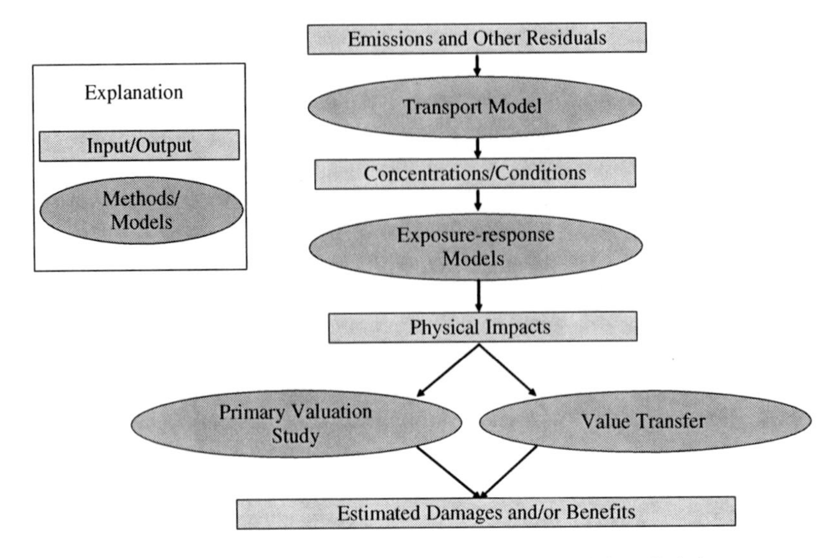

Figure 1. Damage Function Approach Applied to Air or Water Emissions

of WTP for the unit value, and (ii) the choice of an exchange rate, so that values can be measured in a common currency. Still, adjustment for differences in income will not take care of differences in preferences, environmental conditions, and cultural and institutional conditions between countries.

2.2. *Value Function Transfer*

Instead of transferring the value estimates, the analyst could transfer a **value function**. This approach is conceptually more appealing than just transferring unit values because it uses more information about the differences between the policy site and the study site, and the affected populations. The benefit relationship to be transferred from the study site(s) to the policy site could again be estimated using either revealed preference (RP) approaches like travel cost (TC) and hedonic pricing (HP) methods or stated preferences (SP) approaches like the contingent valuation (CV) method and choice experiments (CE). From any of these approaches, a benefit function is estimated:

$$(1) \qquad \mathrm{WTP}_{ij} = f(G_j, H_i)$$

where WTP_{ij} is the willingness-to-pay of household i for the environmental good at site j, Gj are the characteristics of the environmental good at site j, and H_i are characteristics of household i. Alternatively, the benefit function could predict mean WTP for the entire population at site j, based on aggregate measures of H_i.

To implement this approach the analyst would have to find a study in the existing literature with estimates of the parameters of the WTP function. Then the analyst would collect data on the two groups of independent variables G and H at the policy site, put them into equation (1), and calculate households' WTP at the policy site.

The main problem with the benefit function approach is due to the exclusion of relevant variables in the bid or demand functions estimated in a single study. When the estimation is based on observations from a single study of one or a small number of recreational sites or a particular change in environmental quality, a lack of variation in some of the independent variables usually prohibits inclusion of these variables. For domestic benefit transfers researchers tackle this problem by choosing the study site to be as similar as possible to the policy site. The exclusion of methodological variables makes the benefit function approach susceptible to methodological flaws in the original study. In practice researchers tackle this problem by choosing scientifically sound original studies.

Instead of transferring the benefit function from only one valuation study, results from several valuation studies could be combined in a meta-analysis to estimate one common benefit function. **Meta-analysis** has been used to synthesize research findings and improve the quality of literature reviews of valuation studies to come up with adjusted unit values. In a meta-analysis original studies are analyzed as a group, where the results from each study are treated as a single observation into new analysis of the combined data set. This allows us to evaluate the influence of the characteristics of the environmental good, the features of the samples used in each analysis (including characteristics of the population affected by the change in environmental quality), the valuation method used, and the modelling assumptions. The resulting regression equations explaining variations in unit values can then be used together with data collected on the independent variables in the model that describes the policy site to construct an adjusted unit value. The regression from a meta-analysis would look like equation (1), but with one added independent variable C_s = characteristics of the study s (and the dependent variable would be WTP_s = mean willingness-to-pay from study s).

Smith and Kaoru's (1990) and Walsh, Johnson and McKean (1990, 1992) meta-analyses of recreation demand models from both TC and CV studies for the US Forest Service's resource planning program were the first attempts to apply meta-analysis to environmental valuation. Later there have been applications to HP models valuing air quality (Smith and Huang 1995), CV studies of both use and non-use values of water quality improvements (Magnussen 1993), CV studies of groundwater protection (Boyle, Poe and Bergstrom 1994), TC studies of freshwater fishing (Sturtevant, Johnson and Desvousges 1998), CV studies of visibility changes at national parks (Smith and Osborne 1996), CV studies of morbidity using health status indices (Johnson, Fries and Banzhaf 1997), CV studies of endangered species (Loomis and White 1996), CV studies of environmental functions of wetlands (Brouwer et al. 1999), and HP studies of aircraft noise (Schipper, Nijkamp and Rietveld 1998). Only the last two studies are international meta-analyses, including both European and North American studies. All the others, except Magnussen (1993), analyse US studies only.

Many of these meta-analyses of relatively homogenous environmental goods and health effects focus mostly on methodological differences, limiting their usefulness for value transfer.[1] Methodological variables like "payment vehicle", "elicitation format", and "response rates" (as a general indicator of quality of mail surveys) in CV studies, and model assumptions, specifications and estimators in TC and HP studies, are not particularly useful in predicting values for specified change in environmental quality at the policy site. This focus on methodological variables is partly due to the fact that some of these analyses were not constructed for benefit transfer (e.g. Smith and Kaoru 1990, Smith and Huang 1995, and Smith and Osborne 1996), and partly because there were insufficient and/or inadequate information reported in the published studies with regards to characteristics of the study site, the change in environmental quality valued, and income and other socio-economic characteristics of the sampled population. Particularly, the last class of variables would be necessary in international benefit transfer, assuming between-country heterogeneity in preferences for environmental goods and health effects.

In most meta-analyses, secondary information is collected on at least some of these initially omitted site and population characteristics variables or for some proxy for them. These variables makes it possible to value impacts outside the domain of a single valuation study, which is the main advantage of meta-analysis over the benefit function transfer approach. However, the use of secondary data and/or proxy variables introduces added uncertainty, for example using income data for a regional population in lack of income data for fishermen at the study site. On the other hand, secondary data are more readily available at the policy site without having to do a new survey.

Most meta-analyses caution against using them for adjusting unit values due to potential biases from omitted variables and specification/measurement of included variables. To increase the applicability of meta-analysis for benefit transfer, one could select studies that are as similar as possible with regards to methodology, and thus be able to single out the effects of site and population characteristics on the value estimates. However, there are usually so few valuation studies of a specific environmental good or health impact that one cannot do a statistically sound analysis.

3. VALIDITY AND RELIABILITY OF VALUE TRANSFER

Several studies have compared value transfers to the results of new studies conducted at the policy site, to test the validity of value transfer. In Chapter 3, Rosenberger and Phipps review many of these studies. These studies often statistically reject the validity of value transfer. For example, Loomis (1992) argues that between-state benefit transfer in the U.S. (even for identically defined activities) are likely to be inaccurate, after rejecting the hypotheses that the demand equations and average benefits per trips are equal for ocean sport salmon fishing in Oregon versus Washington, and for freshwater steelhead fishing in Oregon versus Idaho.

Likewise, Bergland, Magnussen and Navrud (1995) developed a test procedure, used in several other validity studies, that showed that transferred and original use and non-use value estimates of improved water quality in two very similar and closely located lakes in Norway, were statistically different in both unit value and benefit function transfers.

But, whether a value transfer is valid is a different question from whether it is reliable. Validity requires that the values, or the value functions, generated from the study site be statistically identical to those estimated at the policy site. Reliability requires that the difference between the transferred value estimates and the values estimated at the policy site be small. This distinction is important. While some studies find very large transfer errors, it can be the case that standard statistical tests can reject equality between two value estimates, even when the difference between them is not large. In the Loomis et al. study, the observed transfer errors are 4–39% and 1–18% for unit value and benefit function transfers respectively. In the Bergland et al. study, transfer errors were 25–45% for unit value transfer, and 18–41% for value function transfer.

In Chapter 11, Kristofersson and Navrud propose equivalence testing, used in pharmaceutical research, as a combined test for validity and reliability. By defining acceptable transfer errors prior to conducting the validity test, and using a null hypothesis of inequality rather than equality between transferred and original benefit estimates, they show how results from previous validity tests could be reversed. Pattanayak, Smith and Van Houtven in Chapter 13, go one step further. They interpret value transfer as an identification problem, and propose a preference calibration approach to secure consistency between the transferred estimate and the economic concepts underlying the definition of WTP for environmental quality and quantity changes.

Since the numbers of valuation studies are very unevenly distributed geographically, transfers across countries are increasingly done. However, most validity tests are conducted on transfer within a country, and relatively few studies compare results of identical valuation studies conducted in multiple countries. In Chapter 5, Ready and Navrud report results of a CV study valuing improvements in health conducted in five European countries. Average transfer error when predicting WTP in one country from survey responses collected in other countries was ± 37–39%. In another benefit transfer validity test across countries, Rozan (2004) conducted the same CV survey in Strasbourg, France and the neighbouring city of Kehl, Germany; and asked respondents to state their WTP for a specified improvement in air quality. She found transfer errors of 15–30%. The transfer errors found in these two studies are not markedly larger than those found in studies of within-country transfer.

Are these transfer errors small enough for the purposes of the policy analyst? Transferring values from a study site to a policy site necessarily increases the uncertainty in those values. Whether the resulting level of uncertainty is acceptable to the analyst will depend on several factors. How expensive would it be to conduct a new study at the policy site? What level of uncertainty would the values from a new study have? How critical is precision in the values to the analyst? Some of these issues are discussed and tested by David Barton in Chapter 14, in his empirical study of the value of added information of benefit transfer. Further, is uncertainty over values of practical importance compared to uncertainty in predicted physical and ecological impacts in the analysis? David Brookshire and co-authors (Chapters 2 and 6) explore the issue of the relative importance of value uncertainty in analyses.

4. CONTENT OF THE BOOK

The book consists of four main parts. Part 1 outlines the theory and methods for value transfer. Chapter 2 places value transfer into the broader context of information transfer, and provides a framework for thinking about uncertainty over values versus other sources of uncertainty in a policy analysis. Chapter 3 reviews studies testing for validity and reliability of benefit transfer, and concludes that the evidence from previous empirical tests of value transfers is still inconclusive.

Part II presents applications of benefit transfer to different environmental goods, natural resources and health impacts. Chapter 4 focuses on transfer of aesthetic values of landscape. Chapter 5 looks at transfers of estimates for environmentally induced morbidity. Chapters 6 and 7 consider transfers of water quantity and quality, respectively, while Chapters 8, 9 and 10 look at recreational activities. Most of the applications use CV studies, but Chapters 4 and 7 look closer at the pros and cons of CE in value transfers. In addition to these Stated Preference methods, Chapter 10 also shows how the revealed preference methods like Travel Costs (TC) can be used in value transfers. Chapter 5 shows that there are differences in values across European countries.

Part III presents some new approaches. Chapter 11 introduces equivalence testing, which combines the concepts of statistical validity and policy relevance into one test, and applies it to use and non-use values of fish stocks. Chapter 12 develops a method based on Bayesian updating to improve benefit transfer, and applies it to recreational value of a National Park. Chapter 13 presents a new way of looking at benefit transfer that assures that transferred values are consistent with the theory underlying welfare measures, with an application to use value for water quality.

Part IV is the concluding part, and focuses on the policy utility of environmental value transfer, and on the future research agenda. Chapter 14 shows how Bayesian updating can be used in a policy context to inform decision makers on how benefits from adding more primary data in terms of increased accuracy of the transferred value, compared with the costs of data collection, and demonstrates the approach for a policy decision related to waste water treatment in Costa Rica. Chapter 15 tries to draw some general conclusions, and outlines future challenges for environmental value transfers that will improve the reliability of transfers and their policy use.

Ståle Navrud is associate professor, Department of Economics and Resource Management, Norwegian University of Life Sciences, Ås, Norway. Richard Ready is associate professor, Department of Agricultural Economics and Rural Sociology, The Pennsylvania State University, State College, PA, USA.

5. NOTE

[1] Carson et al. (1996) is an example of a meta analysis of different environmental goods and health effects, which was performed with the sole purpose of comparing results from valuation studies using both stated preference (CV) and revealed preference methods (TC, HP, defensive expenditures and actual market data).

6. REFERENCES

Bergland O, Magnussen K, Navrud S(1995) Benefit Transfer: Testing for Accuracy and Reliability. Discussion Paper #D-03/1995, Department of Economics and Social Sciences, Agricultural University of Norway.

Boyle KJ, Poe GL, Bergstrom JC (1994) What Do We Know About Groundwater Values? Preliminary Implications from a Meta Analysis of Contingent Valuation Studies. American Journal of Agricultural Economics 76 (5):1055–1061.

Brouwer R, Langford IH, Bateman IJ, Turner RK (1999) A Meta-Analysis of Wetland Contingent Valuation Studies. Regional Environmental Change 1 (1):47–57.

Carson R, Flores N, Martin K, Wright J (1996) Contingent valuation and revealed preference methodologies: Comparing the estimates for quasi-public goods. Land Economics 72 (1):80–99.

Desvousges WH, Johnson FR, Banzhaf HS (1998) Environmental Policy Analysis With Limited Information: Principles and Applications of the Transfer Method. Edward Elgar, Northampton, MA.

Downing M, Ozuna T (1996) Testing the Feasibility of Intertemporal Benefits Transfers Within and Across Geographic Regions. Journal of Environmental Economics and Management 30 (3):316–322.

Johnson F, Reed, Fries Erin E, Spencer Banzhaf H (1997) Valuing Morbidity: An Integration of the Willingness-to-Pay and Health-Status Index Literatures. Journal of Health Economics 16:641–665.

Loomis JB (1992) The Evolution of a More Rigorous Approach to Benefit Transfer: Benefit Function Transfer. Water Resources Research 28 (3):701–705.

Loomis J., White D (1996) Economic Benefits of Rare and Endangered Species: Summary and Meta-analysis. Ecological Economics 18 (3):197–206.

Magnussen K (1993) Mini meta analysis of Norwegian water quality improvements valuation studies. Unpublished manuscript, Norwegian Institute for Water Research, Oslo, 29 p.

Rosenberger RS, Loomis JB (2003) Benefit Transfer. In: Champ PA, Boyle KJ, Brown TC (eds) A Primer on Nonmarket Valuation. Kluwer Academic Publishers, Dordrecht, The Netherlands, pp 445–482.

Rozan A (2004) Benefit Transfer: A comparison of WTP for Air Quality between France and Germany. Environmental and Resource Economics 29: 295–306.

Schipper Y, Nijkamp P, Rietveld P (1998) Why do aircraft noise value estimates differ? A meta-analysis. Journal of Air Transport Management 4:117–124.

Smith VK, Huang J (1995) Can Markets Value Air Quality? A Meta-Analysis of Hedonic Property Value Models. Journal of Political Economy 103 (1):209–227.

Smith VK, Kaoru Y (1990) Signals or Noise?: Explaining the Variation in Recreation Benefit Estimates. American Journal of Agricultural Economics 72 (2):419–433.

Smith VK, Osborne L (1996) Do Contingent Valuation Estimates Pass a Scope Test?: A Meta-analysis. Journal of Environmental Economics and Management 31 (3):287–301.

Sturtevant LA, Johnson FR, Desvousges WH (1998) A Meta-analysis of Recreational Fishing. Unpublished Manuscript. Triangle Economic Research, Durham, NC.

Walsh RG, Johnson DM, McKean JR (1990) Nonmarket values from two decades of research on recreation demand. In: Smith VK, Link AN (eds) Advances in applied micro-economics, Vol. 5. JAI Press, Inc., Greenwich, CT, pp 167–193.

Walsh RG, Johnson DM, McKean JR (1992) Benefit Transfer of Outdoor Recreation Demand Studies, 1968–1988. Water Resources Research 28 (3):707–713.

D. BROOKSHIRE AND J. CHERMAK

BENEFIT AND INFORMATIONAL TRANSFERS

1. INTRODUCTION

The use of benefit transfers for policy analysis has increased dramatically during recent decades. This has been, in large part, due to the required use of benefit-cost analysis for many government projects[1]. Furthermore, the complexity of problems has increased in recent years as factors such as population growth and migration, forecast climate changes and increased use of depletable resources make resource management ever more important. These have all contributed to increased data needs. In many cases it is not practical or feasible to gather primary data, resulting in the use of data transfer studies. Desvousges *et al.*, (1998) note that:

"Transfer studies are the bedrock of practical policy analyzes..." and thus adopt a broader definition of the transfer method as "...the use of existing information designed for one specific context to address policy questions in another context." (p1)

They further state:

Strictly speaking, all analysis involves transferring some information, whether the information is data such as census counts, the intuition and tools of economics and other disciplines, or the analysts' prior knowledge, assumptions and language. (p4)

Smith (1992, p 544) notes the focus of a benefit transfer study:

"...benefit transfers focus on measuring (in dollars) how much people affected by some policy will gain from it. They are not forecasts, and they usually do not attempt to predict other exogenous influences on people's behavior. Instead, a predefined set of conditions is assumed to characterize the non-policy variables. Then benefit estimates are derived by focusing on the effects of the conditions assumed to be changed by the policy."

it is Desvousges *et al.*'s observation of information from other than economics, coupled with Smith's observation of assuming a predefined set of conditions that represent the non-policy variables (predefined set of conditions) that provide a basis for this paper. Specifically what is the information we take from other disciplines in our economic analyses? Does this information itself rely on some type of transfer? What predefined conditions, if any, is the policy analyst to assume? Traditionally, within economics, we assume economic conditions such as preferences, utility, and market structure. However, what non-economic, predefined conditions are directly or indirectly presumed, and on what models and related information are those based? Are the pre-defined conditions appropriate for the problem being investigated? If not, does the use of the information bias the result to the extent of

S. Navrud and R. Ready (eds.), Environmental Value Transfer: Issues and Methods, 11–22.

altering the policy prescription? It is these types of questions that go to the heart of the value of information.

While economic transfer studies are widely accepted, we rarely consider the sheer magnitude of the totality of model and data transfers that are actually included in such studies[2]. We explore the use of the informational transfers from across the spectrum of scientific disciplines that are incorporated into analyses.

2. ISSUES IN INTEGRATING BEHAVIORAL AND SCIENCE INFORMATION

Ever increasing populations, regional migration, and changing regional demands on resources, concerns about climate change, as well as changing regulatory environments contribute to an ever increasing need for policy analyses of resource allocations. This places increasing demands on models and data from all branches of the physical, natural, and behavioral sciences that are used in policy decisions.

For example, in semi-arid environments this has resulted in new inquiries into how to manage water resources using data transfers in an integrated setting. Continued population growth challenges the current institutional and regulatory structure in the provision of water resources. Also, studies focusing on climate are pointing to further potential challenges in the expected availability of water resources. The result is the need to evaluate alternative institutional frameworks, further the physical and natural understanding of the environment, and ascertain the magnitude of the impact from such changes within an integrated setting. For behavioral scientists this might include efforts to clarify through an adjudication the water property rights structures coupled with an investigation of more efficient allocation systems such as the practicalities of water banking. Clearly, preferences in more information rich settings would also be explored. For the natural and physical sciences this could include further efforts in the nature of linkages of groundwater systems to surface systems, prediction of snowpack run-off and long-term climatic changes. These effects, again understood in a more information rich setting, would serve as an engine of "reduced uncertainty" in an integrated policy framework. In that there are interactions, or impacts from one branch of scientific knowledge to another, we argue that policy cannot consider these separately, but rather requires the simultaneous consideration of the physical, the natural, and the behavioral sciences and, indeed, we see a need for such integrated research. While economists have focused on the assumptions and integration aspects from the behavioral components, less focus has been placed on the components from other science venues. The impact of the assumptions is, however, of paramount importance. For the policy analyst confronting an immediate policy decision the following questions become relevant:

- Is the current information from the physical, natural, and behavioral sciences of sufficient precision overall to warrant a policy statement or the corollary;
- Is there some specific aspect of the existing models and data that will challenge the applicability, usefulness, and/or appropriateness of the eventual policy statement?

The physical, natural, and behavioral sciences did not, historically, approach broad issues such as water resources management in an integrated framework. Integration has been difficult because it requires scale compatibility, but also consistent modeling goals, as well as adequate computational power to solve the resultant models. As a result the behavioral sciences often stylize the physical or natural setting in such a way as to raise questions about the relevance to the representation of the resource change under consideration[3]. Further, within the behavioral sciences, current thought would state that benefit transfers suffice when an original study is not warranted or is not practical. That is, the benefit transfers are, by definition, more cost effective than an original study and in contrast to our above raised questions[4].

Thus we argue, whenever a transfer is used, the assumption is made (implicitly or explicitly) that the cost incurred from carrying out a primary study would have been greater than the incremental value added from the improved accuracy of a primary study. Do we, as society, make that assumption too lightly? If we consider informational transfers from the broader perspective, what are the implications for policy analysis? Indeed, are transfers the "most economic method" if the possibility of an error in policy occurs? Perhaps, the more appropriate question is: "What is the value of information, or improved information?"

The issue is not "a primary study" (or even a set of similar studies) to improve the resulting policy analysis, but what original component or components of the policy analysis would either make the transfer study results more credible and/or improve the accuracy of the analysis? By raising the issue of credibility, we are suggesting that some required component (either models or data from physical, natural, and/or behavioral sciences) may be so uncertain or nothing appropriate exists, so as to make the result effectively useless. If the uncertainty around the estimate is so large then that aspect alone drives the final policy prediction. In physical or natural science processes this might be represented by the current state-of-the-art of climate modeling as against weather predictions. Within the behavioral sciences this could be the uncertainty surrounding a benefit estimate. An example of this might be the benefit estimate for the preservation of an endangered species.

Focusing on economists for a moment, do we accept the information resulting from the physical, natural and institutional settings too lightly? Alternatively, is the extent of our efforts focused on only the benefit portion of the transfer misplaced? While the evaluation methods for a benefit transfer have progressed substantially over the last decade, rarely has the precision of all of the component parts that are included in a transfer been held to the same level of scrutiny as the benefit portion.

Desvousges *et al.*, (p 183) recognize the problematic aspects of the physical or natural science issues in their analysis of the external costs of additional electric power plants. For instance they state;

"...modeling the physical effects of marginal changes in pollution on visibility levels is a complicated task because visibility is a regional problem that cannot readily be linked to individual pollutant sources (they cite the National Research Council 1993). Moreover, many of the needed parameters for some models (visual range, sulfate levels, nitrate levels, carbon levels and ammonia levels) *are not available over the study area* (italics added for emphasis)."

Hence, do policy analysts utilize and scrutinize the physical or natural science and institutional settings of an informational transfer to the same degree as economic benefit transfers? To some extent the issues of combining information have been explored within a statistical context (Draper *et al.*, 1992). We do not know, however, of systematic efforts to explore the precision of outcomes in an integrated resource management analysis[5].

3. FRAMEWORK FOR VIEWING DATA TRANSFERS

Morgenstern (1973, p 88) presents a systematic relationship between data, historical events, and economic theory, which, while not intended as a commentary on the precision of an informational transfer or the issues associated with integrated policy analysis, seems quite appropriate. Consider Morgenstern's diagram in Figure 1.

Let area A represent the body of data gathered consisting of numerical statistics (quantitative information) and let area B represents other data such as historical events, expectations, behavioral observations, preferences or other qualitative data. Area C is theory. Thus, $A \cup B \cup C$ is the information obtained by combining data, historical influences (and other relevant information), and theory.

It is the intersection of these three areas that yields primary information in an original study area. This intersection also yields the information, including physical, natural, and behavioral, that is transferred from one study area to another. The intersection represents the "zone" of transferability for information from the transfer site. This is the information that following the notion of Desvousges *et al.*, is identified in the

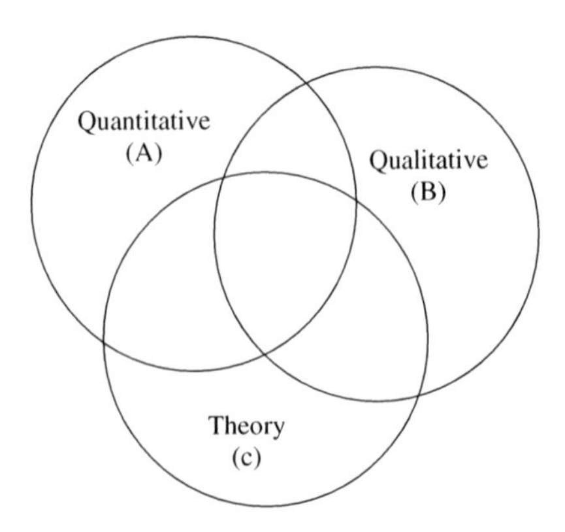

Figure 1. Schematic of an information transfer

"...competent application of transfer methods... [which (sic)]...demands all the advanced technical skills required in original research and more. Further, the...transfer analyst must employ great judgement and creativity both in manipulating available information and in presenting results to decision makers." (p1)

For purposes at hand, we consider this framework to further discuss the complexity of an overall transfer process. Every discipline that provides a component has a theoretical, quantitative and qualitative aspect. Essentially, a theoretical/predictive process, whether it be physical, natural, or behavioral in nature, has an underlying skeleton. In the context of Figure 1 (and moving beyond economics) each component (A, B, or C) is really a vector. As such, the precision of the information transfer depends on a number of aspects of each original component transferred.

The first consideration of a transfer concerns data applicability. How closely do the original data resemble the data in the area of interest? How apropos is the theory employed in the original data area to the problem in the transferred area? If areas A, B, and C were identical across the original study area and the area of interest, then the precision from the informational transfer would be exactly equal to that of a primary study, and thus identical in all respects. As the similarity of the three areas decreases, the precision of the information transfer decreases. Since it is nearly impossible to expect an exact match of the data from any discipline, the relevant question then becomes the degree to which the analysis is biased by an imprecise transfer.

Thus, the second consideration is when inexact physical, natural, and behavioral information transfers are all incorporated as primary components of the analysis. We can imagine each transfer being an intersection of quantitative data, qualitative factors, and appropriate theoretic model. The inclusion of multiple information transfers, all of which to some degree are inexact, could have positive or adverse impacts on the analysis and the recommended resultant policy. The question in this case, is which transfer, if any, has the largest impact on the precision of the analysis. If research dollars were to be spent to improve the analysis, where should the research dollars be spent? The answer, of course, depends on the value of the improved information from each component.

The third consideration is the impact of non-behavioral information transfers on the benefit transfer (from economics) itself. The level of precision of the benefit transfer may be reduced if we consider that there may be imprecision in the quantitative or qualitative factors, or even theoretical framework. For example, in semi-arid climates, the infiltration rate of water through the vadose-zone is very important when considering recharge. Historically, the accepted theoretical relationships have been used, regardless of climate. Recent research (e.g., Walvoord and Phillips, 2001) suggests the relationships may be quite different from the accepted norm and that a new paradigm is necessary for semiarid vadose-zone. Given this, economic analysis that relied on stock estimates using the accepted paradigm would over-estimate recharge to the system, resulting in production levels in excess of the steady state. Resultant policy could be ineffective and unsustainable.

Simply stated, these considerations all point out that any information transfer assumes a predefined set of conditions. This, in essence, assumes a predefined state of nature (usually the most likely or average value or outcome) in which all components have a probability of occurring of 100%. Assuming a predefined range of conditions and the probability of occurrence for each specific state in the range, results in a most likely outcome and a confidence interval around the value. Precision, of course, is enhanced by improved information, which tightens the confidence interval. How much precision is required or desirable? The straightforward answer is that the desired level of precision is that which results in the incremental benefits of the information equaling the incremental costs of gaining the additional level of precision.

4. A STYLIZED RESOURCE POLICY INITIATIVE

To illustrate points from the previous section we present the following stylized example. A government is considering instituting a policy to improve a specific species' probability of survival. In order to make this decision, the relevant factors from the behavioral, physical, and natural sciences have to be considered. The species is located in several locations and the government is considering a program encompassing five separate locations. The decision to implement the policy, or not, will be based on the expected benefits of the program across the $n = 5$ sites. That is;

$$(1) \qquad ENB = \sum_{x=0}^{n} \rho(x)(EB_x - EC_{n-x}),$$

where ENB are expected net benefits, $\rho(x)$ is the probability of x sites being successful, EB_x are the expected net benefits associated with x successful sites, and EC_{n-x} are the expected costs associated with $n-x$ unsuccessful sites, where n are the total number of sites. Thus, all areas from Figure 1 are represented (quantitative, qualitative, and theoretic) in this decision. The policy decision will come from the analytic results that can (abstractly) be represented by $A \cup B \cup C$ in Figure 1. If $ENB < 0$, the policy will not be implemented. If ENB ≥ 0, the policy will be implemented.

Physical and natural sciences supply the information concerning the probability of success of the program at each site. Assume that there are only two outcomes at each site; success or failure of the program. This assumption, of course, is an information transfer in itself. Physical scientists have determined the probability of success, S, which is the same across each location by extrapolating from other habitat studies. Given that there are only two possible outcomes, the probability of zero, one, two, three, four, or all five sites being successful is determined by a binomial probability function. That is;

$$(2) \qquad \rho(x) =_n C_x S^x F^{n-x}$$

Where

$$(3)\qquad {}_nC_x = \frac{n!}{x!(n-x)!}.$$

Furthermore, n is the total number of areas, x is the number of successes, S is the probability of success, and F is the probability of failure, which is equal to $(1-S)$. The distribution, of course, changes as S changes.

Analysis from the behavioral sciences determines the expected gross benefits and expected costs. For simplicity, assume that a high gross benefit value per successful site, B_h, and a low gross benefit value, B_l, have been determined through a benefit transfer, as has the probability of occurrence of the high value, ϕ. Similarly, high and low cost value per site, C_h and C_l, respectively, have been determined, as has the probability of the high cost occurring, ω. Expected gross benefits can be expressed by;

$$(4)\qquad EGB = \phi B_h + (1-\phi)B_l$$

and expected costs by;

$$(5)\qquad EC = \omega C_h + (1-\omega)C_l.$$

ENB can be expressed as;

$$(6)\qquad ENB = \sum_{x=0}^{n}\left[\frac{n!}{x!(n-x)!}S^x F^{n-x}\left(x(\phi B_h + (1-\phi)B_l)\right.\right.$$
$$\left.\left.-n(\omega C_h + (1-\omega)C_1)\right)\right].$$

The value of ENB is obviously impacted not only by the values and costs determined by behavioral analysis, but also by the probabilities associated with high or low values, and the probability of success of the program, as determined by the physical scientists.

For illustrative purposes, assume $B_h = 500$, $B_l = 375$, $C_h = 475$, and $C_l = 375$. As specified before, $n = 5$ and of these five site, 0,1,2,3,4,or 5 could be successful. We first hold the benefits and costs constant, as well as the probability of having high benefits or costs. That is, $\phi = \omega = 0.5$. The expected gross benefit per site is \$437.50 and the expected cost is \$425.00. In this scenario, however, we do allow the probability of success (determined by the physical and naturals sciences) to vary. Table 1 presents the results for probabilities of success (S) ranging from 0.10 to 0.90.

As can be seen from these results, the policy would not be enacted if the probability of success were 0.49 or less, but would be, if it were greater than 0.49. This illustrates the importance of understanding the uncertainty associated with

Table 1. Varying probability of success

Probability of success(es)	Expected net benefits ($)	Resultant policy
0.10	−1693.75	Don't Enact
0.20	−1262.50	Don't Enact
0.30	−831.25	Don't Enact
0.40	−141.40	Don't Enact
0.49	−11.88	Don't Enact
0.495	9.69	Enact
0.50	31.25	Enact
0.60	462.50	Enact
0.70	893.75	Enact
0.80	1325.00	Enact
0.90	1756.25	Enact

the physical component of this model and with having a physical model that is appropriate for the situation at hand[6].

In the second scenario we hold the probability of success constant at $S = 0.5$, but we now vary the probability of occurrence of high gross benefits and costs. Table 2 presents the results for the probability of occurrence (where $\phi = \varpi$) ranging between 0.10 and 0.90.

In this case, the policy would be enacted, regardless of the probability of occurrence since all the expected net benefits are non-negative. Information that more exactly estimated expected net benefits would not alter the policy decision. Thus, the value of more precise information in this regard would be zero. The policy would not be altered, and true benefits would not change. However, the impact of the physical and natural science information in this example would be critical. Depending on information, the policy decision would change. Thus, the value of the information would be the change in benefits less the cost of acquiring the more

Table 2. Varying probability of occurrence

Probability of occurrence(s)	Expected net benefits ($)	Resultant policy
0.1	6.25	Enact
0.2	12.50	Enact
0.3	18.75	Enact
0.4	25.00	Enact
0.5	31.25	Enact
0.6	37.50	Enact
0.7	43.75	Enact
0.8	50.00	Enact
0.9	56.25	Enact

precise information. If the incremental net benefit were equal to or greater than the cost to acquire the more accurate information, the value of the information would be non-negative.

While there are an infinite number of combinations of probabilities we could present, these examples serve to illustrate the main point that the analysis is only as good as the information that is incorporated into the analysis. The transfers from both the physical and the behavioral sciences should be scrutinized for their applicability to the problem at hand.

5. RECENT EVENTS: THE CASE OF ARSENIC IN THE UNITED STATES[7]

There have been efforts (and controversy over such efforts) for many years to reset the arsenic drinking water standards in the United States. This issue came to the forefront in a series of recent decisions by the Clinton and Bush Administrations. The arsenic standard was originally set in 1942 by the U.S. Public Health Service and the standard was re-iterated in 1962 at 50 parts per billion (ppb). As consideration for public health have grown over time with the ever-increasing broader concerns for the environment, the U.S. Congress directed the U.S. Environmental Policy Agency (EPA) to set a new standard. In June of 2000 the EPA proposed a revised standard of 5 ppb[8]. The EPA recognized that in setting the standard, compliance would require a significant investment in infrastructure, which might be financially difficult for small communities[9].

Specifically, in setting the standard;

> ...EPA determined that the "feasible" level is 3 ppb, but that the benefits of this standard would not justify the costs. Consequently, EPA proposed a slightly less stringent standard of 5 ppb. Also, EPA proposed to require non-transient, non-community water systems (e.g., schools only to monitor and report, as opposed to treating) because of cost-benefit considerations and because of the relatively low arsenic occurrence for these water systems. (Tiemann, p3).

In the waning days of the Clinton Administration, the standard was officially set at 5 ppb. While there had been controversy up to that point, matters significantly escalated. In the initial days of the Bush administration, the executive order was set aside. We suggest that the controversy surrounding the setting of the standard embodies all of the issues of an information transfer.

The standard is based upon animal studies and medical studies from several countries. Specifically, Taiwanese medical data suggested that high doses of arsenic could cause cancer. However, the data could not speak to the issues of the low rates of ingestion in the United States relative to those in Taiwan. Furthermore, the issue of the possibility of detoxification of arsenic by the body at low levels remains unknown. These results suggest that little is known about the complete dose response function. For instance, is the function non-linear suggesting the possibility that low doses cause little or no harm or is it linear? As a result the physical science data transferred for the standard setting carried significant uncertainties and possibly little knowledge of the underlying dose response model.

The behavioral side of the equation also carries significant uncertainties. By definition, if the low dose response is not known, it is hard to believe that the morbidity and mortality effects of high dose studies do not carry a significant amount of uncertainty. In considering the data that the EPA used, it was estimated that a 5 ppb standard would:

"...prevent about 20 cases of bladder cancer and about 5 bladder cancer deaths nationwide annually, while a 10 ppb standard would prevent 3 bladder cancer deaths annually. ...and arsenic-related lung cancers and cardiovascular diseases would be reduced as well." (Tiemann 2000, p3).

Surely a defensible behavioral study would have been a large undertaking. Not only would the underlying preference relative risk structure need to be understood, but also consideration would have to be taken for the possibility that some individuals would choose an averting behavior (or already do) within the understanding of the preference structure for morbidity.

In addition the wide range of "values of life" that exist in the literature would have to be reviewed for the appropriateness of those individual models and data that lead to the estimates. For example, Viscusi (1983) estimated the average cost-per life saved values for "loose", "medium", and "tight" arsenic regulation standards to be $1.25, $2.92. and $5.63 million, respectively. At the same time, a survey of the value of life studies finds the implicit value for a human life to be between $0.7 and $6.4 million[10].

The essential point is that when the arsenic policy decision is viewed from the stylized example in the previous section, the policy decision may suffer from a large degree of imprecision in both the behavioural and physical factors. We would note that when viewed from an integrated framework, EPA has acknowledged the need for recommending further research, apart from further presidential decisions. At this point in time, the questions we posited earlier remain germane:

- Is the current information from the physical, natural, and behavioural sciences of sufficient precision overall to warrant a policy statement and/or
- Is there some specific aspect of the existing model (s) and data that will challenge the applicability, usefulness, and/or appropriateness of the eventual policy statement?

If the answer to the second question is yes, then the follow-up questions are: (1) what component, or components, are the components of concern and (2) is the value that will be gained from improved information be at least as great or greater than the cost to obtain the improved information? If benefits are greater than the incremental costs, then the improved information has a value, if not, than it does not.

6. PERSPECTIVES

Tiemann, from whom we directly drew the facts of the arsenic study, concludes with what we would like to view as a variation on the overall theme of our analysis:

"In developing this complex regulation, EPA finds itself in a not uncommon position of working to develop a drinking water standard that protects public health to the extent feasible in the face of time constraints, scientific uncertainty, and incomplete information." (p4)

That is, the problems of uncertainty of data transfers possibly are more pervasive than is commonly acknowledged. We view the arsenic case as just one example of the issues regarding science and behavioral models. A careful examination of the literature might well yield many examples where the problems discussed herein are dominant in the final outcome of the policy analysis. These examples might well range across policy topics or goals, resource types, as well as the extent of well-developed science and behavioral models that might be considered for transfer.

We do not have an answer to the potential problems discussed in this paper. Indeed, if the answers were apparent, the points in the paper would be moot. However, we suggest that as behavioral scientists, we should become less satisfied with the partial approach typically used to date in policy studies and results that take the physical and natural science portion as given. Instead, we will be well served to expand the arena of concern to include all information transfers, rather than the (in some cases) small sub-set of benefit transfers. Further, continued interaction between scientists from the different disciplines to develop hybrid models, rather than simply taking a developed model component and assuming it will work or (even worse) making simplifying assumptions that completely neglect the important nuances of another discipline will help refine policy tools. If, in fact, a system is only as strong as its weakest link, then, by analogy, a policy prescription developed from an inadequate model may suffer similarly.

David Brookshire is professor and Janie Chermak is associate professor, Department of Economics, University of New Mexico, Albequerque, NM, USA. *This material is based on work supported in part by SAHRA (Sustainability of semi-Arid Hydrology and Riparian Areas) under the STC Program of the National Science Foundation, Agreement No. EAR-9876800. The authors would like to thank participants at the EVE Workshop, Lillehammer, Norway, October, 1999 for thoughtful discussions regarding benefit transfers.*

7. NOTES

[1] For example, the use of benefit-cost analysis in Europe, the OECD, the United Nations, and The World Bank has increased dramatically. Similarly, the 1980 US Comprehensive Environmental Response, Compensation, and Liability Act (Superfund) expanded the role of benefit-cost analysis, as did the 1981 US Executive Order that required all new projects be subjected to benefit-cost analysis.

[2] The Devosouges *et al.*, study is remarkable in laying out the relevant terrain for handling the economic portion of a study. As we discuss later, they did not choose to undertake a similar analysis for the physical or natural science underpinnings. While some issues and methods are appropriate for physical, natural, and behavioral model and data transfers, it is not clear that the sets are a complete union.

[3] These statements are not meant to be a criticism of either the physical, natural, and/or behavioral processes but are simply an acknowledgement of how disciplines do not tend to plan their research from the perspective of an integrated framework. Instead the tendency is to ask questions of "science" that

are germane to a single discipline and then possibly later ask the question "How do we make this fit together?".

[4] It is probably the nature of the beast, but we seldom encounter a study that concludes that no policy conclusions can or should be drawn given the modeling and/or data availability.

[5] Later in this volume, Brookshire *et al.*, offer a stylized case study of this process.

[6] This example is, obviously, extremely simple. For simplicity, it is a static representation of a dynamic problem. As such, we do not consider the impact of other complicating and/or decision factors. For example, not only would the time horizon be of interest (chosen terminal time, T versus an infinite horizon problem), but also in an intertemporal problem the choice of discount rate would impact the results, and perhaps the policy decision.

[7] This section draws directly on Tiemann (2000) for the historical background and other facts. The interpretation is ours.

[8] In setting the standard, EPA must effectively set two standards: (1) non-enforceable maximum containment level goal at which there are no know effects and (2) an enforceable standard for which "feasibility" is determined typically based upon cost of compliance. EPA is further required to determine if the benefits of the standard justify the cost. It is important to note that in this process, the National Research Council reported in 1983 "...that studies on U.S. populations failed to confirm the association between arsenic in drinking water and the incidence of cancer seen in Taiwan, and that 25 to 50 micrograms of arsenic a day may be nutritionally required." (Tiemann, p 2).

[9] Feasible in EPA's terminology is dependent on the costs to water systems for 50,000 or more people. In the U.S. this is only 2% of the water systems but does represent approximately 56% of those served by community systems.

[10] Viscusi *et al.*, (2000), Table 20.3 p 673.

8. REFERENCES

Brookshire DS, Chermak JM, DeSimone R (in this Volume) Uncertainty, Benefit Transfers and Physical Models: A Middle Rio Grande Valley Focus," *in this Volume*.

Cameron TA (1992) Issues in Benefit Transfer, paper presented at the 1992 Association of Environmental Resource Economists Workshop, Snowbird, UT USA June 2–5, 1992.

Desvousges WH, Johnson FR, Banzhaf HS (1998) Environmental Policy Analysis With Limited Information Principles and Applications of the Transfer Method. Edward Elgar, Cheltenham, UK (1998).

Draper D, Gaver DP Jr, Goel PK, Greenhouse JB, Hedges LV, Morris CN, Tucker JR, Waternux CM (1992) Combining Information: Statistical Issues and Opportunities for Research. National Academy Press, Washington, DC. (1992).

Morgenstern O (1973) On the Accuracy of Economic Observations. 2nd edn. Princeton University Press, Princeton, NJ. (1973).

Smith VK (1992) On separating defensible benefit transfers from "smoke and mirrors." Water Resources Research 28(3): 541–564 (1992).

Tiemann Mary (2000) RS20672: Arsenic in Drinking Water: Recent Regulatory Developments and Issues, Congressional Research Report, September 14, 2000.

Viscusi WK (1983) Risk by Choice: Regulating Health and Safety in the Workplace. Harvard University Press, Cambridge Massachusetts USA (1983).

Viscusi WK, Vernon JM, Harrington JE Jr (2000) Economics of Regulation and Antitrust. 3rd edn. MIT Press, Cambridge Massachusetts USA (2000).

Walvoord MA, Phillips FM (2001) Shifting Paradigms in Semiarid Vadose-zone Hydrodynamics: Implications for Water and Solute Balance Estimates on the Basin-scale, presented at the SAHRA 1st Annual Meeting, Tucson, AZ USA, February 20–23, 2001.

R. ROSENBERGER AND T. PHIPPS

CORRESPONDENCE AND CONVERGENCE IN BENEFIT TRANSFER ACCURACY: META-ANALYTIC REVIEW OF THE LITERATURE

1. INTRODUCTION

Benefit transfer is the adaptation of existing information or data to new contexts. Benefit transfer has become a practical way to inform decisions when primary data collection is not feasible due to budget and time constraints, or when expected marginal payoffs from primary data collection are small. Primary research is conducted to address valuation needs for a specific resource, in space and time, while benefit transfer uses existing information about similar resources and conditions. Traditionally, the context of primary research is referred to as the study site, and the benefit transfer context is referred to as the policy site. Benefit transfers include two general approaches: value transfers and function transfers. Value transfers are the use of point estimates of value or range of point estimates of value. Function transfers entail the adjustment of a valuation (benefit or demand) function from a study site to characteristics of the policy site. The degree of correspondence between the study site and the policy site determines the validity of a benefit transfer.

Benefit transfer is potentially a very important tool for policy makers since it can be used to estimate the benefits of a study site, based on existing research, for considerably less time and expense than a new primary study (see, for example, *Water Resources Research* 28(3) (1992), and Krupnick (1993) for a discussion of the concept of benefit transfer and Brookshire and Neill (1992) and Desvousges, Johnson and Banzhaf (1998) for reviews of the issues and problems involved with benefit transfer). The primary obstacle to realizing this potential is developing an accepted framework for assessing the magnitude of error, termed generalization error, involved in benefit transfer (Rosenberger and Loomis 2003; Smith and Pattanayak 2002).

Generalization errors arise when estimates from study sites are adapted to policy sites. These errors are inversely related to the degree of correspondence between the study site and the policy site. Validity measures have been used in past studies to test for the accuracy of benefit transfers (Table 1). These measures specify the difference between the known value for a policy site[1] and a transferred value to the policy site. Little research has been conducted on the relationship between these measures and the factors that affect them. These factors include the quality and robustness of the study site data, the methods used in modeling and interpreting the study site data, analysts' judgments regarding the treatment of study site data and questionnaire development, other errors in the original study, and the physical

S. Navrud and R. Ready (eds.), Environmental Value Transfer: Issues and Methods, 23–43.

Table 1. Summary of Benefit Transfer Validity Tests

Reference	Resource/Activity	Value transfer percent error[a]	Function transfer percent error[a]
Loomis (1992)	Recreation	4–39	1–18
Parsons and Kealy (1994)	Water\Recreation	4–34	1–75
Loomis et al. (1995)	Recreation		
Nonlinear Least Squares Model		—	1–475
Heckman Model		—	1–113
Bergland et al. (1995)	Water quality	25–45	18–41
Downing and Ozuna (1996)	Fishing	0–577	—
Kirchhoff et al. (1997)	Rafting	36–56	87–210
	Birdwatching	35–69	2–35
Kirchhoff (1998)	Recreation/Habitat		
Benefit Function Transfer		—	2–475
Meta-analysis Transfer		—	3–7028
Brouwer and Spaninks (1999)	Biodiversity	27–36	22–40
Morrison and Bennett (2000)	Wetlands	4–191	—
Rosenberger and Loomis (2000a)	Recreation	—	0–319
VandenBerg et al. (2001)	Water quality		
Individual Sites		1–239	0–298
Pooled Data		0–105	1–56
Shrestha and Loomis (2001)	International Recreation	—	1–81

Adapted from and expanded on Brouwer (2000).
[a] All percent errors are reported as absolute values.

characteristic, attribute, and market correspondence between the study site and the policy site (Bergland, Magnussen, and Navrud 1995; Boyle and Bergstrom 1992; Brouwer 2000; Desvousges, Naughton and Parsons 1992). Protocols for conducting benefit transfers have been suggested as an attempt to minimize the effect of these factors on benefit transfer error (Rosenberger and Loomis 2001; 2003).

This chapter identifies and discusses an implicit assumption necessary for conducting benefit transfers. The primary assumption is the existence of a meta-valuation function from which values for specific issues can be inferred. The validity or accuracy of benefit transfers depends on the robustness and stability of this valuation function, and the degree of information existing for a specific issue. Secondary assumptions include the ability to capture differences between the study site context and the policy site context through a price vector. This assumption is that the multi-dimensionality of site characteristics is reducible to a single dimension price variable (Downing and Ozuna 1996; Smith, Van Houtven and Pattanayak 2002). A tertiary assumption is that values are stable over time, or vary in a systematic fashion that is captured in a price deflator index (Eiswerth and Shaw 1997).

We posit a meta-valuation function to address the first assumption. The second and third assumptions are indirectly addressed as we provide an overview of the literature on benefit transfer validity testing. This chapter begins with the conceptual development of a meta-valuation function. It then reviews the relevant literature, attempting to pull together the information into a comprehensive conclusion.

Primary research is traditionally reductionistic by collecting data and estimating values for a single site, without examining the broader valuation context. For example, individual site models cannot account for the effect of modeling decisions and site characteristics on site values because these factors are, by default, held constant. The valuation function places these individual studies in a broader context that models how values are related to factors across sites and studies.

2. A META-VALUATION FUNCTION

Figure 1 is a conceptual illustration of the proposed meta-valuation function. First, let us assume there is an underlying or meta-valuation function, $F(V)$. $F(V)$ is a function that links the values of a resource (such as wetlands) or an activity (such as downhill skiing or camping) with characteristics of the markets and sites, across space and over time, $F(V) = [g(A), g(B), g(C)]$ where $g_i(\cdot) = g_i(\cdot)(MK_i, SC_i, SP_i, T_i)$. The meta-valuation function ($F(V)$) is the envelope of a set of study site functions ($g(.)$) that relates site values to characteristics or attributes associated with each site, including market characteristics (MK) physical site characteristics (SC), spatial characteristics (SP) and time (T) Market characteristics may include factors such as individual preferences, socio-economic status (income, age, education, health), socio-cultural characteristics (attitudes, beliefs, dispositions), and socio-political influences (institutions, regulations, citizen participation). Physical site characteristics may include factors such as quality and diversity

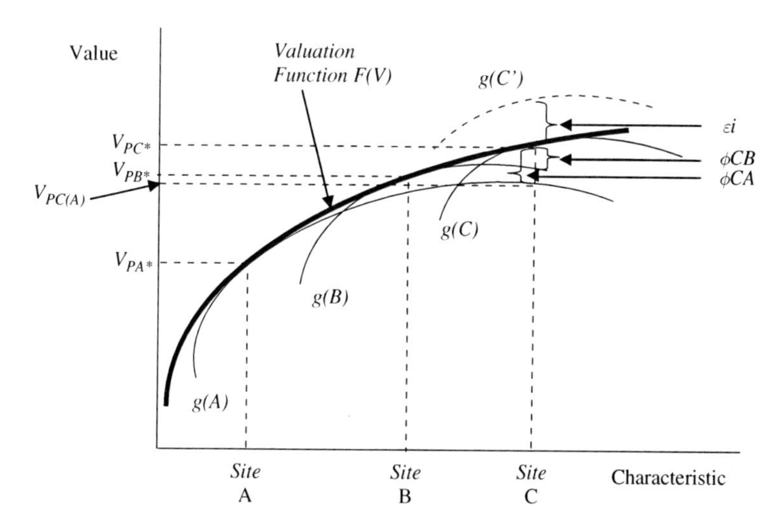

Figure 1. Meta-Valuation Function and Benefit Transfer Error

of the site, resource composition and complexity, and other physically measurable and observable factors. Spatial factors may include distance from point of origin to the site, scope or scale of the site, geographic location of the site, and diversity of the surrounding region. Temporal factors may include such issues as stability of demand and supply and socio-cultural evolution (changes in tastes, values, preferences, knowledge). The degree that any of these sets of factors affects benefit transfer accuracy is an empirical question. Some research illustrates the efficiency gains from calibrating preference functions (Smith and Pattanayak 2002; Smith, Van Houtven and Pattanayak 2002).

We hypothesize that primary research projects attempting to value a resource at a specific place, for a specific market, at a point in time, are randomly sampling from this function. How close the original research project gets to estimating the actual or 'true' value from the meta-valuation function depends on the quality of the research and the assumption that there is one true value for the resource at that place, for these people, at this point in time. Thus, in Figure 1, $g(A)$, $g(B)$ and $g(C)$ are independent site-specific functions estimating the value of the same (or similar) resource at three different locations. Assuming the primary research was conducted properly, *ceteris paribus*, the estimated 'true' value for each site is V_{PA^*}, V_{PB^*} and V_{PC^*}, respectively.

Benefit transfer validity tests typically assume the value estimated using primary research is the 'true' value for a site, or V^P. However, since V^P is unobservable, primary research approximates it, V_{pp}. In terms of notation, let the subscripts be s for a study site and p for a policy site. In benefit transfer applications, the study site values V_{ss} are used to inform the value of a similar, but unstudied, site. That is, V_{ss} is transferred to a different, but similar site j, where site j is the policy site. When the study site measure, V_{ss}, is transferred to the policy site, it becomes a transfer value, V_{ps}.[2]

$$(1) \qquad V_{ps} = V^P + \delta_{ps},$$

where δ_{ps} is the error associated with the transfer of a benefit measure from site i to site j. The empirical tests of the convergent validity, or accuracy in estimating V^P by V_{ps}, presented in Table 1 are based on calculating the percentage difference between V_{ps} and V_{pp} :

$$(2) \qquad \%\Delta V_{ij} = [(V_{ps} - V_{pp})/V_{pp}] * 100$$

when $i \neq j$. Given equations [1] and [2], the convergent validity measures become $\delta_{ps}/V_{pp} * 100$.

There are two sources of variability in δ_{ps}, and thus, errors in benefit transfers: (1) differences in the characteristics of the study site and policy site (ϕ_{ps}); and (2) errors associated with estimating V^P via V_{pp} (ε_i) (Woodward and Wui 2001). Figure 1 illustrates the potential generalization error associated with applying different benefit estimates to other sites. The individual site value distribution $g(C')$ is a deviation from the 'true' distribution of the value for Site C ($g(C)$) by the amount ε_C. This source of error arises from poorly conducted primary research, such as poor sample design, questionnaire development, and other sources for bias (Mitchell and Carson 1989). Dealing with this form of error requires subjective judgments about the quality of primary research, and is beyond the scope of this paper.

The remaining form of error, which is the main thesis of this paper, is the error associated with the correspondence between the study site and policy site. In Figure 1, ϕ_{CA} and ϕ_{CB} are the errors associated with using the distribution of value for Site A or Site B, respectively, to estimate the value for Site C given the differences in the site characteristics. For example, ϕ_{CA} is the error associated with adjusting $g(A)$ by the characteristics of site C to estimate the value for site C, or $V_{PC(A)}$. This error arises in part because $g(A)$ was not developed with the characteristics of Site C in mind. However, the greater the correspondence, or similarity, of the two sites, the smaller the expected error (Boyle and Bergstrom 1992; Desvousges, Naughton and Parsons 1992).

3. SITE CORRESPONDENCE EFFECTS

This chapter focuses on the site correspondence factors affecting the validity of benefit transfers. Other factors such as time and research methodology are important, but they require additional assumptions that are beyond the scope of this chapter. The correspondence between two sites, i and j, (SC_{ij}) is measured as the differences between study site i and policy site j based on observable measures of the distribution of market characteristics (age, gender, and income) and the distribution of physical characteristics (topography and other landscape features, resource qualities, and other measures of the physical attributes of the respective sites). We hypothesize that δ_{ij} is a function of several factors:

$$\delta_{ij} = h\left(\phi\left(SC_{ij}\right), \varepsilon_i\right) \tag{3}$$

where δ_{ij} is the generalization error associated with the transfer of study site measure i to policy site j. ϕ, or the error associated with movements along the valuation function, is defined by SC_{ij}, which is a measure of the correspondence between characteristics of study site i and policy site j. ε_i is the error associated with the study site measure. In [3], we hold ε_i as a stochastic component of the study site.

Thus, the site correspondence model takes on the following form:

$$\%\Delta V_{ij} = h(\%\Delta SC_{ij}). \tag{4}$$

That is, the percentage difference in the value transfer from site i to site $j(\%\Delta V_{ij})$ is a function of the percentage difference in the characteristics of site i and site $j(\%\Delta SC_{ij})$, where SC_{ij} includes characteristics of the sample population or market and physical characteristics of the study sites. Market characteristics can be measured in terms of the demographic profiles of the sample populations for the sites and the physical characteristics can be measured as the physical differences between the sites.

4. EVIDENCE FROM THE LITERATURE

We will provide an overview of the literature listed in Table 1 that tests the convergent validity of benefit transfers. Convergent validity is measured as the degree to which benefit transfer derived estimates of value for a policy site match the known value for the policy site. The known or actual value for the policy site is based on primary research conducted at that site for estimating values. It is assumed that this actual value is correct, that is, there is no error associated with this value. However, if there is error in the context of the primary research, then the validity test outcomes are biased. This measurement without error assumption for known values at the policy site is the major caveat in what follows.

We review the literature by first providing an overview of each of the studies listed in Table 1. This overview will summarize the context and purpose of each study. We then discuss the outcomes of these studies in the context of the goals of this chapter. Namely, we will discuss issues related to the effect of the correspondence between the study sites and policy sites and the convergent validity estimates for each study in Table 1.

4.1. Overview of the Literature

Loomis (1992): The goal of this study was to test the convergent validity of value transfers vs. function transfers. Loomis developed multisite zonal travel cost models for (1) Oregon ocean sportfishing for salmon; (2) Washington ocean sportfishing for salmon; (3) Oregon freshwater steelhead fishing; and (4) Idaho freshwater steelhead fishing. All multisite models included measures of the distance from the point of origin of anglers from the port or site and a measure of fishing quality as the total sport harvest per port or site. The steelhead fishing models included a measure of substitute sites as fishing quality divided by the distance from the point of origin of the anglers to the substitute site.

Tests of benefit transfer included a test of equality of coefficients in the models estimated for each recreation activity. Thus, models (1) and (2) were compared, and models (3) and (4) were compared. Equality of coefficients was rejected at the 0.01 significance level for the first two models, and equality of coefficients was rejected at the 0.05 significance level, but failed to reject equality at the 0.01 level, for the latter two models.

Percent errors associated with transferring value estimates were calculated for the Oregon steelhead fishing data. The multisite zonal travel cost model for Oregon steelhead fishing, which pools the data across 10 rivers studied, was used to estimate the known value for each river or site. A data-splitting technique was used to estimate a multisite benefit transfer function. This technique pools all data except for the data at the policy site (n–1 data pooling technique) and uses the n–1 model to predict the value at the excluded site. Two convergent validity tests were conducted: transferring the point estimates from the n–1 models, and transferring the benefit function to the policy site by adjusting each function to the characteristics of the policy site. All models were specified as described above. The results, provided in Table 1, show that benefit function transfers provided more accurate measures than value transfers. Benefit function transfers resulted in lower percent errors (mean = 6%; median = 5%) than value transfers (mean = 20%; median = 20%)70% of the time.

Parsons and Kealy (1994): The goal of this study was to test the convergent validity of value transfers vs. function transfers. Parsons and Kealy developed random utility travel cost models for lake recreation in Wisconsin. They split their sample between residents of Milwaukee and non-residents of Milwaukee (rest of the state). They also tested a Bayesian update modeling technique using a variety of random subsamples from the Milwaukee subsample. All models were specified using a price and price-low income interaction variables, site characteristics variables (lake size, commercial facilities, lake depth, dissolved oxygen measures, and visibility measure), and spatial factors (location of lake in a northern county or remoteness).

Tests of benefit transfer included equality of coefficients tests and percent error calculations. They included several treatments of the models, including the case of no information on lake choice or user characteristics, information on user characteristics, and behavioral information and user characteristics for three random subsamples from the Milwaukee subsample. Model equality was rejected for the rest of state model and Milwaukee model. However, equality of individual coefficients was accepted for the price-low income interaction variable; lake size; existence of a lake inlet-boater/angler interaction variable, and a dummy variable identifying lakes that had no dissolved oxygen at critical points in time.

Percent error calculations are presented in Table 1. The transfer of the average value estimated from the rest of state model to the Milwaukee policy site resulted in a–34% error. If information regarding the characteristics of the Milwaukee users is incorporated by simulating the rest of state model by adjusting it to the Milwaukee subsample (a function transfer approach), the percent error was –4%. The incorporation of full information on behavior (lake choice and number of trips taken) and user characteristics is tested using the following procedure. First, three subsamples of Milwaukee respondents are drawn ($n = 13, 28$, and 55). Three models are then developed: (a) Milwaukee subsample; (b) pooled Milwaukee subsample-rest of state sample; and (c) a Bayesian update of rest of state model parameters with the parameters estimated in the Milwaukee subsample model. The percent errors associated with using these models to estimate the known value from the full Milwaukee model are provided in Table 2. The general results of this study show that

Table 2. Percent Errors in Full Information Benefit Transfers

Model	N = 13	N = 28	N = 55	Mean absolute
Milwaukee Subsample	16%	75%	−66%	52%
Pooled Subsample+Rest of State	−3	−1	−15	6
Bayesian Update of Rest of State	−1	−3	−10	5

function transfers outperform value transfers, and that part of the increased accuracy of function transfers is through adjusting the function to fit the characteristics of the policy site or through Bayesian updates of the transfer function.

Loomis et al. (1995): The goal of this study was to test function transfers. Loomis et al. estimated several multisite zonal travel cost models for three Army Corps of Engineers districts, including the Sacramento, California district (10 reservoirs), the Little Rock, Arkansas district (8 reservoirs), and the Nashville, Tennessee district (8 reservoirs). Each model included variables measuring travel costs, reservoir size, a measure of substitute sites (reservoir size/distance from origin to substitute site), county median age, and county population. Two methods were used to estimate each model – nonlinear least squares and Heckman two stages. They tested the transferability of the multisite models including equality of coefficients and percent error.

Loomis et al. argue that multisite models increase the ability to capture visitation patterns (a demographic measure) and differences in site characteristics. The results of the coefficient equality test reject equality in all cases, although equality in coefficients between the Little Rock and Nashville models was weak. Although the general indication from Table 1 that the Heckman model outperformed the nonlinear least squares model in cross district comparisons, the nonlinear least squares model outperformed the Heckman model for the Little Rock – Nashville comparisons. In fact, the Little Rock – Nashville transfers were quite accurate (1–25% percent error). This is potentially due to the greater similarity of the eastern districts demographically and geographically.

Bergland, Magnussen and Navrud (1995): The goal of this study was to test the accuracy of benefit transfer under controlled conditions. The authors control for a variety of factors that can affect the accuracy of benefit transfers, including the use of the same survey instrument, valuing the same resource change, and conducting the study in the same time period. As such, the authors have effectively held constant the effects of methodology, quality changes, and temporal factors. They use a double-bounded contingent valuation method in two communities that border two different rivers in Norway. The variables in the final models reported included a measure of water quality change, respondents' use of the waterway for recreation, respondents' availability of alternative recreation sites, and level of education dummy variables.

The test of benefit transfer used by the authors included an equality of willingness-to-pay (WTP) measures between sites, equality of actual and predicted WTP from adjusting the function to the characteristics of the alternative site, equality of the functions, and equality of the samples (or whether the samples came from the same population).

Results of their tests rejected equality of the WTP measures at the two sites, and equality of actual and predicted WTP from adjusting the benefit function. The percent error calculations are provided in Table 1. They also reject equality of the benefit functions and their coefficients. And finally, they reject that the two samples are drawn from the same population and therefore cannot be pooled.

Table 3. Percent of Cases the Function is Transferable

	1987–1988		1987–1989	
	Within bay	Across bay	Within bay	Across bay
Transferable	63	50	50	41
Questionable	25	36	38	39
Nontransferable	12	14	12	20

Source: Downing and Ozuna (1996).

Downing and Ozuna (1996): The goal of this study was to test the transferability of benefit functions and value transfers across sites and time. The authors used a contingent valuation method with dichotomous choice to estimate benefit functions for saltwater fishing trips in eight bay regions (160 total sites) of Texas' Gulf Coast. The only variable included in their benefit functions was the offer amount (bid) in their questionnaire. The authors estimated 24 benefit functions – eight bays by three time periods (1987, 1988, and 1989). They conducted three transfer tests: (1) within a bay and across time periods; (2) across bays and within a time period; and (3) across bays and across time periods.

Their test of function transferability included specifying dummy intercept and dummy-bid slope interaction variables. If both dummy variables are insignificant, then they conclude the function is transferable. If either dummy variable is significant, then they conclude the transferability of the function is questionable. If both dummy variables are significant, then they conclude that the functions are not transferable. Table 3 shows that the majority of functions are transferable or are questionable, with higher probabilities of transferability in within bay and across time period transfers of the functions.

They next tested whether the functions led to statistically similar welfare measures. They conducted this test by using an overlapping confidence interval approach. Over 90% of the welfare estimates were not transferable, which includes the majority of the welfare estimates based on transferable functions. The authors argue that the nonlinearity of the benefit functions and the welfare measures derived from them limit our confidence in being able to transfer functions and/or welfare measures in a benefit transfer context. Even in conditions when the resource/activity is the same.

We calculated percent errors based on the welfare measures provided in Downing and Ozuna (1996). Percent errors were calculated for all possible combinations of the welfare measures, including across bays and within time, within a bay and across time, and across bays and across time periods. Table 4 breaks out the ranges reported in Table 1. Generally speaking, Table 4 shows that there is no discernible pattern, but some bays welfare estimates are more stable across time than other bays.

Kirchhoff, Colby and LaFrance (1997): The goal of this study was to test value transfers vs. function transfers. The authors use a contingent valuation method with payment card to estimate benefit functions for two pairs of sites. The first pair of

Table 4. Percent Errors Associated with Value Transfers in Downing and Ozuna (1996)

	Percent errors across bay and within time period transfers		
	1987	1988	1989
Mean	31	29	92
Median	26	24	64
Range	0–138	2–105	2–577
N	56	56	56

	Percent errors within bay and across time period transfers (bay = a, . . . , h)							
	a	b	c	d	e	f	g	h
Mean	57	91	71	17	17	24	31	31
Median	47	64	59	22	18	30	25	28
Range	26–122	22–219	36–163	2–29	7–28	1–43	18–57	16–57
N	6	6	6	6	6	6	6	6

	Percent errors across bay and across time period transfers					
	1987–1988	1987–1989	1988–1987	1988–1989	1989–1987	1989–1988
Mean	58	90	32	56	50	62
Median	52	52	35	46	45	45
Range	0–224	0–428	0–69	1–235	0–204	1–315
N	56	56	56	56	56	56

sites was concerned with protecting riparian habitat for two nationally renowned birdwatching sites in southern Arizona. The focus of the survey for both sites was the protection of perennial streamflows. However, one site focused on protecting a single species (hummingbirds) with habitat protection being implied; the other site protecting habitat with bird species diversity implied. The second pair of sites was concerned with current instream flow effects on the value of whitewater rafting on two rivers in northern New Mexico. One site was rated medium to difficult for rafting, the other medium at best. Their specified benefit functions included the following variables. For the Arizona models, socioeconomic variables (income, age, education, residency status), whether rafting at the site was the main reason for the current trip, and past visitation levels for the site were included in the models. For the New Mexico models, socioeconomic variables (stated willingness to pay for instream flow levels, trip expenses, income, state residency and gender), physical characteristic (actual flow rate at time of survey), and attitudinal variables (perceptions of flow levels, importance of rafting for site use) were included in the models.

Kirchhoff, Colby and LaFrance tested function and value transferability using overlapping confidence interval approaches. In the function transfer tests, the function for one site was adjusted by the characteristics of the other site (the policy site) in order to predict a value for this site. The results of their tests show that neither functions nor values are transferable between the pairs of sites. However, the Arizona models had fewer rejections of transferability and also lower percent errors in the function transfer than the New Mexico models (Table 1).

Kirchhoff (1998): The goal of this study was to compare the accuracy of benefit function transfers with meta-regression analysis function transfers. The policy site known values and benefit function transfer tests included those reported in (A) Loomis et al. (1995), (B) Loomis (1992), and (C) Kirchhoff, Colby and LaFrance (1997). Model specifications and function transfer tests are discussed above for each of these studies. The meta-regression analysis studies included (I) Smith and Kaoru's (1990) meta-analysis on travel cost method studies for recreation spanning 1970 to 1986. The variables specified in Smith and Kaoru's meta-regression function included the type of site, inclusion of substitute site prices in the original studies, treatment of opportunity costs in the original studies, whether the original studies developed site-specific or regional models, functional form and estimator used, and the year of the original study. The second meta-regression analysis was (II) Walsh et al.'s (1992) meta-analysis on travel cost method and contingent valuation method recreation studies spanning 1968 to 1988. The variable included in the Walsh et al. models (they report a combined model, travel cost method model, and contingent valuation method model) were site quality, specialized vs. generalized recreation activity, household vs. on-site survey, inclusion of out-of-state visitors, travel cost, type of contingent valuation question, region of visitor origin, and type of recreation activity. The third meta-regression analysis study was (III) Loomis and White's (1996) meta-analysis on contingent valuation method studies on threatened and endangered species. The variables included in the Loomis and White's model consisted of percent change in species population, payment frequency, visitor vs. household survey, and type of species valued.

In all cases, the benefit function and meta-analysis function parameters were adjusted by the characteristics of the policy site in order to predict its value. The comparison test across the different applications was based on percent error in the transferred estimate from the known value for a policy site. Table 5 disaggregates the percent error calculations reported in Table 1 for Kirchhoff (1998). The first column provides the comparison made; for example, A (I) is the comparison between the Loomis et al. (1995) benefit function transfer for reservoir recreation and the application of the Smith and Kaoru (1990) meta-analysis.

The results show that (a) benefit function transfers outperformed meta-analysis function transfers for studies A and B; (b) meta-analysis function transfers outperformed benefit function transfers for study (C); and (c) meta-analysis function transfers performed equally well as benefit function transfers for study (D). We interpret these results to be very promising for meta-analysis function transfers. In all cases, the underlying meta-analysis models were not developed for the purpose of benefit transfers (e.g., model specifications), while the benefit functions were developed for the specific issues being investigated.

Table 5. Comparisons and Percent Errors in Kirchhoff (1998)

Comparisons	Percent error	
	BFT	Meta
A (I)	1–475	233–7028
A (II-TCM)	1–475	3–817
B (I)	1–18	5–73
B (II-TCM)	1–18	28–62
C (II-CVM)	87–210	11–81
D (III)	2–35	3–39

Brouwer and Spaninks (1999): The goal of this study was to test value transfers vs. benefit function transfers. The authors estimate benefit functions for two Dutch peat meadows using a contingent valuation method with payment card. There were two surveys for one site, and one survey for the other site. Their benefit functions included variables measuring socioeconomic characteristics (gender, education, household membership in different age classifications, income, and membership in environmental organizations), attitudes (toward the resource, willingness to pay for the resource, knowledge about the site) and whether the survey design included annual or monthly payments (for the split-sample testing for one site). Their tests included (a) equality of sample mean WTP estimates and distributions; (b) equality of benefit function coefficients; (c) equality of benefit functions through a pooled data test; and (d) equality of WTP estimates across sites and by adjusting benefit functions to policy site characteristics.

The results of their tests were (a) rejection of equality of mean WTP in one of two comparisons and rejection of equal distributions of WTP in all cases; (b) rejection of equality of function coefficients in three or four cases; (c) acceptance of equality of functions in pooled model comparison; and (d) rejection of equality of WTP measures derived from adjusting the benefit functions and actual values using an overlapping confidence interval test. We calculate percent errors (Table 1), which show that benefit function transfers outperformed value transfers with regards to minimizing percent error.

Morrison and Bennett (2000): The goal of this study was to test value and function transfers based on multi-attribute stated choice modeling. The authors estimate benefit functions that value water diversions from irrigation to two wetlands in Australia. The attributes included in their stated choice experiment were water rates, irrigation-related employment, wetland size, presence of breeding waterbirds, and presence of endangered and protected species. Their benefit function specifications included variables that measured socioeconomic characteristics (households with children, ownership or paying for home, income, and water rates), physical characteristics (irrigation-related jobs, wetland size, frequency of waterbird breeding, and presence of endangered and protected species), and attitudes (environmental vs. development dispositions, and several measures testing for content validity of their experiment). They used two tests for convergent validity: (1) non-overlapping confidence intervals for estimated implicit prices (similar to an equality of coefficients test) across the two models; and (2) non-overlapping confidence intervals for welfare estimates derived various treatments (level of attributes) of each model. We calculate percent errors associated with the various welfare estimates provided by the authors.

The results of their tests show promise for the transferability of multi-attribute stated choice models. For the implicit prices comparison, six of eight implicit price's confidence intervals overlap. Welfare estimate's confidence intervals for various treatments of attribute levels overlap in five of nine cases (however, fewer overlaps when the convolutions approach to confidence interval estimation is used). The percent errors in value transfers ranged in absolute values from 4 to 191%, depending on the treatment of the attribute levels (Table 1). Table 6 further disaggregates the percent errors. Wetland A had consistently lower values than wetland B, and subsequently would understate the value for wetland B, and vice versa. The mean percent errors are consistent with other value transfers.

Rosenberger and Loomis (2000): The goal of this study was to test the convergent validity of a meta-analysis benefit function for in-sample transfers. Meta-analysis was conducted on outdoor recreation use value estimates spanning the literature from 1967 to 1998 in the US and Canada. In total, 682 estimates from 131 studies covering 21 recreation activities are included in the meta-analysis. Five meta-analysis transfer functions (a National model and four Census Region models) are developed by retaining variables significant at the 0.20 level or better. The variables in the models covered details of each study, including methodological characteristics (contingent valuation vs. travel cost models, elicitation method, survey design, and functional form), time trend, physical characteristics (lake, river, public land, forested land), and activity type. No socioeconomic characteristics are present in the models as these factors were rarely reported in the original studies.

The meta-analysis percent errors were calculated by first calculating the sample's average consumer surplus by activity by region. The meta-analysis transfer functions were then adjusted to the specific characteristics of the activity and region. This adjustment was accomplished by holding the methodological variables at their regional average level and turning on or off the dummy variables specifying characteristics of the activity/region. Percent errors were then calculated between the sample's average value and the meta-analysis predicted value. The results show that

Table 6. Value Transfer Percent Error Summary for Morrison and Bennett (2000)

	Wetland A to B	Wetland B to A
Mean	−32	59
Median	−32	48
Range	−4 to −66	4–191

the National model was more robust than the Census Region models in predicting the sample averages. Use of regional averages for adjusting the explanatory variables in the National model was more accurate than use of the national averages. Each model specification and adjustment performed better for activities and regions with more data (such as big game hunting, fishing, wildlife viewing) than activities or regions with little data (such as boating, swimming, skiing). The authors support this result with a regression test that show percent error declines as sample size increases. The ability to systematically model the effect of study and site differences on value measures demonstrates the ability of meta-analysis in benefit transfers.

VandenBerg, Poe, and Powell (2001): The goal of this study was to test benefit transfer accuracy using multisite models of groundwater quality values. The study encompassed 12 towns, with four each in Massachusetts, Pennsylvania, and New York. Six of the towns had previous experience with contaminated ground water. Willingness-to-pay was calculated from a contingent valuation survey. The benefit functions estimated included variables measuring socioeconomic characteristics (education and income), attitudes/beliefs (perceptions of risk and safety, interest in water quality, trust in government and scientific organizations), and previous contamination experience. The authors conducted three tests, including equality of mean willingness-to-pay across sites, equality of benefit function predicted values with mean values for the sites, and value transfers vs. function transfers per percent errors.

Vandenberg, Poe and Powell developed several transfer models and scenarios. In their first comparison, they transferred values between the towns, resulting in a 46 and 55% rejection of equal mean values using t-tests and overlapping confidence interval tests. Thirty-six percent of the time they rejected transferability of the functions, which result in a 57% rejection rate that predicted values based on benefit function transfer were in a 90% confidence interval for the known value at the site. Forty-eight percent of the time, benefit function transfers resulted in lower percent errors than value transfers.

In their second comparison, they estimated a multisite model using an n–1 data-splitting technique (the omitted site was the policy site). Mean value calculated from the n–1 model was transferred to the policy site, resulting in 33% rejection of equal means and overlapping confidence intervals. Twenty-five percent of the time they rejected transferability of the n–1 model, with only 17% of the n1 model predicted values not being in a 90% confidence interval of the known site value. Seventy-five percent of the time, benefit function transfers resulted in lower percent errors than value transfers.

In their third comparison, they estimated state-specific models (i.e., one model each for Massachusetts, Pennsylvania, and New York using data for towns in that state only) and transferred these models to their related towns. They only reject equality of mean values 25% of the time whether via t-tests or overlapping confidence interval tests. Twenty-five percent of the time they rejected transferability of the state-specific models, with only 17% of the state-specific model predicted values not being in a 90% confidence interval of the known site value. Sixty-seven percent of the time, benefit function transfers resulted in lower percent errors than value transfers.

In their final comparison, they estimated a model based on sites that had contaminated ground water problems in the past. This model resulted in a 42% rejection rate of equal mean values whether via t-test or overlapping confidence interval tests. Seventeen percent of the time they rejected transferability of the contaminated sites model regardless of the test performed. Ninety-two percent of the time, benefit function transfers resulted in lower percent errors than value transfers.

Similar to Rosenberger and Loomis's (2000) finding, VandenBerg, Poe and Powell (2001) found an aggregation effect associated with increasing number of observations and variability of data included in the model. This led to efficiency gains, especially with regards to benefit function transfers. Multisite benefit functions, due to aggregation effects, consistently outperformed value transfers, although value transfer accuracy was improved via aggregating sites according to likeness (e.g., by state or previous contamination).

Shrestha and Loomis (2001): The goal of this study was to test international benefit transfers of a meta-analysis function using out-of-sample studies. The authors apply the meta-analysis transfer function described in Rosenberger and Loomis (2000) to predict average consumer surplus for policy sites in countries other than the United States. They included three sets of data for the Rest of the World (ROW) studies, including set A (all 83 observations), set B (removal of three outliers for a total of 80 observations), and set C (removal of the studies supplying the outlier observations for a total of 77 observations). These dataset were further disaggregated into three groupings: (1) all countries; (2) high income countries; and (3) low income countries. The consumer surplus estimates from the studies were adjusted using the purchaser power parity (PPP) method and the per capita income (PCI) differential method. They tested equality of original study mean values and predicted mean values from adjusting the meta-analysis function to the characteristics of the policy site and the partial correlation between predicted values and original study values.

Shrestha and Loomis found that, in the case of the PPP adjustment, four of nine tests rejected equality of mean estimates with a percent error ranging from 1 to 81% in absolute terms. Using the PCI adjustment, only two of nine times did they reject equality of mean estimates, and a 4 to 46% range in percent error. The results of their correlation tests between original site values and meta-analysis predicted values showed a significant and positive correlation for all treatments of their data. This means that the meta-analysis function predicted the magnitude of value consistently with empirical estimates in the literature. Even though there are important cultural differences between the United States and the Rest of the World (investigating these factors was beyond the scope of the present study), the performance of the meta-analysis function was quite high for international transfers.

4.2. Summarizing the Evidence on Benefit Transfer Accuracy

4.2.1. Site correspondence effects Several of the studies listed in Table 1 support the hypothesis that the greater the correspondence, or similarity, between the study site and the policy site, the smaller the expected error in benefit transfers. Lower transfer errors resulted from in-state transfers than from across-state transfers (Loomis 1992; VandenBerg, Poe and Powell 2001). This is potentially due to lower socioeconomic, sociopolitical, and sociocultural differences for transfers within states, or political regions, than across states. In the Loomis et al. (1995) study, their Arkansas and Tennessee multi-site lake recreation models performed better in benefit transfers between the two regions (percent errors ranging from 1 to 25% with a nonlinear least squares models and 5 to 74% with the Heckman models) than either one when transferred to California (percent errors ranged from 106 to 475% for the nonlinear least squares models and from 1 to 113% for the Heckman models). This suggests that the similarity between the eastern models implicitly accounted for site characteristic effects. VandenBerg, Poe and Powell (2001) show accuracy gains when they transfer values and functions within communities that have experienced groundwater contamination in the past, than transferring across states, within states, or to previously unaffected communities.

Several of the studies in Table 1 also support the hypothesis that generalization errors can be reduced by transferring functions instead of point estimates or values. Benefit functions enable the calibration of the function to differences between the study site for which the function was developed and the policy site to which the function is applied (Loomis 1992; Parsons and Kealy 1994; Bergland et al. 1995; Kirchhoff et al. 1997 (for the birdwatching model only); Brouwer and Spaninks 1999; and VandenBerg, Poe and Powell 2001 (pooled data models)). However, the gains in accuracy may be more a function of the similarity of the sites than the calibration of site characteristics in the function transfers. This is because most of the functions did not include variables measuring the physical differences between the sites, but socio-economic differences between the markets. Many of the physical differences important for calibrating values across sites are unmeasured in the original functions. In part, this is because these characteristics are fixed, or constant, in individual site models, or the researchers assumed these differences are captured in the price coefficient (Downing and Ozuna 1996).

In those cases where variables measuring the physical characteristics of the sites are included in the model, they are inadequately so. For example, Loomis (1992) included a single fishing quality variables measured as total sport harvest at the sites. Parsons and Kealy (1994) included several measures of physical characteristics of their sites, including size and location of the lake, accessibility to the lake, clarity and depth of the lake, dissolved oxygen content, and the presence of developed facilities important to boaters and anglers. Loomis et al. (1995) included a single measurement of the size of reservoirs as they attempted to transfer their models between California, Tennessee, and Arkansas. Bergland et al. (1995), after arguing for experimental designs intended to minimize generalization errors through consistent application of survey method, resource change, and within the same time period, only include the water quality class change of the valuation context and whether the waterway was used for recreation. Kirchhoff, Colby and LaFrance (1997) include the actual flow rate of the rivers in New Mexico at the time of the survey, but do not include any measures of physical characteristics of the riparian areas valued in their Arizona application. However, the Arizona transfer was more accurate than the New Mexico model. Morrison and Bennett (2000) include several measures of physical characteristics in their multi-attribute stated choice models. However, the percent errors calculated from results reported in their paper are based on their setting each function at the same hypothetical levels for each site. They did not adjust one site model to the characteristics of the other site.

Several of the studies listed in Table 1 included variables measuring market characteristics in their transfer models. For example, Loomis et al. (1995) included measures of median age and population of the origin counties in their multisite zonal travel cost models. The median age and population of the policy site (a county in this case) are used to adjust, or calibrate, the study site models to better fit the characteristics of the policy site. Bergland et al. (1995) include variables measuring education and age of their sample respondents. These measures are then calibrated to the contexts of the other site. Downing and Ozuna (1996), although they do not include any variables in their models besides the offer amount in their experiment, assume this 'price' coefficient captures differences in market characteristics of their samples. Kirchhoff, Colby and LaFrance (1997) include several market characteristic variables in their Arizona birdwatching model, potentially resulting the greater accuracy in their function transfers compared to value transfers. Brouwer and Spaninks (1999) include several socioeconomic measures and attitudinal measures in their models based on respondent characteristics. Although they did not include any measures of physical characteristics in their models, their function transfers outperformed value transfers. This is potentially due to the calibration of their models by market characteristic differences between the sites. Morrison and Bennett (2000) included several measures of the market characteristics of their samples, however none of these variables directly enter into their tests of transferability. VandenBerg, Poe and Powell (2001) include education and income in their models, along with several variables measuring perceived risk, safety, and attitudes of the respondents.

4.2.2. Meta-analysis and the meta-valuation function Three of the studies listed in Table 1 test meta-analysis functions for benefit transfer purposes. Theoretically, we discussed how meta-analysis may enable us to estimate the meta-valuation function. Recall that the meta-valuation function is an envelope of a set of site or regional valuation functions. If this function exists and we can estimate it using meta-analysis, and its calibration results in unique points on the multi-dimensional surface, then it would provide us with the means to accurately predict values for policy sites. However, the estimation of this function depends largely on available data. In cases where information is limited, Rosenberger and Loomis (2000) show benefit transfers based on a meta-analysis function result in inaccurate predictions of value. In cases where we have more information, they show meta-analysis functions result in very accurate predictions of value. Parsons and Kealy (1994), using multisite models, show the increased accuracy of function transfers with increasing levels and types of information.

An exogenous factor affecting our ability to use meta-analytic techniques to estimate the meta-valuation function include access to information. As shown above and noted in Rosenberger and Loomis (2000), empirical valuation studies do a poor job at recording and reporting characteristics of their study sites, including physical characteristics of the sites and characteristics of the sample population. In all of the meta-analyses tested in Kirchhoff (1998), Rosenberger and Loomis (2000) and Shrestha and Loomis (2001), none of them included market characteristics of the underlying samples in the original studies. As shown above, both physical differences and market differences between the study sites and policy sites are important factors to be accounted for in conducting accurate benefit transfers.

Multisite models share several of the advantages of meta-analysis functions. Namely, they include more information across populations and sites. Thus, those studies that tested function transfers based on multisite models (Downing and Ozuna 1996; Loomis 1992; Loomis et al. 1995; Parsons and Kealy 1994) performed reasonably well, especially given the level of specification of their functions.

5. DISCUSSION AND CONCLUSIONS

Several implicit assumptions are made whenever benefit transfers are undertaken. We argue that the primary assumption is the existence of a meta-valuation function from which policy site estimates are drawn. In applications of benefit transfer, quite often it is further assumed that the multi-dimensionality of this meta-valuation function can be reduced to a price vector and is stable across time. This chapter investigates these assumptions by developing a conceptual model of the meta-valuation function and reviews the empirical literature on benefit transfer validity testing.

In general, evidence presented in this chapter supports the need for primary research to target the development of benefit transfer models (Bergland et al. 1995), or at least be more sensitive to potential use of primary research outcomes in benefit transfer applications. This sensitivity to benefit transfers would entail gathering and reporting data that may not be of immediate use in the primary research, but essential to developing valid and reliable benefit transfers. For example, insignificant covariate effects or explanatory variables that define the context of a study site (and are therefore constant) are often not reported in primary research. However, in order to develop a meta-valuation function, these covariate effects and measures of other variables are necessary for the estimation of the meta-valuation function and its calibration to policy site contexts.

We expect research on meta-analyses and their use in benefit transfers to continue, especially as our body of knowledge for certain resources and issues expands.

It will be interesting to see if a general pattern in the effects of different site characteristics emerges. This pattern, if it existed, could have a tremendous effect on how and what kind of data is collected in non-market valuation surveys. The literature has just begun addressing spatial issues related to non-market valuation of resources (Bateman et al. 2002) and a couple of studies investigate the use of Geographic Information Systems (GIS) in benefit transfer contexts (Bateman, Lovett and Brainard Bateman1999; Eade and Moran 1996). Eiswerth and Shaw (1997) discuss some of the implications of using the Consumer Price Index to update, or inflate, values to account for temporal changes. Temporal issues are important given that our body of knowledge accumulates over time, during which state-of-the-art in nonmarket valuation, tastes and preferences, knowledge and understanding, political and cultural contexts, and attitudes and beliefs may evolve.

Given no conclusions can be drawn from past empirical tests of benefit transfers with certainty, what are resource managers and policy makers to do? One of the distinct advantages of benefit transfer over primary research is the time and resource savings. But if the costs of benefit transfer (in terms of valuation accuracy) are high, then there may be no recourse beyond initiating primary research. We recommend resource managers and policy makers consider the correspondence between their policy site and the available information from study sites. Lacking a specified meta-valuation function that takes into account the differences across the candidate study sites, resource managers and policy makers should choose candidate study sites by the degree to which they correspond, or are similar, to their policy site needs across physical attributes of the site and the affected market. By doing so, the literature and evidence provided in this study suggest that generalization errors may be minimized.

We also strongly suggest that funding agencies and other interested parties should continue to support (and fund) primary research. Primary research provides the information required to estimate the meta-valuation function through multisite modeling and meta-analysis. Without the addition of new information, the confirmation of existing information, and growth in our body of knowledge pertaining to the values of nonmarket goods, our ability to develop robust and valid benefit transfer functions is limited.

Randall Rosenberger is assistant professor, Department of Forest Resources, Oregon State University, Corvallis, OR, USA. Timothy Phipps is Professor, Division of Resource Management, West Virginia University, Morgantown, WV, USA.

6. NOTES

[1] The known ('true' or actual) value for a policy site is derived from an original study designed to estimate a value for this site.

[2] Although we speak of value in the singular, the reader should recognize that benefit transfers quite often use multiple values to infer a value and its distribution for a policy site.

7. REFERENCES

Bateman IJ, Lovett AA, Brainard JS (1999) Developing a Methodology for Benefit Transfers Using Geographical Information Systems: Modelling Demand for Woodland Recreation. Regional Studies 33(3):191–205.

Bateman IJ, Jones AP, Lovett AA, Lake IR, Day BH (2002) Applying Geographical Information Systems (GIS) to Environmental and Resource Economics. Environmental and Resource Economics 22(1–2):219–269.

Bergland O, Magnussen K, Navrud S (1995) *Benefit Transfer: Testing for Accuracy and Reliability.* Discussion Paper #D-03/1995. Norway: Department of Economics and Social Sciences 21p.

Boyle KJ, Bergstrom JC (1992) Benefit Transfer Studies: Myths, Pragmatism, and Idealism. Water Resources Research 28(3):657–663.

Brookshire DS, Neill HR (1992) Benefit Transfers: Conceptual and Empirical Issues. Water Resources Research 28(3):651–655.

Brouwer R (2000) Environmental Value Transfer: State of the Art and Future Prospects. Ecological Economics 32(1):137–152.

Brouwer R, Spaninks FA (1999) The Validity of Environmental Benefits Transfer: Further Empirical Testing. Environmental and Resource Economics 14:95–117.

Desvousges WH, Johnson FR, Banzhaf HS (1998) *Environmental Policy Analysis with Limited Information: Principles and Applications of the Transfer Method.* Mass: Edward Elgar.

Desvousges WH, Naughton MC, Parsons GR (1992) Benefit Transfer: Conceptual Problems in Estimating Water Quality Benefits Using Existing Studies. *Water Resources Research* 28(3):675–683.

Downing M, Ozuna T Jr (1996) Testing the Reliability of the Benefit Function Transfer Approach. Journal of Environmental Economics and Management 30(3):316–322.

Eade JDO, Moran D (1996) Spatial Economic Valuation: Benefits Transfer using Geographical Information Systems. Journal of Environmental Management 48(2):97–110.

Eiswerth ME, Shaw WD (1997) Adjusting Benefits Transfer Values for Inflation. Water Resources Research 33(10):2381–2385.

Kirchhoff S (1998) *Benefit Function Transfer vs. Meta-Analysis as Policy-Making Tools: A Comparison.* Paper presented at the workshop on Meta-Analysis and Benefit Transfer: State of the Art and Prospects, Tinbergen Institute, Amsterdam, April 6–7, 1998.

Kirchhoff S, Colby BG, LaFrance JT (1997) Evaluating the Performance of Benefit Transfer: An Empirical Inquiry. Journal of Environmental Economics and Management 33(1):75–93.

Krupnick AJ (1993) Benefit Transfers and Valuation of Environmental Improvements. Resources Winter(110):1–6.

Loomis JB (1992) The Evolution of a More Rigorous Approach to Benefit Transfer: Benefit Function Transfer. Water Resources Research 28(3):701–705.

Loomis J, White D (1996) Economic Benefits of Rare and Endangered Species: Summary and Meta-Analysis. Ecological Economics 18(3):197–206.

Loomis JB, Roach B, Ward F, Ready R (1995) Testing the Transferability of Recreation Demand Models Across Regions: A Study of Corps of Engineers Reservoirs. Water Resources Research 31(3):721–730.

Mitchell RC, Carson RT (1989) *Using Surveys to Value Public Goods: The Contingent Valuation Method.* Washington, DC: Resources for the Future.

Morrison M, Bennett J (2000) Choice Modelling, Non-Use Values and Benefit Transfers. Economic Analysis and Policy 30(1):13–32.

Parsons GR, Kealy MJ (1994) Benefits Transfer in a Random Utility Model of Recreation. Water Resources Research 30(8):2477–2484.

Rosenberger RS, Loomis JB (2000) Using Meta-Analysis for Benefit Transfer: In-Sample Convergent Validity Tests of an Outdoor Recreation Database. Water Resources Research 36(4):1097–1107.

——(2001) *Benefit Transfer of Outdoor Recreation Use Values: A Technical Document Supporting the Forest Service Strategic Plan (2000 Revision).* General Technical Report RMRS-GTR-72. Fort Collins, CO: U.S. Department of Agriculture, Forest Service, Rocky Mountain Research Station. 59 p. (http://www.fs.fed.us/rm/pubs/rmrs_gtr72.html).

—— (2003) Benefit Transfer. In Champ P, Boyle K, Brown T, eds, *A Primer on Non-Market Valuation*. Under contract with Kluwer Academic Publishers; Dordrecht, The Netherlands.

Shrestha RK, Loomis JB (2001) Testing a Meta-Analysis Model for Benefit Transfer in International Outdoor Recreation. Ecological Economics 39(1):67–83.

Smith VK, Van Houtven G, Pattanayak SK (2002) Benefit Transfer via Preference Calibration: 'Prudential Algebra' for Policy. Land Economics 78(1):132–152.

Smith VK, Pattanayak SK (2002) Is Meta-Analysis a Noah's Ark for Non-Market Valution? Environmental and Resource Economics 22(1–2):271–296.

Smith VK, Kaoru Y (1990) Signals or Noise?: Explaining the Variation in Recreation Benefit Estimates. American Journal of Agricultural Economics 72(2):419–433.

VandenBerg TP, Poe GL, Powell JR (2001) Assessing the Accuracy of Benefits Transfers: Evidence from a Multi-Site Contingent Valuation Study of Groundwater Quality. In Bergstrom JC, Boyle KJ, Poe GL, eds, *The Economic Value of Water Quality*. Mass: Edward Elgar.

Walsh RG, Johnson DM, McKean JR (1992) Benefit Transfer of Outdoor Recreation Demand Studies: 1968-1998. Water Resources Research 28(3):707–713.

Woodward RT, Wui Y-S (2001) The Economic Value of Wetland Services: A Meta-Analysis. Ecological Economics 37(2):257–270.

J.M.L. SANTOS

TRANSFERRING LANDSCAPE VALUES: HOW AND HOW ACCURATELY?

1. INTRODUCTION

Market failure provides a rationale for public intervention in the management of landscape change: to ensure that non-market benefits we derive from rural landscapes (i.e. aesthetics, recreation and ecological function) are kept at socially adequate levels. A variety of policies, including planning controls and agri-environmental schemes, have been designed and implemented for this purpose.

The potential significance of the costs of landscape policy, including the social opportunity cost of environmentally sensitive management as well as policy administration costs, suggests that the landscape benefits actually delivered by policy should be directly valued and weighted against policy costs.

This direct valuation and cost-benefit approach is not yet systematically used in the evaluation of landscape policy. Nonetheless, a number of valuation studies have been carried out in this policy context (e.g. Willis et al. 1993a; Hanley et al. 1996 and Santos 1998), some of which commissioned by public agencies running agri-environmental schemes and included in official evaluation reports (European Commission 1998). Contingent valuation (CV) is the non-market valuation technique most extensively used in these studies and the one best fitting of the typical valuation problem facing policy analysts in this context (Willis et al. 1993b).

However, carrying out an original CV study for every single policy decision is not possible in practice. It would not match the time and budget constraints of most policy-evaluation exercises. Transferring valuation information from previous studies sounds as an attractive alternative, in that, being faster and cheaper, it would allow for a more systematic use of valuation and cost-benefit analysis.

Benefit transfer is the application, with the necessary adjustments, of valuation information from an original study, or from multiple studies, to a different context, where such information is required to evaluate a new policy. Several issues involved in transferring valuation information may affect the quality of the transfer. Desvousges et al. (1998) offer a comprehensive review of these issues. Here, we are interested in the following:

- how to select the original studies for a transfer (source-studies);
- whether to transfer a single best benefit estimate or multiple estimates, possibly from multiple studies;
- how to use valuation information from multiple studies in order to produce the benefit estimate to be transferred; and

S. Navrud and R. Ready (eds.), Environmental Value Transfer: Issues and Methods, 45–75.

- whether to transfer (1) an unadjusted benefit estimate; (2) an estimate adjusted for differences between the original study and the policy context; (3) a source-study-specific valuation model; or (4) a meta-model accounting for the inter-study variation in benefit estimates across the relevant valuation literature.

Some of these issues have already been addressed using rather controlled 'transfer experiments' (see e.g. Loomis 1992; Downing and Ozuna 1996), where original benefit estimates at the policy site were compared to estimates transferred from other sites, with both types of estimates produced under very standardised methodological conditions. It is usually impossible to ensure such standardisation in most practical benefit transfers carried out for policy evaluation purposes.

Moreover, most of these past 'transfer experiments' only test for convergent validity, i.e. for whether differences between transfer estimates and original benefit estimates are statistically insignificant. They do not investigate whether those differences are large enough to imply divergent policy recommendations, i.e.: whether transfer error leads, in practice, to wrong decisions. This second type of test is known as an importance test, which requires information on policy cost, usually not available in those more experimental studies.

This chapter is focused on a real policy-evaluation case, that of the Pennine Dales Environmentally Sensitive Area (ESA) scheme, for which we had a large amount of available cost and benefit information, as well as many candidate source-studies for transfer (Santos 1998). This focus on a real-world policy setting, with available cost information, enabled us both (1) to address the issues raised above in not so standardised ways, which is more relevant for real-world benefit transfers, and (2) to carry out importance as well as convergent validity tests.

The Pennine Dales comprise some of the most spectacular upland scenery in England. The interesting patterns of colour and texture in the Dales landscapes, which are so appreciated by visitors, were shaped by farming practices that evolved over the centuries. Many cherished landscape features, such as stone walls, flower-rich meadows and broadleaved woods, are threatened by current changes in farming practice, comprising a mix of abandonment of some features (walls and field barns) and intensification of meadows use for forage production. The Pennine Dales ESA scheme is a voluntary scheme which offers farmers management agreements aimed at conserving the cherished landscape attributes of the Dales.

The cost-benefit problem in this case was to compare aggregate landscape benefits for visitors with policy costs, including farm income foregone, and additional work for wall and barn repair and better wood management; policy-administration costs were to be included as well.

Within this cost-benefit frame, a CV survey of visitors to the Pennine Dales ESA was undertaken in 1995, in order to estimate visitors' willingness-to-pay (WTP) for the ESA scheme on a per-household per-year basis. 422 visitors were successfully interviewed, producing usable questionnaires. Respondents were asked to choose between (1) the continuance of a specified policy scheme, at a given tax-rise cost, and (2) giving up the scheme altogether with no tax increase. This implies a discrete-choice (DC) approach to the elicitation of WTP.

Different policy schemes were offered to different respondents, which represented diverse mixes of three basic measures: P1, or the conservation of existing stone walls and field barns; P2, or the conservation of flower-rich hay meadows; P3, or the conservation of remaining small broadleaved woods.

The DC (would/wouldn't pay) answers were analysed through censored logistic regression (Cameron 1988). Among the independent variables in the model were the presence/absence of each programme ($Pi = 1/0$, for $i = 1, 2$ or 3) and the respondent's first-choice programme (FIRST $Pi = 1/0$, for $i = 1, 2$ or 3). The estimated model is presented in Table 1.

This model allows us to predict average WTP conditional on any combination of the independent variables. For our current purposes, we are interested in WTP for the most complete policy mix (i.e.: Pi = 1 for i = 1, 2 and 3) by the average individual in the sample (i.e. sample averages for all other independent variables). This value of the most complete policy mix for the average individual is estimated at £112.19 per household per year. Matrix-algebraic manipulation of the variance-covariance matrix of the censored WTP model (Cameron 1991) enabled us to build a 95% confidence interval for this WTP estimate: [£101.08; £123.30].

In what follows, this original benefit estimate is supposed to represent the 'correct' estimate of the landscape benefit of this ESA scheme for visitors. Assuming this

Table 1. Censored-regression model of WTP for the Pennine Dales ESA scheme

Variable	Parameter estimate	t-ratio	Label
INTERCEPT	–47.53	–2.898	Intercept
P1	22.05	3.650	Program 1: stone walls and field barns
P2	13.27	2.465	Program 2: flower-rich meadows
P3	19.36	3.691	Program 3: broad-leaved woodland
P1*FIRSTP1	10.97	1.530	Programme 1 when first in preferences
P2*FIRSTP2	37.85	4.769	Programme 2 when first in preferences
P3*FIRSTP3	19.57	2.292	Programme 3 when first in preferences
INCOME	0.00079	3.952	Household income before taxes (£)
DAYTRIP	14.53	2.715	Day trip
ACTBIRD	–26.36	–2.980	Birdwatching in the Dales
ACTOTHER	–13.00	–2.490	Other activities in the Dales
LANDQUAL	10.95	2.641	Quality/uniqueness of the landscape (4-point scale)
ATTPROD	15.00	2.659	Selected env. friendly products
ATTCHAR	8.95	1.794	Gave money to env. or cons.charities
MEMBNT	18.93	3.579	Member of National Trust
MEMBRSPC	25.51	2.455	Member of RSPCA
MEMBAC	34.05	2.441	Member of angling club
MEMBOTHE	30.69	2.970	Member of other env. group
CATB	–14.43	–2.251	Retired
CATD	34.11	2.652	Employer or managerial profession
FEMALE	13.04	2.346	Sex (female =1)
κ	43.86	20.476	Dispersion parameter

original benefit estimate is (as usual) unknown, the next section will look for sources of valuation information from past studies in the literature that would be relevant for predicting this benefit without the need to carry out a full CV survey at the policy site.

2. SELECTING THE CANDIDATE STUDIES FOR TRANSFER

Which original studies could we use as sources of valuation information to be transferred to our policy case-study? This question led us to undertake a review of the relevant CV literature, using the, broad, selection criterion of keeping all studies providing at least one CV estimate of WTP for the conservation of agricultural landscapes. This search considered the following sources: (1) past literature reviews of valuation studies;[1] (2) recent issues of journals of environmental and agricultural economics; (3) references in collective books on farming and the countryside;[2] (4) conference papers; (5) bibliographical references included in all landscape studies that have been found; (6) some prominent practitioners, who have added some references to our list (namely, research reports) and confirmed its completeness.

As it is usually impossible to ensure complete coverage of the literature, selection criteria and sources should always be fully reported, as above. Only this will enable readers to assess how representative and complete the final list of studies is, while ensuring this list is replicable using the reported criteria and sources (Glass et al. 1981).

A final list of 19 studies was arrived at, which is probably biased towards works written in English and already published in the Spring of 1997.[3] These 19 studies were quite uneven with respect to research quality. Rough indicators of quality of survey design, sampling procedures and CV methodology vary widely across studies (cf. Santos 1997), which meant that different studies were characterised by rather different levels of: sampling and non-response biases; understanding and acceptance of the CV scenarios by respondents; and statistical reliability of WTP estimates. Yet, no a priori exclusion of studies on quality grounds was made: the effect of research quality on results should be empirically assessed, as part of the literature review itself, not prejudged at its very beginning (Glass 1976; Glass et al. 1981).

Table 2 presents the studies retained for analysis. Each study includes one or more surveys, which referred to different populations or used different CV formats. A total of 36 surveys are reported in the 19 studies. In some cases, each survey respondent valued different landscape changes, which raises the number of results to 49. Taking into account different estimators (i.e.: levels of α-trimming of open-ended data, truncation of DC data), the total number of estimates rises to 64.[4]

Table 2 shows the wide variation in WTP estimates within the 19 CV studies of agricultural landscapes, with a minimum of £0.95 and a maximum of £233.7. Variation in estimates is large not only across studies but also within each study.[5]

Table 2. CV studies of changes in agricultural landscapes

Study	Landscape change and population	Surveys	Estimates	WTP range
Pruckner (1995)	Nationwide conservation of Austrian agricultural landscapes, valued by Summer tourists (mainly foreign people)	1	2	2.67–4.47
Drake (1992 and 1993)	Nationwide preservation of Swedish agricultural landscapes from conversion into forestry, valued by national residents	1	1	77.42
Stenger and Colson (1996)	Nationwide conservation of French agricultural landscapes; enhancement of the 'bocage' landscape in the Loire-Atlantique 'departement', and in one 'canton' in it; valued by the departement's residents	1	3	4.99–77.65
Halstead (1984)	Preventing urban development of the farmland area closest to respondent's home; valued by residents in three Massachusetts towns (USA)	1	3	45.80–233.72
Bergstrom et al. (1985); Dillman & Bergstrom (1991)	Preserving public amenity benefits of prime agricultural land from development in Greenville County (South Carolina, USA), valued by the County residents	1	1	9.32
Bateman et al. (1992 and 1993)	Preservation of the Norfolk Broads landscape from flooding by the sea; valued by visitors; local and UK residents (OE and DC format)	3	7	23.98–160.08
Willis and Garrod (1991 and 1992)	Conserving the Yorkshire Dales (UK) 'today's landscape', as opposed to the 'abandoned landscape'; valued by local residents and visitors	2	2	29.12–29.74
Willis et al. (1993a)	Maintaining 2 ESAs (South Downs and Somerset Levels and Moors) and the ESA scheme in England as a whole, valued by visitors; local and national residents (OE and DC format)	6	17	2.18–103.97
Santos (1997)	Conserving today's agricultural landscapes in the Peneda-Gerês NP (Portugal); valued by visitors	1	2	47.40–56.48
Willis (1982); and Willis & Whitby (1985)	Preserving Tyneside Green Belt (UK) from urban development; valued by residents in two communities surrounded by Green Belt land	2	2	43.33–79.13
Willis (1990)	Preventing detrimental landscape changes from current trends in farming practices in three SSSI (Skipwith Common, Derwent Ings, and Upper	1	3	0.95–2.02

(Continued)

Table 2. (Continued)

Study	Landscape change and population	Surveys	Estimates	WTP range
	Teesdale; all in the North East of England); valued by residents up to 200 Km from sites			
Hanley et al. (1991)	Conservation of typical lowland heath in Avon Forest Park (Dorset, UK); valued by visitors	1	1	11.78
Hanley and Craig (1991)	Preservation of the Flow Country of Caithness and Sutherland (Northern Scotland) from afforestation; valued by Scottish residents	1	1	20.33
Campos and Riera (1996)	Conservation of the 'dehesa' landscape (traditional Iberian agro-forestry farming system, combining holm and cork oaks with grazing and cereal fields) in the Monfragüe NP (Spain); valued by NP visitors	1	1	37.20
Hanley et al. (1996)	Breadalbane, and Machair ESA schemes (in Scotland); valued by visitors; local and national residents (using both OE and DC formats)	8	12	9.19–100.35[a]
Gourlay (1995)	Loch Lomond and Stewartry ESA schemes (in Scotland); valued by local residents	2	2	13.78–21.83
Bullock and Kay (1996)	Landscape improvement from reduced grazing pressure in the Southern Uplands ESA, Scotland; valued by visitors and national residents	2	2	70.66–84.99
Beasley et al. (1986)	Preserving remaining farmland in the Old Colony and Homestead areas of south Central Alaska (USA); valued by residents in 2 communities adjacent to existing farmland	1	2	76.64–145.21
Total		36	64	0.95–233.72[a]

Note: a. Excluding a figure of £387.07, resulting from a DC CVM estimate based on a quite small sample and exhibiting an extremely wide confidence interval (cf. Hanley et al. 1996).

3. A META-ANALYSIS OF THE RELEVANT VALUATION LITERATURE

How to take stock of all this valuation information from multiple source-studies in producing the WTP estimate to be transferred to our policy case-study? A meta-analysis of that valuation information is used here to deal with this problem. Later, an alternative procedure is also considered: simply taking WTP averages across studies.

Meta-analysis emerged as an alternative to the classic, rhetoric, approach to literature review in the social sciences (Glass 1976; Glass et al. 1981; Wolf 1986).

In these sciences, many studies on the same research problem often produce different answers, with such contradictory findings typically accumulating over long periods of research practice. This divergence in findings is due to three types of factors: (1) different sampling procedures, types of tests or variable definitions, as well as other method differences; (2) differences between studies' participants, with respect to class, age, education and other socio-economic variables; and (3) sampling variation.

There is an important task for literature reviews in interpreting and integrating this type of divergent findings. For this purpose, leading authors often adopt an idiosyncratic, authoritarian and impressionistic approach. Authors such as Glass (1976), Glass et al. (1981) or Wolf (1986) recommend an alternative approach, which is based on the idea that to understand and make sense of the variety of empirical findings on the same research problem, one should take into account the three types of factors that explain this variety. Accounting for the role of so many factors across large numbers of studies is generally well beyond the capacity of the human mind. Hence, the same statistical techniques which assist analysts in extracting information from individual studies' data should also be recommended to assist the reviewer in making sense of a complex collection of studies' findings and characteristics.

Several meta-analyses of non-market valuation studies have been published to date. In a pioneer paper, Smith and Kaoru (1990) reviewed the literature of travel cost recreation studies carried out between 1970 and 1986 in the USA. Walsh et al. (1992) reviewed both travel cost and CV studies of recreation benefits. These meta-analyses, as well as more recent ones (cf. e.g. Loomis and White 1996; Smith and Huang 1995), have at least two points in common.

First, the individual studies reviewed are not studying the same cause-effect link (as in typical meta-analyses, e.g. of the effect of class size on student achievement). Yet, they all provide measurements of WTP for particular environmental commodities.

Second, the purpose of meta-analyses of valuation studies is not to summarise their findings with respect to a common research problem, but to address a new problem: that of accounting for the variation in WTP estimates across studies. This is done by analysing the inter-study co-variation of WTP estimates, on the one hand, and, on the other, site attributes, population characteristics and method options characterising each study. Multiple regression techniques are the natural choice for this task.

In this chapter, the variation in mean WTP estimates across studies is explored using inter-study models of this type, that is meta-models. Variables characterising the studies with respect to (1) the landscape change (or programme) under valuation, (2) the surveyed population, and (3) the method applied are used as the predictors.

Depending on whether or not each of these variables should count for valuation, the analysis is gauging either theoretical validity (conformity with expectations for variables that should count) or evidence of unreliability (significant effects of variables that shouldn't count). Thus, in addition to taking stock of differences across studies with respect to landscapes and surveyed populations (some of which should count for valuation) our analysis enables us to assess the effect of particular methodological options that vary across studies, such as the elicitation format (e.g. OE vs DC elicitation format). The problem with these methodological options is that, when there is no consensus about the preferred option for the particular valuation problem, difficult reliability problems are raised if WTP is found to be sensitive to different options. Problems are also raised when using the model for benefit transfer purposes, as in this chapter, because this implies selecting the preferred methodological alternative.

Good modelling practice requires reporting all independent variables initially considered for inclusion in the model, even if later dropped on grounds of statistical significance. This enables the reader to check whether a particular variable is not in the final model because it was not statistically significant or, simply, because it was not considered in the first place. For such list of initially considered variables, as well as problems in defining and measuring these variables for the limited information reported in the case-studies in the literature see Santos (1998).[6]

The log transformation of the dependent variable (WTP) yielded the best results, both in terms of goodness-of-fit and significance of individual predictors. Hence, the semi-log (dependent) functional form was selected, except as regards the income variable, for which the log-linear form performed better. Given the selected functional form, the parameter estimates are interpreted as elasticity, in the case of income (log-linear), or percentage-effect on WTP of marginal changes, for the other continuous variables (semi-log). With respect to dummy variables (semi-log), the formula put forward by Halvorsen and Palmquist (1980) was used to calculate the percentage-effect on WTP of a change from 0 to 1 in the value of each sdummy.

According to Wolf (1986), explicitly weighting individual observations to take into account research quality is a current practice in meta-analysis. In the valuation literature reviewed here, small sample sizes were observed to lead to low precision of WTP estimates (Santos 1997). So, weighting individual observations according to sample size would be a good option. However, weighted-least-squares (WLS) estimators were not significantly different from OLS ones, and did not improve statistical significance. Hence, only the OLS results are presented here.

Three outlier observations were identified, for which the residuals were consistently larger than twice the estimated standard deviation of the random term, whatever the specification assayed. Outliers were high-variance WTP estimates, mainly associated with small sample size and/or the DC CV format. They were dropped, as they would be given too much weight if included in an un-weighted (OLS) model.[7]

Hypothesis testing in a OLS setting is subject to bias in the presence of heteroskedasticity and autocorrelation. It is probable that several forms of heteroskedasticity are present in this data set, which includes studies with different levels of research quality. Furthermore, the fact that several WTP estimates were drawn from each individual study gives a panel structure to our data, which creates potential for correlation among residuals of individual observations from the same study. Hence, resort was made to the technique proposed by Newey and West (1987) for robust estimation of the parameter variance–covariance matrix in the presence of many known forms of heteroskedasticity and autocorrelation. Newey–West-based t-ratios for the parameter estimates do not lead to significantly different conclusions, and thus, for simplicity, we report only the usual OLS t-ratios.

Despite a relatively small sample size (N = 61), 14 of the 16 variables in our first model in Table 3 (meta-model 1) have statistically significant coefficients at the 5-per cent significance level. Some detailed scrutiny will also reveal that all parameter estimates have the expected signs. This is taken as evidence of theoretical validity for the policy and population variables, which should count for valuation, and evidence of unreliability for those method variables that shouldn't count for valuation. For these latter, serious transfer problems are raised if we have no

Table 3. Regression of log (WTP) on variables describing the landscape change (policy), the population, and the method

	Meta-model 1			Meta-model 2		
Independent variables	Coeff.	t-ratio	Adjust. factor[a]	Coeff.	t-ratio	Adjust. factor[a]
Constant	−198.9	−2.873		−2.081	−0.724	
• Landscape type, landscape change and substitution effects (policy variables):						
Small site	−0.576	−1.705	−0.438			
ESA or NP	0.669	2.795	0.952	0.909	3.091	1.482
Absolutely Unique Site	0.559	1.695	0.748			
Nationwide scope	1.312	6.959	2.713			
Change in character	0.346	1.045	0.414			
Change to Moderate/heavy development	1.084	2.457	1.956			
Format involves sequential or embedded valuation	−0.348	−1.987	−0.294	−0.210	−1.314	−0.190
• Surveyed population (population variables):						
NIMBY situations	1.358	3.709	2.888			
Majority non-users	−0.676	−4.919	−0.491	−0.358	−2.193	−0.301
Majority Foreign	−2.336	−6.008	−0.903			
log(GDP per capita)	0.593	2.070		0.457	1.578	
• Survey and estimation effects (method variables):						
DC and IB format	0.948	6.441	1.580	1.193	6.242	2.296
Level of trimming (OE)	−0.077	−3.009		−0.188	−3.157	
Truncation (DC data)	−0.203	−0.720	−0.183			
'Tokens' technique	−1.291	−4.005	−0.725	−1.816	−5.617	−0.837
Year of survey	0.098	2.844				
Sample Size		61			32	
R^2		0.933			0.892	
Outliers Excluded		3			1	

Note: a. For dummy variables, this column provides the percent effect on WTP of a ceteris paribus change from 0 to 1 in each variable; figures in this column were calculated from regression coefficients according to a formula put forward by Halvorsen and Palmquist (1980). The parameter estimates for continuous dependent variables should be interpreted as usual, that is: (1) as an elasticity when the specification is log-linear (income, in our case); (2) as percentage effect on WTP of a small change in the independent variable when the specification is semi-log (level of trimming, year).

certainties about the best methodological option. A full discussion of the effects of all these variables on WTP, as regards their implications for theoretical validity and reliability of CV landscape benefit estimates can be found in Santos (1998).

Another meta-model was estimated from a smaller subset of the CV benefit estimates in the reviewed literature on agricultural landscapes. This subset comprises only those WTP estimates for the cases better matching the type of policy under evaluation: an ESA scheme. This subset was secured by dropping all WTP estimates for which at least one of the following criteria (in terms of the independent variables) applied:

- conservation or recreational site = 1;
- nation-wide scope = 1;
- considerable change in character = 1; or
- moderate/heavy development = 1.

Excluding these observations, not all parameters in meta-model 1 could be estimated. The model estimated from this (more homogeneous and closer to our policy's context) subset of WTP estimates is meta-model 2, also presented in Table 3. Later in this chapter we will compare our two meta-models as regards their performance as predictive devices used for benefit transfer purposes.

4. HOW TO TRANSFER LANDSCAPE VALUES?

In order to address, in an applied way, the issues involved in transferring WTP information that were raised in the introduction, a series of different benefit-transfer exercises were carried out, by considering different source-studies in the reviewed literature and different transfer procedures. These exercises illustrate how landscape values can be transferred in a practical policy evaluation setting.

To assess the different transfer sources and procedures against the three criteria also referred to in the introduction (i.e.: convergent-validity, importance and the value-of-information criteria), we were required to produce not only point transfer estimates but also interval estimates.

To classify the different transfer exercises that follow, the first criterion is whether (1) a single best study or (2) multiple studies were transferred.

For the single-best-study approach, two best-studies were selected: one matching the policy case (Pennine Dales ESA scheme) as regards surveyed population and the particular policy, but not as regards the method used in the Pennine Dales original survey; another, methodologically similar, but different as regards population and policy. For both source studies, three transfer procedures were considered: (a) transferring an unadjusted scalar WTP estimate reported by the source-study; (b) transferring such a scalar estimate after adjustment for CV format or income difference; and (c) transferring a valuation function reported by the source-study, to adjust for differences in the independent variables in this function. Procedure (c) was only possible for one of the selected studies. Thus, a total of 5 benefit transfers were carried out following the single-best-study approach.

For multiple-study transfers, two different procedures were considered: (a) taking averages of reported WTP estimates for all ESA-like policy schemes; and (b) using the two meta-models referred to above to predict WTP estimates for the Pennine Dales ESA setting. Both in (a) and (b), the DC and open-ended (OE) elicitation formats were dealt with separately, which produced 6 benefit-transfer estimates for the multiple-study approach.

5. TRANSFERRING A SINGLE BEST STUDY

As above mentioned, two alternative best-studies were selected: one fitting well the policy and population relevant for the Pennine Dales ESA, but fitting poorly the type of method used in the Pennine Dales ESA survey (OE, as opposed to the DC format); another, which fits poorly the relevant policy and population, but uses exactly the same methods as in that survey. This choice of source-studies is aimed at assessing the relative importance of adjusting for real value differences (i.e.: due to different landscapes, policies and people), as compared to methodological differences.

In real-world cases, both differences will typically occur simultaneously, which will generally lead to less accurate transfers. Thus, the level of accuracy achieved here for the single-best-study approach to benefit transfer can be interpreted as providing an upper bound for accuracy in most real-world transfer problems.

5.1. Similar Population, Similar Policy, Different Method

Willis and Garrod (1991) carried out a CV survey of 300 visitors to the Yorkshire Dales National Park (NP) in 1990,[8] exactly five years before the Pennine Dales ESA survey. Willis and Garrod's survey provides an excellent source-study for the transfer problem we are analysing here. They asked all respondents how much they would be willing to pay to preserve 'today's' landscape as opposed to the 'abandoned' landscape (hence, they used the open-ended, OE, CV format).

Table 4 presents the differences between 'today's' and the 'abandoned' landscape scenarios, as presented to respondents, including differences between the pictures used to convey these alternative future states of landscape. Note that the qualitative landscape change implied by these two alternative states is similar to that valued, five years later, in the Pennine Dales ESA survey. Moreover, the pictures (water-colour paintings) representing the policy-off and policy-on states of landscape are almost identical between the two surveys.

The only significant differences between these surveys' scenarios are that the former (Yorkshire Dales NP survey) underlines the impacts on the local economy and refers to threats to heather moors, whilst the latter (Pennine Dales ESA survey) does not. In fact, the ESA scheme is focussed on landscape effects alone and does not protect heather moors on the fells.

Another difference refers to the geographic extent of the landscape good: in the Yorkshire Dales NP survey, respondents valued the landscape change at stake as occurring all over this NP, including both dales (valleys) and fells (upland ridges),

Table 4. 'Today's' and 'abandoned' landscape scenarios in the Yorkshire Dales questionnaire

TODAY'S LANDSCAPE
'This typical Dales scene supports a community earning a living from farming or tourism. It is the product of an agricultural system that was, until recently, supported by subsidies intended to increase food production and maintain a healthy rural economy. Some meadows are cut for hay, while others are used for silage production. Some walls and field barns remain in good order, but many derelict ones are replaced by fences or modern sheds. Many broad-leaved woodlands are being damaged by stock and some heather moorland is deteriorating due to over-grazing.'
(Picture: shows a flower-rich meadow in the foreground; a field barn in good condition; some well-kept stonewalls as well as some derelict ones, and some modern fences; a small broad-leaved wood and field trees are present in the middle and background.)
ABANDONED LANDSCAPE
'In this landscape future farming subsidies have been taken away, leaving upland farmers to compete with better farms in the lowlands. In this situation, many owner-farmers would sell-up, while tenanted farms would be taken back into estates and their buildings used for alternative purposes. The few remaining farmers would keep smaller flocks on improved land, but outlying meadows and pastures would be abandoned. With no money to maintain field barns, walls and woodlands, they would decay and become derelict. To survive some farmers would turn to farm-based tourism or forestry.'
(Picture: the most striking differences, as compared to the picture for 'today's' landscape, are the absence of flowers in the foreground meadow; the field barn ruined, without roof; only derelict walls and fences; the broad-leaved wood reduced to a few trees, and almost no field trees.)

Source: Adapted from Willis and Garrod (1991's) appendix 1.

where the heather moors are. On the other hand, in the Pennine Dales ESA survey, respondents were asked to value that change as occurring all over the (broader) Pennine Dales,[9] but excluding fell land, where the ESA scheme does not apply. Hence, the former survey valued a change that is more inclusive in one respect (includes fell land) but less in other (does not include the northern half of the Pennines).

The surveyed population was the same in both cases, i.e.: only visitors to the Yorkshire Dales NP were interviewed in both cases. Thus, we can expect that the fact of the ESA scenario including dales outside the NP (while the NP survey scenario does not) has not caused large value differences. For visitors to the NP, changes taking place outside the NP might have a rather marginal value (at least use value).

Still concerning survey population, there is the possibility that rising income and concern for landscape degradation might have led to increase WTP for landscape conservation over the 5-year period separating the two surveys.

Taking stock of all these comparisons between surveys, we would expect elicited WTP values to be not very different between surveys. This enhances the prospects for a successful benefit transfer, perhaps adjusted for income and cultural change. Indeed, such an appropriate source-study for transfer (as Willis and Garrod's is for the current transfer problem) is very rarely available in practice for most benefit transfer problems.

There is, however, an important methodological difference between Willis and Garrod's benefit estimate and the original estimate from the Pennine Dales ESA survey: the first used the OE format whilst the latter used the DC format. An adjustment of the former would, therefore, be justified if the DC format is judged superior in methodological terms.

5.1.1. Transferring an unadjusted scalar The sample average of WTP to preserve today's landscape is £24.56 per household per year (Willis and Garrod 1991). Adjusting for price changes between 1990 and 1995 yields an estimated £29,04.[10] Using estimated standard deviation and sample size, reported by the authors, a 95-per cent confidence interval was built for the population mean of WTP, which, after price adjustments, is: [£23,30; £34,79].

These point and interval benefit estimates can be transferred without any other adjustment to deliver the benefit estimate that is required to evaluate the ESA scheme. Given the similarity between the context of Willis and Garrod's original study and that of the ESA scheme, as regards both the relevant population and the valued policy, this simple transfer procedure is supposed to deliver a sufficiently accurate benefit estimate for policy evaluation. Note that this procedure has the advantage of not requiring any information in addition to the original study used as the source for the transfer (i.e.: Willis and Garrod's).

5.1.2. Transferring an adjusted scalar As above mentioned, Willis and Garrod's OE benefit estimate needs to be adjusted for CV format if the DC format is judged superior in methodological terms. This requires more information on past CV studies of landscape changes, so that the appropriate adjustment factor can be derived. Additional information of relatively good quality is available in this case: that resulting from the previously discussed meta-analyses of the literature of CV studies of landscape changes. To derive an appropriate adjustment factor for CV format, we used meta-model 2, as this includes estimates for ESA-like policies alone, which more closely match the relevant policy setting.

Note that estimating meta-model 2 does not require the estimates from the Pennine Dales ESA survey, which we are trying to predict here. Thus, an adjustment factor was derived which is independent from the (generally unknown) difference between the source-study's estimate and the true value one is trying to predict — i.e.: that adjustment factor is only based on the analysis of available literature, as in any practical transfer exercise.

From meta-model 2 (Table 3), we know that the estimated increase in WTP resulting from using the DC as compared to OE format is 229.6 per cent Adjusting

Willis and Garrod's (1991) estimate (after adjustment for price changes) based on this figure yielded an estimated WTP for a DC CV study carried out under similar circumstances which is £95,73, with a 95%-confidence interval of [£76,79; £114,66].

A similar adjustment could easily be made for the income change in the population between 1990 and 1995, by using the income-elasticity parameter in meta-model 2 and the estimated income difference between the two survey samples. However, average household income in the first survey's sample is not known.

5.1.3. Transferring a valuation function Income, other socio-economic variables, cultural attitudes and recreational uses of the landscape could have changed over the 1990–95 period, which could have led to changes in WTP. A possible way to take these changes into account is deriving adjustment factors from meta-analysis, as referred to in the previous section for the CV-format and income variables. Another way is to use a WTP model provided by the original study (usually for purposes of theoretical validity testing) and transferring this whole functional relationship, instead of the final scalar benefit estimate. This allows one to take account of differences between the original and transfer contexts as regards the values of the independent variables in the WTP model.

Willis and Garrod (1991) provided such a WTP model with seven independent variables: (1) income, (2) number of kids, (3) rating given to the Dales landscape in a 4-point scale, (4) whether did engage in walking, or (5) cycling, (6) whether knew Dales is a national park, and (7) number of days spent in the Dales over previous 12 months. We had measurements of variables (1), (2), (5) and (7) from our 1995 Pennine Dales ESA survey. Yet, we had no measurement of variable (6) for 1995, and 1995 measurements of variables (3) and (4) were based on a different variable definition. For these three variables, the only possible solution was using sample averages from the 1990 survey. However these were not known,[11] which precluded the transfer of the model.

Note that if the transfer of the model could be successfully completed, the final estimate should still be adjusted for CV format if the DC is judged superior, as the transferred functional relationship is based on OE data.

5.2. Different Population, Different Policy, Similar Method

Santos (1997) carried out, in 1996, a CV survey of 704 visitors to the Peneda-Gerês National Park (NP), the only NP in Portugal. This survey was designed to value the landscape benefits for visitors of national agri-environmental measures currently in application within the NP.

This study used a similar survey instrument and was run in approximately the same way (by the same researcher, using the same procedures for interviewer selection and training, the same type of questionnaire, pilots, and sampling procedures) as the Pennine Dales ESA survey, whose results we are trying to predict here. In particular, both surveys used the DC format for the valuation question. Moreover, WTP estimates were secured, in both cases, from similarly specified models, estimated with the same econometric techniques (e.g., censored logistic regression, using the same type of stepwise procedures for variable selection) and data-correction procedures (e.g. for protest answers).

The conservation scheme valued in the Portuguese survey also exhibited some similarities with the one valued in the Pennine Dales ESA survey. Three basic measures were also designed, with respondents valuing all possible bundles of these programmes. There were even some interesting parallels between the particular three programmes for the Portuguese study and those for the Dales ESA study (Table 5).

Table 5. Comparison of the scenarios valued in the Pennine Dales ESA and Peneda-Gerês NP surveys

	Pennine Dales ESA survey	Peneda-Gerês NP survey
Programme 1	P1. STONE WALLS AND BARNS Will ensure the conservation of the currently existing dry stone walls and barns, by providing farmers with funds for repair of walls and barns.	P1. TRADITIONAL TERRACED FARMING Agricultural area typically occupies the lower slopes of the NP with small terraced fields. P1 will ensure the conservation of this landscape, by providing farmers with payments for keeping traditional farming practices.
Programme 2	P2. HAY MEADOWS Will ensure the conservation of the meadows at the current levels of flower diversity and habitat conditions for breeding birds, by paying farmers to maintain traditional meadow management practices.	P2. TRADITIONAL IRRIGATED MEADOWS Meadows introduce a green note, sometimes reinforced by the presence of hedge trees, in an otherwise dry landscape dominated by rocks on the uplands. They are also an important habitat for wildflowers. P2 will ensure the continued management of currently existing meadows, using agri-environmental contracts with farmers.
Programme 3	P3. SMALL WOODLAND Will ensure the conservation of currently existing small broad-leaved woodlands, by paying farmers to adopt better management practices.	P3. SMALL OAK WOODS IN FARMS This is an important ecological resource in the Park, crucial for many animal and plant species, as these oak woods represent the only remains of a broader ecosystem in the past. P3 aims at conserving existing wood area by contracting practices with farmers, which will promote the natural regeneration of trees.

Source: Santos (1997).

There are, however, important differences between the studies' contexts, which concern, for example, the relevance of agricultural elements in the landscape. Agricultural area represents a larger share of the land and is more visible a landscape attribute in the Dales than in the Peneda–Gerês. This stronger weight of agricultural landscape elements in the Dales was clearly perceived by visitors, who, in turn, had stronger preferences for agricultural elements in the Dales (walls, barns and meadows) and for more 'natural' elements (oak woods) in the Peneda–Gerês. As the conservation schemes in both areas were mainly targeted at agricultural attributes, this difference in landscapes and preferences for landscapes led us to expect that WTP for the conservation scheme in the Peneda–Gerês was lower than the value we were trying to predict for the Dales.

Differences between visitors to each area also led us to expect lower WTP for conservation in the Peneda-Gerês. In fact, visitors to the Peneda-Gerês had lower levels of income, environmental concern, environmental group membership, familiarity with the area and frequency of visits; they also engaged less often in landscape-dependent activities such as long walks during their visits (Santos 1997). All of these factors were revealed to be strong predictors of WTP, at least in the Portuguese study, which we are using here as the transfer source.

Summarily, we had a source-study using almost the same method as in the Pennine Dales ESA survey, but that did not match so well the transfer setting as regards the policy and the population. Differences in these two factors led to the judgement that the Peneda-Gerês NP WTP figure should be adjusted, somewhat upwards, to secure a more accurate benefit transfer for the Pennine-Dales-ESA policy context.

5.2.1. Transferring an unadjusted scalar If we assume that the required adjustment is, nevertheless, small and can be ignored, we can simply take the per-household benefit estimate arrived at in the Portuguese study and transfer it directly for the Pennine Dales ESA case. So, using the exchange rate of PTE for £s in 1996 (the year of the Portuguese survey), and adjusting back to 1995£s using the UK price index, we transferred the benefit of £56.48 per visiting household per year for the Pennine Dales ESA policy, with a 95-per cent confidence interval of [£52.11; £60.86].

5.2.2. Transferring an adjusted scalar From the several differences between the Peneda-Gerês NP and the Pennine Dales ESA cases, the only that is controlled for by our meta-model 2 (data on ESA-like landscape changes alone) is the income difference. 'Income' in this meta-model is GNP per capita at constant prices (1990 US$) for the country and year of the survey. This is because the sample average of household income was not available for all studies. This GNP was $6472 for Portugal 1996, and $17968 for the UK 1996. Thus, using meta-model 2, we predicted that WTP in the Pennine Dales case would be approximately 1.595 times that in the Peneda-Gerês case. Multiplying the unadjusted scalar benefit transfer by this factor yielded a benefit transfer adjusted for income differences of £90.11, with a 95-per cent confidence interval of [£83.13; £97.08].

Note that the income difference between the Peneda-Gerês NP visitors and the Pennine Dales ESA visitors is actually smaller than the difference between the corresponding countries' GNPs. However, as argued elsewhere (Santos 1997), using the meta-analytical income-elasticity estimate (which requires using the variable defined in the same way as in the meta-model) has advantages for predicting purposes, as it may account for GNP-correlated differences between countries as regards cultural attitudes towards landscape conservation.

Furthermore the use of a meta-analysis of existing studies combined with available statistical GNP data has the practical advantage of avoiding the need for a survey of the visitors to the Pennine Dales to determine their average household income.

5.2.3. Transferring a valuation function The transfer just described only adjusts for the income difference. This was the only possibility of adjustment based on our meta-analysis of ESA-like landscape policies. However, as previously discussed, there are many other differences between the two relevant contexts, as regards the policy and the population. Some of these differences could be adjusted for by transferring not the scalar benefit estimate resulting from the Portuguese study, but the whole valuation function provided by this source study. Assuming that this functional relationship holds for the Pennine Dales context, we need to evaluate the independent variables in the Dales case and, then, use the model and these evaluations to predict WTP for the Pennine Dales ESA policy.

Note that this procedure, while allowing that adjustments are made for a large number of context-differences, has further information requirements: it needs the values of the relevant independent variables for the Dales case, which requires, in practice, carrying out new data collection operations, such as a simplified survey of visitors. As we actually had this information from the Pennine Dales ESA survey, this is not a problem here. But it will often be in practice.

The WTP model from the Peneda-Gerês case that is relevant for our present transfer purposes is presented in Table 6. As the contents of the policy measures P1, P2 and P3 and the preferences for these measures were markedly different between cases, no adjustments were made for the first five independent variables (plus the intercept).

Likewise, no adjustment was made for RURALRES and RURALCHI. The fact of living in a rural area (or having been born in such type of area) is a negative WTP predictor in the Portuguese case and has been shown to be a positive predictor in some Northern European cases. It was argued elsewhere (Santos 1997) that this is understandable, as, basically, many of those living in rural areas live there by 'obligation' in Portugal and 'by option' in the UK. And if this is by option, the beauty of the rural landscape will definitely be a factor for some.

For all other independent variables, except two, we had the average value in the Pennine Dales ESA sample, using the same variable definition, which is an obvious advantage of having two very similar studies for the Dales and Peneda-Gerês cases. For the two variables for which we had not average

Table 6. WTP model from the Peneda-Gerês study used for benefit transfer

Variable	Parameter estimate	t-ratio	*Label*
INTERCEPT	−4665	−2.838	Intercept
P1	3396	5.929	Program 1: traditional farming - terraces
P2	2185	4.374	Programme 2: irrigated hay meadows
P3	2640	4.184	Programme 3: oak woods
P1*FIRSTP1	4018	5.307	Programme 1 when first in preferences
P3*FIRSTP3	4540	6.186	Programme 3 when first in preferences
LANDQUAL	762	1.747	Quality/ uniqueness of the landscape (4-point scale)
DAYTRIP	2821	3.730	Day trip
DAYS	85	2.708	Days in the NP over the last 12 months
FIRSTIME	−2119	−3.689	First visit ever made to the area
WALKING	2253	3.275	Walking more than 3 Km
WILDWATC	5076	4.180	Looking for wildlife species
VILLAGES	1417	2.108	Visiting traditional villages
OTHERACT	−2973	−1.751	Other recreational activities
ENVCAMP	5189	4.209	Actively campaigns for environment
HIGHED	−1402	−2.471	Went on to higher education
RETIRED	−6602	−4.449	Retired
SELFEMP	2569	2.455	Self employed
INCOME	0.00091	9.448	Household annual pretax income (esc.)
RURALRES	−2624	−4.131	Rural residence
RURALCHI	−3994	−4.899	Urban residence but rural childhood
FOREIGN	−3536	−2.805	Non-Portuguese
κ	6250	24.047	Dispersion parameter

Source: Santos (1998).

sample values for the Pennines, that is WILDWATCH and FOREIGN, we used the following proxies; BIRDWATCH and 'living more than 400 miles away' respectively.

Transferring the WTP model yielded a benefit estimate of PTE 16 864 for the Pennine Dales ESA, which converted to 1995 £s gives an estimated £67.52 per household per year.[12] The variance-covariance matrix of the model and the Dales independent-variable vector were then combined to analytically compute the variance of our benefit transfer, according to simple matrix algebraic procedures proposed by Cameron (1991) for censored logistic regression. In this way the 95-per cent confidence interval for our model-based benefit transfer was estimated as [£61.74; £73.31].

It is interesting to note that, when adjusting for differences between the Peneda-Gerês and Pennine Dales cases, some of the independent variables in the WTP model lead to much stronger adjustments than others, and that these adjustments often take different signs across variables. For example, the variables INCOME, WALKING, DAYTRIP, ENVCAMP and FIRSTIME implied upward adjustments between 13 and 48% (this last for income); on the other hand, RETIRED implied a downward adjustment of 37%. Adding up, the several adjustments partly offset each other, which leads to an overall upward adjustment of only 19,5%.

6. TRANSFERRING MULTIPLE STUDIES

If, as it is often the case, there is no single best study, as no single study best matches the policy evaluation problem over the policy, population *and* method dimensions altogether, then it is possibly better to transfer multiple studies rather than a single study. Transferring the average, or other more complex model, of multiple studies conveys more information and enables one to avoid the unpredictable effects on the benefit transfer of a more or less arbitrary selection of one single study, plus more or less arbitrary adjustments to the selected estimate.

6.1. Transferring Averages of Similar Studies

A simple option to transfer multiple studies is to take the average of a series of benefit estimates, whose contexts are judged sufficiently close to each other and to that of the policy to be evaluated. This average is then transferred with or without adjustments.

If some studies match better the policy, others the population, and others the method, the probability of the average of all studies being close to the real value (to be predicted with the transfer) is higher than the corresponding probability for any individual study chosen arbitrarily and taken in isolation.

Alternatively, we may view the several estimates in the literature that sufficiently match our policy evaluation problem as independent samples from the same WTP distribution. In this case, it would be advisable to use their average rather than one single estimate, as the former is a more precise estimator of the central tendency of WTP in the population.

Of course, the appropriate procedure would be using a weighed average (inverse of standard deviations, or sample sizes, used as weighs) or even 'random-effect averages' if we cannot assume that the same WTP distribution characterises all studies; but, instead, that each study is a sample from a particular WTP distribution which is, itself, extracted from a same underlying common distribution (Desvousges et al. 1998).

In this chapter, we use simple arithmetic averages as a first approach. For this purpose, we used only those benefit estimates in the reviewed literature of landscape CV studies that could be considered close enough to the policy evaluation problem at stake, that is the evaluation of the Pennine Dales ESA scheme. Thus, we excluded from our initial set of 64 estimates all those estimates that matched at least one of these conditions:

- correspond to small sites, which people tend to perceive as having many substitutes, even at the regional level;
- involve nation-wide landscape conservation programmes;
- involve considerable or very considerable, and abrupt, landscape changes, completely modifying landscape character; such as massive afforestation and

urban development of the land; these changes do not match the kind of gradual, progressive, changes that characterise the ESA scheme;
- involve elicitation mechanisms implying substitution effects, such as embedded or sequential values, as we wanted to evaluate the ESA scheme as the next addition to the status quo.

Also excluded was an extreme outlier estimate, which was more than threefold the next highest estimate. This selection procedure left us with only 20 out of our initial 64 estimates.

Next we had to decide whether to average across all OE and DC estimates or to produce separate average estimates for each CV format. The latter option was selected, which implied that, once averages were taken, a decision had to be made about the best CV format, hence the best average estimate to use for the transfer.

There were 13 OE estimates selected by the criteria above, with an average WTP estimate of £20.67 per household per year (median: £21.32; 95-per cent confidence interval of [£16.05; £25.28]; all figures after conversion to 1995 £s).

There were 7 DC estimates retained by the selection criteria above, with an average WTP estimate of £68.46 per household per year (median: £69.00; 95-per cent confidence interval of [£54.15; £82.77]). This would, in principle, provide the best prediction for the original Pennine Dales ESA survey's estimate, which was based on the DC format as well.

6.2. Transferring Meta-Model Predictions

If we consider that there is not a sufficiently homogeneous set of benefit estimates in the literature, which can be taken as a set of independent samplings from the same WTP distribution (or, at least, from a mother distribution of WTP distributions), then the average becomes a meaningless operator.

However, there are still ways to transfer multiple studies: one is using a probabilistic model of the effects of policy-differences, population-differences, and method-differences on estimated WTP. This model is then interpreted as a model of the underlying process that is supposed to have generated all the benefit estimates in the literature — that is as a meta-model. Thus, it can be used to predict the benefit estimate that would be arrived at by a new study for which we know the values of the appropriate independent variables, that is it can be used to predict average WTP conditional on these independent variables. This only requires assuming that the same basic underlying meta-model applies.[13]

There are two possible strategies to use meta-models of this type for benefit transfer purposes: (1) one is using a model such as meta-model 1, estimated from a vast sample, including estimates for a broad range of different landscape changes, policies and populations; and (2) the other is using models such as meta-model 2, estimated from smaller, more homogeneous, sub-samples including only landscape changes, policies and populations of a particular type, better matching the policy evaluation problem. The first strategy has the advantage that, as more things vary (and vary more) across estimates, it is possible to estimate the effects on WTP of more factors, hence leading to models accounting for more independent variables. It has, however, the limitation that model parameters estimated for this broader class of studies are probably not accurate estimates of the same parameters for the more homogenous subset of studies that are of interest to us. Strategy (2) circumvents this problem, but at the cost of accounting for less factors: note the fact that meta-model 1 has 16 independent variables while meta-model 2 has only 7 variables (table 3).

The combined effect of these limitations and advantages on the accuracy of benefit transfers is, in general, undetermined and needs to be empirically assessed. Hence, in this chapter, we use both strategies for comparative and assessment purposes.

6.2.1. Using the meta-model estimated from all available CV landscape studies To predict the WTP value to be transferred to the Pennine Dales ESA case based on meta-model 1, we simply evaluated the vector of 16 independent variables (data vector) characterising this case.[14] Note that these independent variables could be evaluated without any field data collection at the Pennine Dales (the policy site). So the rough definition of policy, population and method variables in our meta-model (most of them dummies), though causing some error-in-variables problems, presents a clear practical advantage: no additional information is required to predict with the model. Only the original studies in the literature and the meta-analysis of these studies are required to carry out a transfer.

According to the procedure used for transferring averages of several studies, we also produced here a separate benefit transfer (model prediction) for each of the two main CV elicitation formats.

As with the transfer of the Peneda-Gerês WTP model, we also used the variance-covariance matrix of the meta-model, combined with the data vector, to produce the variance estimates used to build confidence intervals for the benefit transfers.

For the OE format the transfer estimate is £33.83 per household per year (after conversion from 1996 to 1995 prices); the 95-per cent confidence interval is [£26.39; £43.38]. For the DC format, the benefit transfer estimate is £87.30 per household per year, with a 95-per cent confidence interval of [£64.34; £118.46].[15]

6.2.2. Using the meta-model based on CV estimates of ESA-like policies only To predict the WTP figure to be transferred to our policy case, based on meta-model 2, we simply combined this model with the vector of independent variables characterising the Pennine Dales ESA scheme. Again, we produced here separate benefit transfers for the two CV formats and built confidence intervals for the benefit transfers.

Hence, for the OE format, the benefit transfer estimate is £26.72 per household per year, with a 95-per cent confidence interval of [£20.75; £34.40]. For the DC format, it is £88.06 per household per year, and the 95-per cent confidence interval is [£61.75; £125.59].

7. HOW TRANSFERABLE ARE LANDSCAPE VALUES?

In this section, the several sources and procedures for benefit transfer considered so far in this chapter are assessed against three criteria. First, convergent validity; second, the practical importance of transfer errors; and, third, the value of the information an original study at the policy site would add to that available from a transfer.

7.1. Convergent Validity

All eleven transfers estimated in this chapter are presented in Table 7 in terms of both point estimates and 95 per cent confidence intervals. Assuming that 'true' WTP is known with certainty and equal to the average benefit estimate from the Pennine Dales ESA original study, point and interval predictions for benefit transfers are also represented as percentages of that 'true' value. This makes easier (1) to assess the per cent deviation of each transfer from the 'true' value (transfer error); and (2) to check whether this 'true' value is inside the 95 per cent confidence intervals for the several transfers.

Although we have more confidence in a (good) original study than in a transfer, it is impossible to know if the former is actually closer to true WTP than transfers, as we have no access to true WTP. So, to be accurate, what we are evaluating here is whether the transfer estimates and the original estimate at the policy site, both supposedly measuring the same value concept, actually converge, i.e.: we are assessing convergent validity.

Table 7. Point and interval benefit-transfer estimates

Type of transfer		in 1995£s			% of Original	
	LL	Mean	UL	LL	Mean	UL
Single-best-study approach						
Similar site, similar policy, different method						
Unadjusted scalar	23,30	29,04	34, 79	20,8	25,9	31, 0
Adjusted scalar	76,79	95,73	114, 66	68,5	85,3	102, 2
Different site, different policy, similar method						
Unadjusted scalar	52,11	56,48	60, 86	46,4	50,3	54, 2
Adjusted scalar	83,13	90,11	97, 08	74,1	80,3	86, 5
Valuation function	61,74	67,52	73, 31	55,0	60,2	65, 3
Transferring multiple studies						
Transferring multiple-study averages						
Average of OE estimates	16,05	20,67	25, 28	14,3	18,4	22, 5
Average of DC estimates	54,15	68,46	82, 77	48,3	61,0	73, 8
Transferring a meta-analytical model						
Meta-model 1 - OE format	26,39	33,83	43, 38	23,5	30,2	38, 7
Meta-model 1 - DC format	64,34	87,30	118, 46	57,4	77,8	105, 6
Meta-model 2 - OE format	20,75	26,72	34, 40	18,5	23,8	30, 7
Meta-model 2 - DC format	61,75	88,06	125, 59	55,0	78,5	111, 9

Notes: LL are lower limits of 95 per cent confidence intervals and UL, upper limits.

From Table 7 it is possible to conclude that the most accurate transfers, and the only ones whose confidence intervals include 'true' WTP, are: (1) the scalar transfer from the Willis and Garrod's (1991) Yorkshire Dales study *adjusted for the DC format*; and (2) both of the two transfers from meta-analytic models (i.e. using meta-models 1 and 2) *when predicting for the DC format*. Transfer error is, in these cases, in the 15–22% range.

Note that the scalar transfer from the Santos' (1997) Peneda-Gerês source-study adjusted for the income-difference alone clearly outperforms the transfer of the WTP model from the same source-study. This model would allow for finer adjustments of many other factors in addition to income. Thus, this result suggests the superior performance of meta-analytic-based adjustment factors — drawing on inter-study variation of WTP estimates — as compared to the transfer of study-specific models. Probably, many study-specific models of intra-study variation of WTP are simply not transferable for the inter-study context.

The worse transfer procedures were all of those implying an OE elicitation format. These are: (1) the unadjusted scalar transfer from the Willis and Garrod's (1991) Yorkshire Dales study; (2) the average of all OE estimates of ESA-like policies; and (3) both of the two transfers of meta-analytic models (i.e. using meta-models 1 and 2) *when predicting for the OE format*. Note, however, that all these transfers produced benefit estimates rather close to each other, with most confidence intervals overlapping. So, while performing badly to predict a DC estimate, OE-based transfers are quite good transfer-sources for each other. This points to a potentially embarrassing problem: transfers provide quite reasonable predictions of true WTP for landscape conservation, but these predictions strongly depend on selecting the 'right' elicitation format, i.e.: the one yielding 'true' WTP. Note that this is not a 'transfer problem' but one involving CV as a whole, because the 'right' elicitation format is also a problem for original CV studies. This is not an absolute limitation of CV if we have reasons to believe that one of the formats is preferable, e.g. on incentive compatibility grounds (Carson et al. 1999).

Earlier proponents of meta-analyses of non-market valuation studies were rather reluctant about using meta-models as direct predictive devices. Thus, another important implication of these results is the good performance of both meta-models when directly used for predicting WTP values for the particular policy context at hand. They provided two of the three best transfers for the Pennine Dales ESA scheme – the third being derived from a CV study carried out in the same area 5 years earlier, a type of source-study that is usually unavailable in practice.

To reinforce this positive assessment of meta-models used as direct predictive devices, we mention that elsewhere (Santos 1998), meta-model 1 was used to predict all of the 66 landscape benefit estimates in the reviewed literature. This was done by, first, estimating the model with all observations from a particular survey dropped from the data; second, using the estimated model to predict the particular benefit estimates that had been dropped. Comparing these predictions with the dropped estimates, we concluded that 44 per cent of the estimates could have been predicted without the original survey, with an error of up to 30 per cent. If one accepts a deviation of up to 50 per cent (still within the bounds set by Cummings et al. 1986 and Mitchell and Carson 1989), 75 per cent of the estimates could have been accurately predicted without the original survey.

Summarily, our transfer results support, in this particular case, the following conclusions:

- first, all transfers based on DC data, predicting for the DC format, or, when based on OE data, adjusted for the DC format using meta-analytic adjustment factors performed rather well in predicting a DC original estimate;
- second, if one aims at a really small and non-significant (in statistical terms) error, the only options are either transferring from a study of practically the same policy and same population, or using meta-model predictions (always adjusting for the elicitation format);
- third, a source-study of a not-very-different policy, different population and using similar method might provide a rather accurate transfer if adjusted for income using a meta-analytic-based factor;
- the next best options are either transferring a valuation function from a study of a not-very-different policy with a different population but using similar method, or transferring a simple average of multiple not-very-different studies (using the same CV format);
- transferring a scalar from a source-study of a not-very-different policy, different population and using similar method, without adjustments, performed slightly worse;
- predicting with meta-models estimated from a smaller data set, which better matches the policy context, such as meta-model 2, does not outperform predictions from a more general meta-model of all landscape changes, such as meta-model 1.

7.2. Practical Importance of Transfer Error

What is practically important as regards transfer error is not whether it is statistically significant, but whether it really makes a difference for policymakers, that is whether it leads to wrong policy recommendations. How large transfer error must be, in order to be of practical importance, depends on the degree of precision required by the evaluation problem. Different such problems, e.g. screening studies to prepare a full cost-benefit analysis as compared to a study aimed at determining compensation required in a court case, have rather different precision requirements (Desvousges et al. 1998).

Fortunately, we have two types of information allowing us to determine the level of precision required by the evaluation of the Pennine Dales ESA scheme: (1) a good estimate of the visiting population, over which to aggregate our per-household benefit estimates; and (2) two ballpark estimates of the social cost of this scheme, including administrative costs (Santos 1998). Dividing policy costs by the number of visiting households, we secure two figures that are directly comparable to the benefit estimates in Table 7,that is a lower bound of £15.50 per visiting household per year, and an upper bound of £28.37.

Table 8 presents the benefit/cost ratios secured by dividing all of our (original and transfer) benefit estimates for the Pennine Dales ESA scheme by these two social-cost ballparks. For each benefit estimate, we have six benefit/cost ratios, that is three benefit figures (mean WTP and the 2 limits of the confidence interval) for each of the two cost ballparks. Particularly important is the ratio corresponding to the lower limit of the confidence interval for benefits. If this is larger than one, policy makers should proceed with the scheme if they are prepared to accept a 2.5 per cent probability of making the wrong decision. Assuming this is acceptable, only benefit/cost ratios based on lower limits of confidence intervals are discussed in what follows.

Table 8. Point and interval estimates of benefit/cost ratios for the original and transfer benefit estimates

Type of transfer		Benefit/Cost ratio		
		LL	Mean	UL
ORIGINAL benefit estimate	LB	6,52	7,24	7,96
	UB	3,56	3,95	4,35
Single-best-study approach				
Similar site, similar policy, different method				
Unadjusted scalar	LB	1,50	1,87	2,24
	UB	0,82	1,02	1,23
Adjusted scalar	LB	4,96	6,18	7,40
	UB	2,71	3,37	4,04
Different site, different policy, similar method				
Unadjusted scalar	LB	3,36	3,64	3,93
	UB	1,84	1,99	2,14
Adjusted scalar	LB	5,36	5,81	6,26
	UB	2,93	3,18	3,42
Valuation function	LB	3,98	4,36	4,73
	UB	2,18	2,38	2,58
Transferring multiple studies				
Transferring multiple-study averages				
Average of OE estimates	LB	1,04	1,33	1,63
	UB	0,57	0,73	0,89
Average of DC estimates	LB	3,49	4,42	5,34
	UB	1,91	2,41	2,92
Transferring a meta-analytical model				
Meta-model 1 - OE format	LB	1,70	2,18	2,80
	UB	0,93	1,19	1,53
Meta-model 1 - DC format	LB	4,15	5,63	7,64
	UB	2,27	3,08	4,18
Meta-model 2 - OE format	LB	1,34	1,72	2,22
	UB	0,73	0,94	1,21
Meta-model 2 - DC format	LB	3,98	5,68	8,10
	UB	2,18	3,10	4,43

Notes: LL are lower limits of 95 per cent confidence intervals and UL, upper limits; LB is the lower bound ballpark for social cost and UL, the upper bound.

Using the original estimates of the Pennine Dales ESA survey, the ESA scheme clearly passes the cost-benefit test. The same happens if we use any of the benefit transfers based on the DC format, adjusted for this format or predicting for this format. Different conclusions are drawn if we use the benefit transfers based on the OE format or predicting for this format: using upper ballparks for costs, all benefit/cost ratios now drop below 1.00.

Again, we have the uncomfortable feeling that, in this case, the only difference that would actually matter in practice is the difference between the two CV formats – the most important single factor to explain differences between transfers. Note, however, that, in cases where benefits are much closer to costs than in the Pennine Dales ESA, adjusting for WTP differences much smaller than that due to different CV formats (e.g. WTP differences due to different recreation activities or landscape changes) will certainly matter in practice.

7.3. The Value-of-Information Test

The ultimate test for a transfer (with given accuracy) is whether the added information that would be acquired through an original valuation study at the policy site (or a better transfer) would justify carrying out such an original (or better transfer) study. This test is conveniently framed as a simple question (cf. e.g. Desvousges et al. 1998): does the difference between the expected value of a policy based on an original (or better transfer) study and the expected value of a policy based on the particular transfer procedure exceeds the added costs of carrying out the original (or better transfer) study as compared to those of the particular transfer study at stake?

In this final section, we compute the expected net-benefit values of policies for the Pennine Dales ESA based on three types of benefit estimates:

- the original Pennine Dales ESA study's estimate;
- the unadjusted scalar transferred from the Peneda-Gerês study (a reasonable transfer, but far from best; in fact, it is the worse using the DC format), thereafter called type I transfer;
- the transfer based on the prediction of meta-model 1 for the OE format (a bad transfer, though not the worse), thereafter called type II transfer.

In this exercise, we make some simplifying assumptions: (1) 'true' WTP is equal to the sample average from the original study; (2) policymakers decide according to benefit/cost ratios based on average benefit estimates; and (3) only upper bounds for costs are relevant for both actual net-benefit calculation and policy decision-making.

With all these assumptions, the 'true' net benefit of keeping the Pennine Dales ESA scheme is estimated as circa £14 000 000 per year. The 'right' decision obviously is continuing with the policy. What are the probabilities of replicated original studies, type I and type II transfers leading to decide not to go ahead with the policy (wrong decision)? This decision would require that the benefit estimate on a per-household basis drops below £28.37.

What is the probability of this happening with a replicated original study? Assuming normality, this probability is that of the standard normal variable being smaller than -14.8, that is practically zero. Therefore, replicating an original study will always lead to the right decision, and thus the expected value of the net benefits is circa £14 000 000 per year.

The required probability for a type I transfer is also very close to zero, which happens here with all transfers associated with the DC format. Hence, the insignificant difference between the expected net benefit of a decision based on this transfer and that of a decision based on an original study does not justify carrying out an original study.

What about this probability for a type II transfer? For this type of transfer, there is a 8.4 per cent probability of a decision of not going ahead with the policy. Thus, a policy based on this transfer will have an expected net benefit of $0.916 \times$ £14 000 000 = £12 827 000 per year. The difference for a policy based on an original study (or a type I transfer) is £1 173 000 per year, which offsets any reasonable cost estimate for an original study (and, a fortiori, for a type I transfer if possible). Thus, carrying out an original study (or a type-I transfer if possible) is justified in this case.

Note that, of course, all of these results rely on the strong assumption that the 'true' WTP value coincides with the average benefit observed in the original study. Adopting a Bayesian approach (see e.g. Atkinson et al. 1992) allows the analyst to ignore what the 'true' WTP value is; by assuming all studies (transfers and the original one) are sampling from the same mother distribution, Barton (1999) uses a sequential Bayesian procedure to update previous valuation information as new information becomes available. This enables the analyst to select the optimal level of information collection.

Furthermore, the above recommendations about whether to carry out an original study or a transfer are only valid for the present context, in which benefits (except if using the OE format) systematically offset costs by several times. In other contexts, carrying out an original study may be justified in terms of the expected value of the gain of information so provided for the decision-maker. Therefore, more applied research in this crucial area, under different cost-benefit contexts, is needed.

José Manuel Lima E Santos is Professor, Department of Agricultural Economics and Rural Sociology, Technical University of Lisbon, Portugal.

8. NOTES

[1] Namely, Mitchell and Carson's (1989) Appendix A, Bonnieux et al. (1992), Römer and Pommerehne (1992), Mäntymaa et al. (1992), Hoevenagel et al. (1992), Navrud and Strand (1992), Johansson and Kriström (1992), Turner et al. (1992), Bennett (1992), Shechter (1992) and Bateman et al. (1994).

[2] Such as Dubgaard & Nielsen (1989), Hanley (1991), and Whitby (1994).

[3] This list possibly misses dissertation material, conference papers and recent research still waiting for publication at that date. Nine of the studies in our final list are journal articles; two are published research reports; three, unpublished research reports; one, a conference paper; and four report on recent research, still unpublished. The studies have been carried out in seven different countries (Sweden, UK, Austria, France, Spain, Portugal and the USA) and include: studies commissioned by the UK's Ministry of Agriculture Fisheries and Food; research projects funded by the EU or by national conservation agencies; and research independently carried out at several universities.

[4] Most of these WTP estimates have been originally elicited on a per-year per-household basis; only three required an adjustment to this basis (Santos 1997). All WTP estimates have been converted into 1996 £, using the 1996 consumer-price index for the respective country and an appropriate exchange rate against the sterling. Price indices for the seven countries were built from ONS (1997a) and UN (1995); average exchange rates for the years of 1994–1996 (ONS 1997b) were used, to avoid that an excessive weight was given to the increase in the value of the sterling over the last months of 1996.

[5] From an analytical point of view, variation in WTP estimates is a desirable fact if it can be associated with variation in some independent variables. Indeed, an important condition for using regression analysis as an analytical tool is that dependent and independent variables vary.

[6] For example not all studies reported sample averages for household income, which led us to take GDP per capita as a proxy.

[7] Note this is not ad hoc exclusion on quality grounds: a clear relationship between sample size and sampling variance was empirically established for this set of studies by Santos (1997), and these observations corresponded to particularly low-precision estimates.

[8] As well as a similar survey of residents, which is not relevant for the purposes of this chapter, and thus is not commented here. Cf. also Willis and Garrod (1992).

[9] I.e.: including not only the dales in Yorkshire, within the NP, but also those in the North Pennines (Cumbria, Durham and Northumberland counties), outside the NP.

[10] The price indexes used thereafter for adjusting current values for the inflation were based on the consumer price index for the respective country (ONS 1997a and UN 1995).

[11] Our thanks to Ken Willis and Guy Garrod for useful information on many other aspects of their questionnaire.

[12] Values in currencies other than £s were converted into 1996 £s using the appropriate exchange rate against the sterling (average of 1994/96 rates; ONS 1997b); conversion into 1995 £s used the ratio of 1996 to 1995 consumer price indexes (ONS 1997a).

[13] Of course, as with averages, random-effects regression models would be a more convenient description than the equal-effects ones used here (a fixed-effects model was impossible to estimate with so few observations from some of the studies). These more sophisticated approaches will be the subject of further research on these transfer issues.

[14] This data vector is: conservation or recreational site = 0; ESA or NP = 1; unique = 0; Nation-wide scope = 0; considerable change in character = 0; moderate/heavy development = 0; substitution effects = 0; NIMBY = 0; mainly non-users = 0; foreignvisitors = 0; log(income) = 9.80; DC and IB elicitation formats = 0 or 1 (depending on the CV format we are predicting for); level of trimming = 0; truncation of DC data = 0; tokens technique = 0; year = 1995.

[15] Note that these predictions were obtained by raising e to the predicted log (WTP) and thus represent medians, not means (obtaining means for a log-normal distribution such as this would require us to multiply by the moment-generating function of the normal distribution.). However, as the WTP data for the regression model are, by definition, sample averages of WTP, what we have here is a median of averages. That is: is a prediction of the WTP sample average that will be exceeded in half of the studies to which the meta-model applies.

9. REFERENCES

Atkinson SE, Crocker TD Shogren JF (1992) Bayesian Exchangeability, Benefit Transfer and Research Efficiency. Water Resources Research 28(3):715–722

Barton, David (1999) The Quick, the Cheap and the Dirty. Benefit Transfer Approaches to the Non-Market Valuation of Coastal Water Quality in Costa Rica. PhD Thesis submitted to the Agricultural University of Norway

Bateman I., Langford I, Willis K, Turner R, Garrod G (1993). The Impacts of Changing Willingness to Pay Question Format in Contingent Valuation Studies: An Analysis of Open-Ended, Iterative Bidding and Dichotomous Choice Formats. CSERGE Working Paper GEC 93–05. Norwich: University of East Anglia

Bateman I, Willis K,Garrod G (1994) Consistency between Contingent Valuation Estimates: A Comparison of Two Studies of UK National Parks. Regional Studies 28(5):457–474

Bateman I, Willis K, Garrod G, Doktor P, Langford I,Turner R (1992) Recreation and Environmental Preservation Value of the Norfolk Broads: A Contingent Valuation Study. CSERGE, University of East Anglia (Unpublished Research Report)

Beasley S, Workman W,Williams N (1986) Estimating Amenity Values of Urban Fringe Farmland: A Contingent Valuation Approach. Growth and Change 17:70–78

Bennett J (1992) Starting to Value the Environment: The Australian Experience. In: Navrud S (ed.) Pricing the European Environment. Oxford University Press, Oxford, 247–257

Bergstrom J, Dillman B, Stoll J (1985) Public Environmental Amenity Benefits of Private Land: the Case of Prime Agricultural Land. Southern Journal of Agricultural Economics 17:139–149

Bonnieux F, Desaigues B, Vermersch D (1992) France. In: Navrud S (ed.) Pricing the European Environment. Oxford University Press, Oxford, 45–64

Boyle K, Bergstrom J (1992) Benefit Transfer Studies: Myths, Pragmatism, and Idealism. Water Resources Research 28(3):657–663

Bullock C, Kay J (1996) Preservation and Change in the Upland Agricultural Landscape – Valuing the Benefits of Changes Arising from Grazing Extensification. Aberdeen: MLURI (Internal Paper quoted by Hanley et al. 1996)

Cameron T (1991) Interval Estimates for Non-Market Resource Values from Referendum Contingent Valuation Surveys. Land Economics 67(4):413–421

Cameron T (1988) A new paradigm for valuing non-market goods using referendum data: maximum likelihood estimation by censored logistic regression. Journal of Environmental Economics and Management 15:355–379

Campos P, Riera P (1996) Social Returns of the Forests: Analysis Applied to Iberian Dehesas and Montados. In: Pearce D (ed.) The Measurement and Achievement of Sustainable Development. Research Report (Project CT 94–367, DG XII Environmental Programme)

Carson, Richard T, Theodore Groves, Mark J. Machina (1999) Incentive and Informational Properties of Preference Questions. Plenary Address to the 9th Annual Conference of the European Association of Resource and Environmental Economists, Olso, Norway (June 1999)

Cummings R, Brookshire D, Schulze W (eds) (1986) Valuing Environmental Goods. An Assessment of the Contingent Valuation Method. Rowman & Allanheld, Totowa

Desvousges WH, Johnson FR, Banzhaf HS (1998) Environmental Policy with Limited Information. Principles and Applications of the Transfer Method. Edward Elgar, Cheltenham

Dillman B, Bergstrom J (1991) Measuring Environmental Amenity Benefits of Agricultural Land. In: Hanley N (ed.) Farming and the Countryside: An Economic Analysis of External Costs and Benefits. CAB International, Wallingford, UK, 250–271

Downing M, Ozuna T Jr (1996) Testing the Reliability of the Benefit Function Transfer Approach. Journal of Environmental Economics and Management 30(3):316–322

Drake L (1992) The Non-market Value of the Swedish Agricultural Landscape. European Review of Agricultural Economics 19:351–364

Drake L (1993) Relations among Environmental Effects and their Implications for Efficiency of Policy Instruments – an Economic Analysis Applied to Swedish Agriculture. (Published dissertation) Upsala: Dept. of Economics: Swedish University of Agricultural Sciences

Dubgaard A, Nielsen A (eds) (1989) Economic Aspects of Environmental Regulations in Agriculture. Wissenschaftsverlag Vauk, Kiel

European Commission (1998) State of Application of Regulation (EEC) No. 2078/92: Evaluation of Agri-Environment Programmes. DGVI Commission Working Document VI/7655/98

Glass G (1976) Primary, Secondary, and Meta-Analysis of Research. Educational Researcher 5:3–8

Glass G, McGaw B, Smith M (1981) Meta-analysis in Social Research. Sage Publications, Beverly Hills, California

Gourlay D (1995) Quoted in Hanley et al. (1996)

Halstead J (1984) Measuring the Nonmarket Value of Massachusetts Agricultural Land: A Case Study. Journal of Northeastern Agricultural Economic Council 13:12–19

Halvorsen R, Palmquist R (1980) The Interpretation of Dummy Variables in Semilogarithmic Equations. The American Economic Review 70(3):474–475

Hanley N (ed.) (1991) Farming and the Countryside: An Economic Analysis of External Costs and Benefits. CAB International, Wallingford, UK

Hanley N, Craig S (1991) Wilderness Development Decisions and the Krutilla–Fisher Model: the Case of Scotland's 'Flow Country'. Ecological Economics 4:145–164

Hanley N, Munro A, Jamieson D (1991) Environmental Economics, Sustainable Development and Nature Conservation. Report to the Nature Conservancy Council, Peterborough, England

Hanley N, Simpson I, Parsisson D, Macmillan D, Bullock C, Crabtree B (1996) Valuation of the Conservation Benefits of Environmentally Sensitive Areas. A Report for Scottish Office Agriculture, Environment & Fisheries Department. Macaulay Land Use Research Institute Aberdeen (Economics and Policy Series no 2)

Hoevenagel R, Kuik O, Oosterhuis F (1992) The Netherlands. In: Navrud S (ed.) Pricing the European Environment. Oxford University Press, Oxford, 100–107

Johansson P-O, Kriström B (1992) Sweden. In: Navrud S (ed.) Pricing the European Environment. Oxford University Press, Oxford, 136–149

Loomis J (1992) The Evolution of a More Rigorous Approach to Benefit Transfer: Benefit Function Transfer. Water Resources Research 28(3):701–705

Loomis J, White DS (1996) Economic Benefits of Rare and Endangered Species: Summary and Meta-Analysis. Ecological Economics 18:197–206

Mäntymaa E, Ovaskainen V, Sievänen T (1992) Finland. In: Navrud S (ed.) Pricing the European Environment. Oxford University Press, Oxford, 84–99

Mitchell R, Carson R (1989) Using Surveys to Value Public Goods: the Contingent Valuation Method. Resources for the Future, Washington DC

Navrud S, Strand J (1992) Norway. In: Navrud S (ed.) Pricing the European Environment. Oxford University Press, Oxford, 108–135

Newey W, West K (1987) A Simple, Positive Semi-definite, Heteroskedasticity and Autocorrelation Consistent Covariance Matrix. Econometrica 55(3):703–708

ONS – Office for National Statistics (1997a) Retail Prices Index – November 1996. The Stationery Office, London

ONS – Office for National Statistics (1997b) Financial Statistics. No. 418 (Feb. 1997). London: The Stationery Office

PA Cambridge Economic Consultants (1992) Yorkshire Dales Visitor Study 1991. Study carried out on behalf of Yorkshire and Humberside Tourist Board, Yorkshire Dales National Park Committee, and Craven District Council

Pruckner G (1995) Agricultural Landscape Cultivation in Austria: An Application of the CVM. European Review of Agricultural Economics 22:173–190

Römer A, Pommerehne W (1992) Germany and Switzerland. In: Navrud S (ed.) Pricing the European Environment. Oxford University Press, Oxford, 65–83

Santos JML (1997) Valuation and Cost–Benefit Analysis of Multi-Attribute Environmental Changes. Upland Agricultural Landscapes in England and Portugal. PhD thesis. University of Newcastle upon Tyne, Newcastle

Santos JML (1998) The Economic Valuation of Landscape Change. Theory and Policies for Land Use and Conservation. Edward Elgar Publish, Cheltenham

Shechter M (1992) Israel – An Early Starter in Environmental Pricing. In: Navrud S (ed.) Pricing the European Environment. Oxford University Press, Oxford, 258–273

Smith VK, Huang J (1995) Can Markets Value Air Quality? A Meta-Analysis of Hedonic Property Value Models. Journal of Political Economy 103(1):209–27

Smith VK, Kaoru Y (1990). Signals or Noise? Explaining the Variation in Recreation Benefit Estimates. American Journal of Agricultural Economics 72(2):419–433

Stenger A, Colson F (1996) Interpretation of an Application of the Contingent Valuation Method to Agricultural Landscapes: the Problem of Embedding Effects. Paper presented at the 7th Annual Conference of the European Association of Environmental and Resource Economists: Lisbon, June 27–29th

Turner R, Bateman I, Pearce D (1992) United Kingdom. In: S. Navrud (ed.) Pricing the European Environment. Oxford University Press, Oxford, 150–176

UN – United Nations (1995) Statistical Yearbook (40th Issue). United Nations New York

UN – United Nations (1997) Monthly Bulletin of Statistics 51 (1: Jan.)

Walsh R, Johnson D, McKean J (1992) Benefit Transfer of Outdoor Recreation Demand Studies, 1968–1988. Water Resources Research 28(3):707–713

Whitby M (ed) (1994) Incentives for Countryside Management. The Case of Environmentally Sensitive Areas. CAB International, Wallingford, UK

Willis KG (1982) Green Belts: An Economic Appraisal of a Physical Planning Policy. Planning Outlook 25(2):62–69

Willis KG (1990) Valuing Non-market Wildlife Commodities: An Evaluation and Comparison of Benefits and Costs. Applied Economics 22:13–30

Willis KG, Garrod G (1991) Landscape Values: A Contingent Valuation Approach and Case Study of the Yorkshire Dales National Park. Countryside Change Working Paper 21. University of Newcastle upon Tyne. Dept. of Agricultural Economics and Food Marketing

Willis KG, Garrod G (1992) Assessing the Value of Future Landscapes. Landscape and Urban Planning 23:17–32

Willis KG, Garrod G, Saunders C (1993a). Valuation of the South Downs and Somerset Levels and Moors Environmentally Sensitive Area Landscapes by the General Public. Research Report to the MAFF. Newcastle: Centre for Rural Economy, University of Newcastle upon Tyne

Willis KG, Garrod G, Saunders C, Whitby M (1993b) Assessing Methodologies to Value the Benefits of Environmentally Sensitive Areas. Countryside Change Initiative Working Paper 39. Department of Agricultural Economics and Food Marketing: University of Newcastle upon Tyne

Willis KG, Whitby M (1985) The Value of Green Belt Land. Journal of Rural Studies 1(2):147–162

Wolf F (1986) Meta-analysis. Quantitative Methods for Research Synthesis. Sage Publications, Beverly Hills

R. READY AND S. NAVRUD

MORBIDITY VALUE TRANSFER

1. INTRODUCTION

In many environmental regulation contexts, an important category of impacts from regulation is impacts on public health. These can include impacts on both rates of mortality and rates of morbidity in the affected population. Indeed, for regulations aimed at improving air or water quality, health benefits can be the dominant category of impacts in a regulatory impact analysis. For example, in a prospective cost-benefit analysis of the 1990 Clean Air Act Amendments (US EPA 1999), decreases in mortality and morbidity from improved air quality constituted over 95% of the total estimated benefit.

When compared to valuation of other environmental goods such as outdoor recreation, scenic quality, wilderness, and wildlife populations, the approach typically used to value improvements in health resulting from improved environmental quality is somewhat unique, in that it relies heavily on unit values and value transfer.[1] The typical approach when valuing environmental health improvements from a proposed action is to follow the damage function approach, which is discussed in the first chapter of this volume. First, projected changes in exposure to pollutants are combined with established exposure-response relationships. This type of analysis gives predictions of how many ill health outcomes would be avoided as a result of the action.[2] These improvements in public health are then valued by multiplying the number of each type of ill health outcome avoided by a constant value specific to each outcome.[3]

The focus of this chapter is on the third step in this approach, multiplication of the number of ill health outcomes to be avoided by an outcome-specific unit value per incidence. Three categories of value are generally considered: (1) the social costs of providing medical treatment to the victim of the ill health outcome; (2) lost labor productivity resulting from the ill health outcome; and (3) the pain, discomfort, and inconvenience suffered by the victim. Per-incidence estimates of the first category of these costs are assembled from hospital records, records of visits to doctors' offices, records of prescription medication use, and surveys of victims of their out-of-pocket health care costs. Per-incidence estimates of lost productivity are usually based on the hourly wages paid to the victim, relying on the theoretical assertion that wages should reflect the marginal value of the victim's labor to his or her employer.

Estimation of the third category of value, the pain, discomfort, and inconvenience suffered by the victim, is more problematic, because there are few market prices or financial records that will reveal this value. Instead, the usual approach is to use stated preference techniques such as contingent valuation or stated choice

S. Navrud and R. Ready (eds.), Environmental Value Transfer: Issues and Methods, 77–88.

approaches to estimate the victim's willingness to pay (WTP) to avoid an ill health outcome.[4]

What makes public health valuation unique among situations where nonmarket valuation techniques are applied is the implicit assumption that all cases of an ill health episode have the same value. In particular, it is usually assumed that the value of an ill health outcome does not depend on (1) the cause of that ill health outcome (so that, for example, a day suffering from itchy eyes and a stuffy nose caused by air pollution is valued the same as a similar episode caused by contaminated water at a swimming beach, (2) whether individuals in the population will avoid at most one incidence of an ill health outcome, or whether some individuals will avoid more than one (so that, for example, the value to an individual of avoiding 7 incidences of ill health is 7 times the value of avoiding one incidence), and (3) the health status of the individuals who will enjoy improved health (so that the value of avoiding an incidence of ill health to a person with chronic health problems is the same as the value to a person who rarely experiences ill health).

In contrast, for most other environmental goods, it is generally believed that the context of the good is critically important in determining its value. The marginal value of improving water quality in a lake depends on how many lakes will be protected. Oil pollution from a tanker spill is valued differently from oil pollution originating from natural seeps. In public health valuation, the issues of context and scale are typically assumed away.

The purpose of this chapter is to review available evidence on the validity of using constant per-episode and per-case values when valuing changes in public health due to changes in environmental quality. A second issue that will be explored is the validity of transferring health values estimated in one geographic region to an analysis conducted in another region. Relatively few environmental health valuation studies have been conducted, especially outside the U.S. Health values are routinely transferred between countries, with little guidance on how values might differ due to differences in health status, socioeconomic conditions, or culture.

2. VALUING ONE EPISODE VERSUS MANY EPISODES

At least for less-serious ill health outcomes, it is common practice in stated preference studies valuing health to value a discrete, marginal change in the number of episodes or cases of ill health that the respondent will experience, rather than valuing a change in risk of ill health. This approach is clearly unrealistic – future health cannot be guaranteed. Further, a risk-free treatment that focuses on health outcomes, rather than on risks, does not allow consideration of potential changes in defensive actions that the respondent might take, such as limiting activity during periods of poor air quality. On the other hand, valuing changes in risk imposes difficulties on both the respondent and the researcher. For this reason, most morbidity valuation studies have measured WTP to avoid, with certainty, one or more specific episodes or cases of ill health.

While exposure-response studies may tell us how many fewer hospital admissions and minor symptom days will occur as a result of an improvement in environmental quality, they usually do not predict how these avoided outcomes will be distributed within the affected population. For many environmental health issues, there is an at-risk subpopulation that suffers a disproportionate share of the total number of ill health outcomes. For example, asthma attacks are concentrated among those who have asthma. The health improvement that results from an improvement in environmental quality will likewise be concentrated within the susceptible subpopulation. Individual sufferers who benefit may avoid more than one episode or case as a result of the policy action. Does the value of avoiding a single episode of ill health depend on how many episodes the individual will avoid?

The evidence is that it does. Tolley et al. (1994) valued avoidance of one additional day of suffering from seven different symptoms that can be caused by pollution, and avoidance of 30 days of suffering from the same symptoms. While WTP to avoid 30 additional days was uniformly higher than WTP to avoid one day, the ratio of the two was only between 7 and 10 to 1. Similarly, Navrud (2001) found that avoidance of WTP to avoid 14 additional days of minor symptoms was 3 to 5 times as large as WTP to avoid 1 additional day. These results would seem to suggest that the marginal benefit from avoiding a symptom day decreases as the number of symptom days avoided increases. Johnson et al (2000), in a study that used paired comparison and stated preference techniques, apply a transformation to duration of illness that assumes diminishing marginal value. However, in a second pair of surveys, Tolley et al. found that WTP to avoid 20 days of severe angina was over three times as large as WTP to avoid 10 days, implying increasing marginal value of duration (or, equivalently, decreasing marginal value of health).

It is not clear, from theoretical grounds, whether WTP to avoid additional symptom days should increase at a less than or greater than proportional rate as the number of additional days to be avoided increases. If health, measured as the number of days in a year the individual does not experience symptoms, is a normal good with decreasing marginal utility, then marginal WTP to avoid an ill health outcome should increase as the number of ill health outcomes the individual will experience increases. However, if health is viewed as something that you either have or do not have, then the marginal disutility of additional ill health may be low, once health status drops below some threshold.

To date, the empirical evidence on whether marginal WTP to avoid ill health increases or decreases as the duration of the ill health increases is mixed, though results consistent with declining marginal disutility of ill health are more common than results consistent with declining marginal utility of good health. Complicating these results is the possibility that the elicitation methods used may be unable to reliably measure how value changes as the scope of the health improvement changes. At a minimum, the evidence to date suggests that it is inappropriate to assume that marginal WTP per outcome avoided is constant regardless of the number of outcomes avoided by each individual.

3. DO PEOPLE WITH POOR HEALTH VALUE HEALTH DIFFERENTLY?

Related to the issue of how many outcomes an individual avoids, is the issue of who in the population avoids the ill health outcomes. If it tends to be persons with poorer health who benefit most from improvements in environmental quality, then it is of interest to know whether marginal WTP to avoid one ill health outcome varies with the individual's health status.

Tolley et al. report conflicting results as to whether health status affects WTP to avoid days of ill health. WTP to avoid one day of minor symptoms was generally positively related to the number of days the respondent experienced those symptoms within the past 12 months, and was negatively related to overall indicators of health. However, WTP to avoid 30 days of minor symptoms or to avoid 10 or 20 days of angina were not related to health status.

Dickie et al. (1987) found that WTP to avoid one day of nine different symptoms that can be caused by ozone exposure was not sensitive to how often respondents experienced the symptoms, or whether respondents were respiratory impaired.

Johnson et al (2000) found that WTP to avoid episodes of respiratory and cardiac ill health was higher for respondents who had been diagnosed with cardiovascular or respiratory conditions, or other serious illness.

Ready et al (2004a) found that WTP to avoid five different episodes of respiratory ill health was significantly positively related to frequency of respiratory symptoms in 6 of 19 regressions, while a significant negative relationship was found in only 1 of the 19 regressions. Further, WTP to avoid the episodes was significantly higher among respondents diagnosed with either asthma or respiratory allergies in 6 of 20 regressions, while no significant negative relationships were found.

To summarize, there are several instances where WTP to avoid ill health outcomes was higher for respondents who suffered from that type of outcome more frequently, or respondents with poorer health measured more generally, while there are very few results that showed the opposite result. We conclude that a weak negative relationship probably exists between health status and WTP to avoid ill health for most ill health outcomes.

4. DOES THE CAUSE OF THE ILL HEALTH MATTER?

Many ill health outcomes that are caused by one type of pollution could also be caused by other types of pollution as well. Nausea, for example, can be caused by air pollution, contaminated drinking water, contaminated swimming beaches, food-borne disease, or by person-to-person transmission of disease. Does the value of avoiding an ill health outcome depend on the cause of that outcome?

Few studies have examined this issue directly. Most environmental health valuation studies are deliberately vague about the cause of the prospective ill health, or the mechanism by which their health would be improved. The fear is that if respondents were told that the health improvement would be delivered by an improvement in environmental quality, they would include in their WTP values the co-benefits (improvements in visibility, ecological services, etc.) that would logically result from the environmental quality improvement, making determination of a value-per-day or a value-per-episode difficult. Indeed, when WTP values measured without reference to the cause of the ill health are compared to WTP values for the same health improvement brought about by an improvement in environmental quality, the latter are found to be larger than the former (Rozan and Willinger 1998).

Ready et al. (2004a) attempted to isolate the impact of the cause of ill health on its value, without confounding the value with consideration of how the improved health would be delivered.[5] In five European countries, WTP to avoid six specific episodes of ill health was measured. Some of these episodes could be caused either by poor air quality or by swimming at contaminated beaches. A split sample design was used, where some respondents were told the cause of the prospective ill health (air pollution or contaminated water) and others were not told the cause. Neither group was told how the ill health would be avoided. Rather, as is common in this literature, respondents were told that by paying a specified sum, they could avoid one episode with certainty. Out of 11 possible tests, none showed a significant difference (at the 10% level) in WTP between the two samples. This result gives some comfort that the common practice of applying per-incidence values, without consideration of the specific cause of the ill health outcome, is valid.

5. TESTING MORBIDITY VALUE TRANSFER AMONG COUNTRIES

Most environmental health valuation studies done to date have been conducted in the United States, though several European studies have been completed more recently. Is it valid to take WTP values for avoided ill health outcomes estimated in one country and use them to value health improvements in a different country (so called unit value transfer)? What types of adjustments should one make when making inter-country value transfers?

The issue that has received the most attention when making inter-country transfers is differences in wealth. If health is a normal good, then WTP for improvements in health should increase with wealth. Indeed, most empirical studies find that within samples of respondents, WTP is positively related to the respondent's income. When using health values estimated in one country (the study country) in a policy analysis in a second country (the target or policy country), it is logical, then, to suppose that WTP should be adjusted to reflect differences in mean income between the two countries. This is of particular importance when transferring unit values estimated in a developed country to a policy analysis in a less-developed country.

There are two common approaches to adjusting WTP values to account for differences in income. First, unit values (WTP to avoid a single incidence of a specific health outcome) from the study country can be adjusted by assuming a constant income elasticity of WTP. A constant income elasticity of 1 would mean that the ratio of WTP to income is the same in the two countries. While an assumed income elasticity of 1 may be intuitive, empirical evidence is that the income elasticity of WTP tends to be positive, but less than one. The second approach is to use value functions estimated in the study country to predict WTP in the target country. This approach, called value function transfer, accounts for differences in not only income, but any other characteristic that was measured for each respondent in the original study, and is measurable in the target country. The value function transfer approach relies on the assumption that the two countries share a common value function.

To test whether any of the three transfer methods (unit value transfer, unit value transfer with adjustment for income differences, or value function transfer) is valid, it is necessary to measure WTP for the same health improvement in two different countries. Alberini et al. (1997) measured WTP to avoid an episode of acute respiratory illness in Taiwan, and compared values for specific ill health outcomes to values previously estimated in the U.S. They transferred U.S.-estimated unit values adjusting for income differences between the U.S. and Taiwan, assuming an income elasticity of 1 or of 0.4, and compared the transferred values to WTP values estimated in Taiwan. They also transferred a value function estimated in Taiwan to predict WTP in the U.S., and compared those predictions to values previously estimated in the U.S. They could not conclusively state whether one of the three approaches outperformed the others, in part because variation in estimated U.S. WTP values was about as large as variation between Taiwan and the U.S. A further complication is that the U.S. studies used different survey instruments, and did not value exactly the same episodes of ill health as were valued in Taiwan.

Similarly, Chestnut et al. (1997) compared WTP to avoid one respiratory illness day estimated in Bangkok, Thailand, with estimates from previous studies conducted in the U.S. They found that, even though average income in Bangkok is about one-quarter that in the U.S., mean WTP was roughly equal in the two countries. Again, interpretation of this result is complicated by the fact that the U.S. and Bangkok studies used different survey instruments.

Ideally, a validity test of value transfer between countries should use the same survey instrument, and value the same outcomes, in both countries. Ready et al. (2004b), estimated WTP to avoid episodes of ill health using the same contingent valuation survey instrument in five different European countries, the Netherlands, Norway, England, Portugal and Spain. The six different episodes valued included two different mild symptom days, a minor restricted activity day, a work-loss day, a bed day, an emergency room visit, and a hospital admission. Table 1 presents brief synopses of the six episode descriptions.[6]

The survey instrument was similar in form to that used by Tolley et al (1994). Respondents were first asked questions about their health status, then asked to rank the episodes in order of severity, then asked their WTP to avoid each episode. Split samples in which the episodes were valued in different order showed no evidence of ordering effects (Ready et al 2004a).

One issue when comparing WTP values from several countries is determining the appropriate exchange rate. Ready et al. (2004b) argue that local currencies should be converted to a common currency using a purchasing-power-parity (PPP) adjusted exchange rate. In the context of the contingent valuation survey, improved health

Table 1. Ill-Health Episode Descriptions

Episode name	Epidemiological end point	Description
EYES (E)	1 Mild Symptom Day	One Day with mildly red, watering, itchy eyes. A Runny nose with sneezing spells. Patient is not restricted in their normal activities.
COUGH (Co)	1 Minor Restricted Activity Day	One day with persistent phlegmy cough, some tightness in the chest, and some breathing difficulties. Patient cannot engage in strenuous activity, but can work and do ordinary daily activities.
STOMACH (S)	1 Work-Loss Day	One Day of persistent nausea and headache, with occasional vomiting. Some stomach pain and cramp. Diarrhea at least twice during the day. Patient is unable to go to work or leave the home, but domestic chores are possible.
BED (B)	3 Bed Days	Three days with flu-like symptoms including persistent phlegmy cough with occasional coughing fits, fever, headache and tiredness. Symptoms are serious enough that patient must stay home in bed for the three days.
CASUALTY (Ca)	Emergency Room Visit for COPD and Asthma	A visit to a hospital casualty department, for oxygen and medicines to assist breathing problems caused by respiratory distress. Symptoms include a persistent phlegmy cough with occasional coughing fits, gasping breathing even when at rest, fever, headache and tiredness. Patient spends 4 hours in casualty followed by 5 days at home in bed.
HOSPITAL (H)	Hospital Admission for, COPD, pneumonia, respiratory disease and asthma	Admission to a hospital for treatment of respiratory distress. Symptoms include persistent phlegmy cough, with occasional coughing fits, gasping breath, fever, headache and tiredness. Patient stays in the hospital receiving treatment for three days, followed by 5 days home in bed.

Note: COPD = Chronic Obstructive Pulmonary Disease

is a market good – it is something that gives positive utility that the respondent can choose to buy at a price. The choice whether to purchase the good depends on the respondent's income, the price of improved health, and on the price of other market goods available to the respondent. If two people have identical underlying preferences, but one faces prices that are uniformly α percent higher than those faced by the other, then their behavior will be identical only if their incomes and the price of improved health also differ by the same proportion.[7] Following this reasoning, all income and WTP values were converted to British Pounds using PPP-adjusted exchange rates.

Mean WTP values for each episode for each country, converted to British Pounds, are shown in Figure 1. As would be expected, WTP is higher for the episodes that are more serious and last longer. The three episodes that only last one day, COUGH, EYES, and STOMACH, have the lowest mean WTP values in every country. Comparing results across countries, Norway and Spain have consistently high WTP compared to the other three countries, while England and the Netherlands have consistently low WTP. Portugal tends to have intermediate WTP values, except for EYES, where it has the highest. These apparent differences are in many cases statistically significant. Pairwise tests of equality show that England has significantly lower WTP than other countries in 14 out of 21 tests, while Spain has significantly higher WTP than other countries in 10 out of 20 tests.

These results are somewhat counterintuitive, given differences in income among the countries. Spain and Portugal have much lower mean real incomes than the three Northern European countries, yet these two countries generally have intermediate to high WTP values relative to the other countries. However, several other differences exist among the countries that have relevance for health valuation (education, family size, current health status). To control for these differences, value functions were

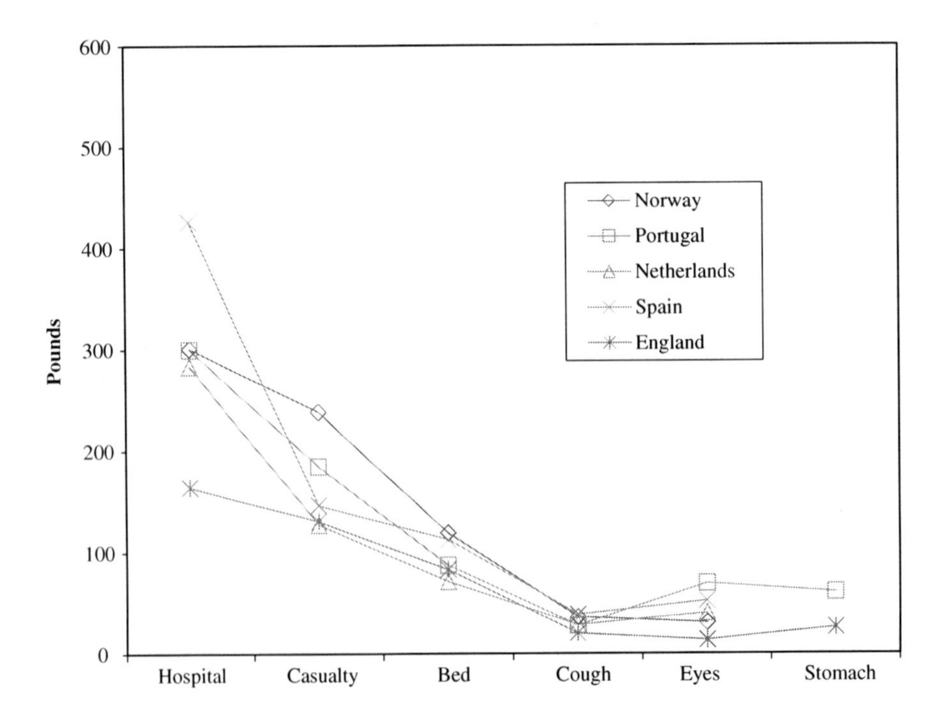

Figure 1. WTP to avoid illness episodes (value per episode).

estimated for each country, for each episode (for regression results, see CSERGE, 1998). Explanatory variables used in the regressions included respondent's income, education level, sex, age, whether there are children in the household, and measures of the respondent's health status and recent experience with symptoms included in the episode descriptions.

Using these value functions, it is possible to construct a WTP estimate for each country for a "standardized" respondent – one who is identical in all measurable characteristics. Figure 2 shows this predicted WTP for each episode for each country, for a respondent with characteristics equal to the mean level of all five countries. Here, the pattern of results is more clear. WTP for the standardized respondent is consistently higher in Spain and Portugal than in the Northern European countries. WTP for episodes in Portugal and Spain is significantly higher than WTP in the Netherlands, Norway and England in 23 of 31 possible pairwise tests. Differences within each of the two groups were small. WTP in Spain differed significantly from WTP in Portugal in only 1 of 5 tests, while WTP differed among the Netherlands, Norway, and England in only 1 of 15 tests.

Even though these results show that unit value transfer and value function transfer are not *statistically* valid between pairs of countries, it is still of interest to know the size of potential transfer errors that might result if transfers were conducted. This can be explored by looking at the percent transfer error resulting from a transfer,

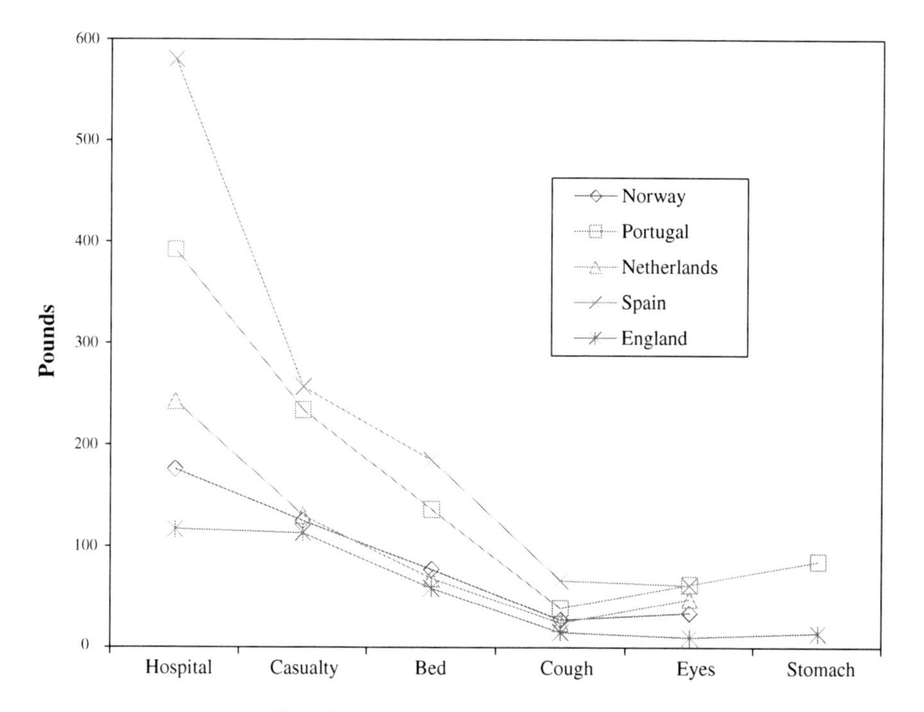

Figure 2. WTP for a "standardized" individual

defined as the absolute value of the difference between the transferred value estimate and the estimate measured in the target country, divided by the estimate measured in the target country. For each target country, for each episode, mean WTP and a value function were estimated from the pooled data set including all other countries. These were then used to conduct a unit value transfer, a unit value transfer adjusted for differences in mean income (assuming an income elasticity of 1), and two value function transfers. In the first value function transfer, the explanatory variables in the transferred value function were set equal to the sample mean for the target country. In the second value function transfer, the transferred value function was used to estimate a WTP for each respondent in the target country sample, and the WTP estimates were averaged.[8]

Table 2 shows average absolute percent transfer error resulting from each transfer approach, averaged over 20 transfer exercises. Interestingly, the performance of the transfer was not substantially improved by taking into account differences between the source countries and the target country. Adjustment for differences in real income gave a slight improvement in performance, over a simple unit value transfer. Value function transfer, with the supposed advantage that it accounts for all measurable differences between the source countries and the target country, actual performed worse than the two unit value transfer approaches.

The average transfer errors in Table 2 include not only the error due to transfer between countries, but also error due to sampling variation both in the study countries and the target country. To give some perspective, a Monte Carlo simulation showed that if the same study was done twice in the same country, the two resulting values would differ by, on average, 16%. The 38% expected transfer error from the unit value transfer approach should be assessed relative to this background level of random sampling error.

Two consistent results emerge from the three studies examined here. First, despite expectations based on economic theory, adjustment of values for differences in measurable characteristics does not necessarily improve value transfer. Second, while value transfer and value function transfer may be statistically invalid, they may generate transferred estimates that are reliable enough for policy analyses. Indeed, the errors associated with value transfer may not be much larger than the

Table 2. Performance of value transfer methods

Transfer method	Mean transfer error
Unit Value Transfer	0.382
Unit Value Transfer with adjustment for income differences	0.377
Value Function Transfer – evaluated for mean individual in target country	0.384
Value Function Transfer – evaluated for all individuals in target country, then averaged	0.419

sampling errors that would result if a new study was conducted in the target country, or than differences in values that result from using different survey instruments in the same country.

6. CONCLUSIONS

A review of the health valuation literature examined the assumptions, commonly relied upon in environmental policy analyses, that the value of avoiding an ill health outcome is independent of (1) the individual's health state, (2) how many fewer such outcomes the individual will experience, and (3) the cause of the outcome. The evidence is that the health state of the individual that will experience the improvement does matter. Respondents with poorer health state are often willing to pay more to avoid a specific ill health outcome than respondents with better health. How many ill health outcomes an individual will avoid also matters. In most studies, marginal WTP to avoid an ill health outcome decreases as the number of ill health outcomes being valued increases. However, if the ill health outcome is well defined to the respondent, it appears that WTP to avoid the outcome is not dependent on the cause of the ill health outcome.

These results suggest that a increased attention to the distribution of health benefits within the affected population is warranted. If the health benefits from improved environmental quality accrue to a subpopulation that is in poorer health, then we need to measure health values that are specific to that subpopulation. If an individual in that subpopulation will experience several fewer ill health outcomes (rather than several individuals each experiencing one fewer outcome) then we need measure WTP to avoid multiple outcomes, rather than rely on values to avoid individual outcomes.

Studies that have investigated the validity of health value transfer and value function transfer between countries show that unit value transfer, while not statistically valid, may provide value estimates that are "good enough" for many policy analyses. A somewhat surprising result is that value function transfer is not necessarily preferred to unit value transfer. Finally, when considering the validity of inter-country value transfer, the potential transfer error should be viewed in the context of the sampling error that would occur if a new study were conducted in the target country. While a new study will usually generate estimates that are more valid than a value transfer, the difference in reliability is not great as might be thought.

Richard Ready is associate professor, Department of Agricultural Economics and Rural Sociology, The Pennsylvania State University, University Park, PA. Ståle Navrud is associate professor, Department of Economics and Social Sciences, Agricultural University of Norway, Ås, Norway. *Some of the results reported in this chapter are from research supported by the European Union's Environment and Climate Research Programme: Theme 4 – Human Dimensions of Environmental Change (contract no. ENV4-CT96-0234).*

7. NOTES

[1] Current practice in valuing early deaths is the focus of the preceding chapter in this book. This chapter focuses on valuation of morbidity outcomes.
[2] Ill health outcomes (called endpoints in the environmental epidemiology literature) might be an episode of ill health such as a minor symptom day or a hospital admission, or it might be a case of a disease, such as chronic bronchitis or asthma.
[3] For a more in-depth discussion of the use of exposure-response functions to value morbidity, see Desvousges et al. (1998), Chapter 5.
[4] For a review of empirical estimates of ill health caused by pollution, see US EPA 1999, Appendix H.
[5] See also CSERGE (1998).
[6] STOMACH was valued only in England and Portugal. In Spain, the CASUALTY episode lasted only 3 days, and the HOSPITAL episode lasted only 6 days.
[7] This follows from the homogeneity properties of the indirect utility function and the expenditure function.
[8] Data and WTP values for Spain were not included in this exercise, because of differences in the descriptions of the episodes, and differences in how health experience variables were measured. Because responses are available for only two countries, transfer tests were not conducted for STOMACH.

8. REFERENCES

Alberini Anna, Maureen Cropper, Tsu-Tan Fu, Alan Krupnick, Jin-Tan Liu, Daigee Shaw, Winston Harrington. (1997) Valuing Health Effects of Air Pollution in Developing Countries: The Case of Taiwan. Journal of Environmental Economics and Management 34 (2):107–126.

Chestnut Lauraine G, Bart D. Ostro, Nuntarven Vichit-Vadakan (1997) Transferability of Air Pollution Control Health Benefits Estimates from the United States to Developing Countries: Evidence from the Bangkok Study. American Journal of Agricultural Economics 79 (5):1630–1635.

CSERGE (1998) Benefits Transfer and the Economic Valuation of Environmental Damage in the European Union: With Special Reference to Health. Final Report to the DG-XII, European Commission, contract ENV4-CT96-0227.

Desvousges William H, Reed Johnson F, Spencer Banzhaf H (1998). Environmental Policy Analysis with Limited Information.Cheltenham, England: Edward Elgar.

Dickie M, Gerking S, Brookshire D, Coursey D, Schulze W, Coulson A and Tashkin D (1987) Reconciling Averting Behavior and Contingent Valuation Benefit Estimates of Reducing Symptoms of Ozone Exposure: Environmental Protection Agency, Washington, D.C.

Johnson F Reed, Melissa Ruby Banzhaf, William H Desvousges (2000) Willingness to Pay for Improved Respiratory and Cardiovascular Health: A Multiple-Format, Stated-Preference Approach. Health Economics 9:295–317.

Navrud Ståle (2001) Valuing Health Impacts from Air Pollution in Europe. Environmental and Resource Economics 20 (4):305–329.

Ready Richard, Ståle Navrud, Brett Day, Richard Dubourg, Fernando Machado, Susana Mourato, Frank Spanninks and Maria Xosé Vázquez Rodriquez. 2004a. "Contingent Valuation of Ill Health Caused by Pollution: Testing for Context and Ordering Effects" *Portuguese Economics Journal* 3 (2):145–156.

Ready Richard, Ståle Navrud, Brett Day, Richard Dubourg, Fernando Machado, Susana Mourato, Frank Spanninks and Maria Xosé Vázquez Rodriquez (2004b) Benefit Transfer in Europe: How Reliable Are Transfers Between Countries? Environmental and Resource Economics 29 (1):67–82.

Rozan Anne, Marc Willinger (1998) Willingness to Pay and Knowledge of the Health Damage Origin. Working Paper, Research Laboratory in Theoretical and Applied Economics (BETA), Louis Pasteur University, Strasbourg, France.

Tolley George, Donald Kenkel and Robert Fabian (1994) Valuing Health for Policy: An Economic Approach Chicago, IL: University of Chicago Press.

US EPA. 1999. The Benefits and Costs of the Clean Air Act 1990 to 2010: EPA Report to Congress Washington, D.C.

D. BROOKSHIRE, J. CHERMAK AND R. DESIMONE

UNCERTAINTY, BENEFIT TRANSFERS AND PHYSICAL MODELS: A MIDDLE RIO GRANDE VALLEY FOCUS

1. INTRODUCTION

Climate issues, institutional changes, population growth, and environmental concerns are requiring policy analyses into new arenas of complex decision-making for water allocations in the desert Southwest. In addition, increased focus on water quality standards, as well as Native American water rights claims and traditional cultural uses further exacerbate resolution. This has lead to a process of continuing re-evaluation of water policies.

The increasing complexity of the water issues has necessitated the development of models at a micro-policy level in order to capture difficult institutional nuances and representations of preference differences across stakeholder groups. More often than not, adequate local micro-data are not available in all settings for modeling and policy decisions. This constrains both the robustness of the model as well as the viability of the policy analysis. In order to circumvent the problem, data and benefit transfers are used in the modeling and policy analysis.

Policy models incorporate physical science and economic data such as preference data. Policy analysis thus can require many types of data transfers when local (primary) data are not available. As such, this "transferred" data can have a great deal of underlying uncertainty. For the physical sciences, the data (as in the case of climate) are usually interpolated from large-scale models. For preference data, the adequacy of benefit transfers depends on the uncertainty of the transferred component, relative to the actual.

Benefit transfer studies have been conducted over the years. Generally, transfers in environmental economics have consisted of transferring a preference-based value from one study to another. These conceptual foundations received renewed interest in the early 1990s in the US. For example, a special issue of *Water Resources Research* (1992), the Association of Environmental and Resource Economists proceedings volume from a benefit transfers workshop and more recently Desvousges *et al.* (1998) are devoted to benefit transfer issues.

The use of benefit transfer data in environmental studies has been grownning. A sampling of these efforts would include Harrison *et al.* (1993), Rowe *et al.* (1994), Lee *et al.* (1995), Douglas and Johnson (1993), Kirchhoff *et al.* (1997), Smith and Kaoru (1990), Feather and Hellerstein (1997) and Kirchhoff (1998).

Benefit transfers or data transfers have been used to varying degrees of success in other areas. For example, data transfers have been used to construct time-of-use (TOU) prices for electricity in areas where no time-of-use data existed. The data from different geographic locations are pooled and used to construct TOU

S. Navrud and R. Ready (eds.), Environmental Value Transfer: Issues and Methods, 89–109.

tariffs for other areas. Examples of this research and the applicability of the results include Kohler and Mitchell (1984), Caves *et al.* (1984), Aigner and Leamer (1984), and Patrick (1990). The transferability in TOU tariffs is heavily dependent on comparable weather patterns, as well as generation capacity. The transfers we consider in this research are closely related to these types of data transfers as well as the traditional value transfers.

Benefit transfers can only provide a precise replicate of the underlying structure when the original study area and the new area are identical, which is highly improbable. When there are characteristic differences between the original study area and the new area, the precision of the data are diminished, which results in increased uncertainty.[1] The degree of uncertainty should be inversely related to the similarity of the sites as well as the similarity of preference structures.[2]

In addition to benefit transfers, physical science data transfers are also incorporated into policy models. This results in an interesting problem for much policy analysis currently taking place. For example, consider the problem of climate change. While there appears to be a growing consensus that, in fact, climate change is occurring, forecasts across competing scientific models are inconsistent. A basic criticism of the climate model is that the scale currently used is too large to adequately forecast local phenomena. As such, environmental economic policy is potentially based on uncertain physical science model predictions, coupled with the uncertainty of the benefit transfer data. Thus, we have a substantive push in both the physical and social sciences to refine the precision of the functions and parameters used in the analyses.

This paper assesses the effects of the relative uncertainty of benefit transfer methods, uncertainty of climate data and alternative population projections on policy decisions. Our motivation stems from the need to address the relative importance of more accurate data for policy analysis, from the physical sciences as well as from demography and economics. For instance, while there has been substantial effort directed at climate assessments in the US and in the arena of benefit transfers, seldom do the uncertainty concerns get addressed in a single framework.

The objective of this research is to evaluate, via a case study, what transfers deserve the most attention from the perspective of reducing uncertainty. That is, is the uncertainty of the benefit transfer larger or smaller than that of the physical science models? We do this by seeking to answer two questions:

- How much does the surrounding uncertainty of the benefit transfer, climate information, and other forecast information impact policy decisions in reallocation issues? and,
- Where should research efforts be focused in order to improve analyses on which policy decisions are based?

While the first question addresses the precision of benefit and data transfers, the second focuses on the "value of information." Models are inherently plagued by uncertainties. Risk and uncertainty can be attributed, for example, to imprecision in data, overly restrictive assumptions, or imprecise model specification (this is certainly not a complete listing). Modeling problems are augmented by similar risks and uncertainties from the physical sciences. How precisely do the current climate models predict future states of nature? Given the uncertainty in the models, where can we make our largest gains from improved information? Where do we make the largest gains in increased precision (per unit of effort) from improved information?

We focus on the problem of water reallocation in the desert southwest of the US and explore the implications of reallocation using both benefit transfer methods and data from a climate model with a high degree of uncertainty. We develop a hybrid economic-engineering model of water consumption in the Middle Rio Grande Valley, located in New Mexico, using local data and current climatic conditions to establish a baseline for the relative uncertainty analyses. Specifically, the model allows the exploration of the effects of differing data sets on the optimal net benefits of allocating water both spatially and across uses. Using primary (local) data and forecasts we run a base case simulation. We then run a series of simulations that systematically replaces the local willingness to pay (WTP) for a safe minimum standard (SMS) level of water in the river, with two benefit transfer values from the extant literature. The values are minimums and maximums. The results allow us to calculate a range of net benefits and thus, bound the relative uncertainty associated with the WTP benefit transfer values. We also run a series of simulations that incorporate the impacts on the system of the minimum and maximum climatic change forecasts. This again, allows us to place bounds on the uncertainty in net benefits associated with the climate forecasts. Finally, we run a series of simulations and bound the relative uncertainty associated with natural stream flow variation, minimum in-stream flow requirements, sustainable groundwater exploitation, and population.

Our results indicate that, while there is uncertainty associated with the climate models, the uncertainties associated with the benefit transfer data and the population forecasts are far greater. We also find the range of uncertainty associated with in-stream flow requirements (for preservation of an endangered species) is larger than either those of climate or the benefit transfer. Furthermore, we find allocation substitutions occur spatially and sectorally. Agriculture is the marginal use in most cases. These results suggest future policy should consider spatial as well as sectoral aspects. Furthermore, future research efforts that are directed towards improving the precision of components, other than climate models, may be the most efficient use of research dollars.

2. THE RIO GRANDE VALLEY

The Rio Grande is the fifth largest river in North America flowing 3033 kilometers (1885 miles) from southern Colorado to southern Texas where it empties into the Gulf of Mexico (See Figure 1). Within New Mexico, the river extends for 756 kilometers (470 miles) north to south. Its discharge area in the state is 83,416 square kilometers (32,207 square miles.) The Middle Rio Grande (MRG) flows approximately 262 kilometers (163 miles) with a drainage area of 64,128 square kilometers (24,760 square miles.)

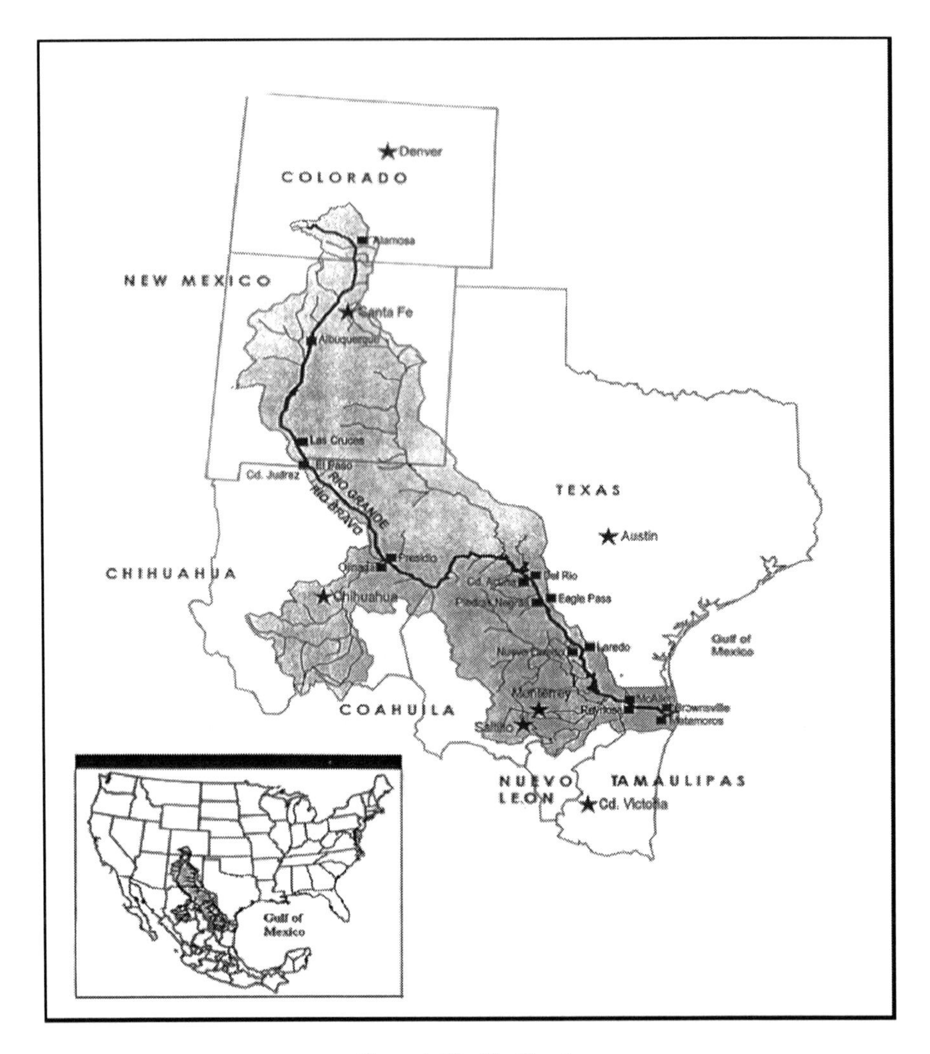

Figure 1. The Rio Grande

The MRG valley is culturally diverse with strong influences from both the Native American and Hispanic cultures. The Pueblo Indians in central New Mexico have diverted water from the river for centuries for crop irrigation. The eight Pueblos in the MRG hold "native rights" to river water and continue to use it for irrigation. The Spanish settlers adopted similar irrigation techniques when they settled in the area.[3] Allocations have been further complicated by a series of diversions and dams (both federal and state) aimed at flood control and irrigation. The growing Albuquerque urban area will continue to place new demands on surface water from the river. Coupled with these more traditional demands are the more recent demands for species and habitat preservation.

Water rights in the MRG are regulated by state statutes and water permits which, compared to other sections of the Rio Grande, is a rather inexact system. While the system historically has solved most problems, the problems and disputes are becoming increasingly contentious. At the heart of the problem is the lack of an adjudication of water rights and the probability that water rights claimed on paper exceed the actual water available. Given the lack of clearly defined rights, the model presented below does not impose water right restrictions on use. Rather, the model is void of water rights and appropriates the water to the highest value use in order to maximize the social benefits from its use.

The MRG can be divided into four distinct reaches separated by agricultural diversion dams. The first diversion occurs at Cochiti dam (designed primarily for flood control) at the Northern end of the MRG. Subsequent diversion dams from north to south are the Angostura, the Isleta and the San Acacia. Elephant Butte Reservoir (designed for long term storage) marks the southern end of the MRG.

Multiple flow gauges with long periods of record are available in each reach. This is an important consideration from the modeling perspective because these SMS mandated flow requirements impose a constraint on diversion and consumption in each reach. These constraints are discussed in further detail at the end of this section. Figure 2 presents a map of the MRG valley and identifies the four reaches. The reaches coincide fairly closely with county lines, and so we distinguish the reaches by the associated county: Sandoval, Bernalillo, Valencia, and Socorro. While Sandoval, Valencia and Socorro are primarily rural areas, Bernalillo is a combination of urban and rural.

Precipitation in the basin is limited and variable, with most areas receiving between 17.5 and 37.5 centimeters (7–15 inches) annually. The majority of flow in the river is derived from snowfall in the mountains of Northern New Mexico and Southern Colorado and rainfall in the late summer. The natural flows are extremely variable from year to year as can be seen in Figure 3.[4]

Primarily, water consumption, in all but Bernalillo County, is for agricultural use. Bernalillo County is the home to Albuquerque, the largest urban area in the State. Not only does this reach consume water for agriculture, there is a substantial urban consumption component. In addition, the US Fish and Wildlife Service (FWS) has designated as critical habitat a portion of the MRG to increase the probability of survival of the Rio Grande Silvery Minnow (RGSM). This designation has resulted in a required minimum flow level. The standard is based on a flow of 5.67 cubic meters per second (m^3/sec) (200 cubic feet per second (cfs)) at the San Acacia Gauge.[5] Finally in all reaches there is the potential to increase the in-stream flow above the SMS in order to restore and protect the eco-system of the river.[6]

Water demand in the MRG can be disaggregated into three basic components:

- Urban: Water consumed by the public water supply system for residential, industrial and municipal consumption. This includes all water diverted from the river for public use that is not returned to the stream.
- Agricultural: Water diverted for agriculture that is not returned to the stream.

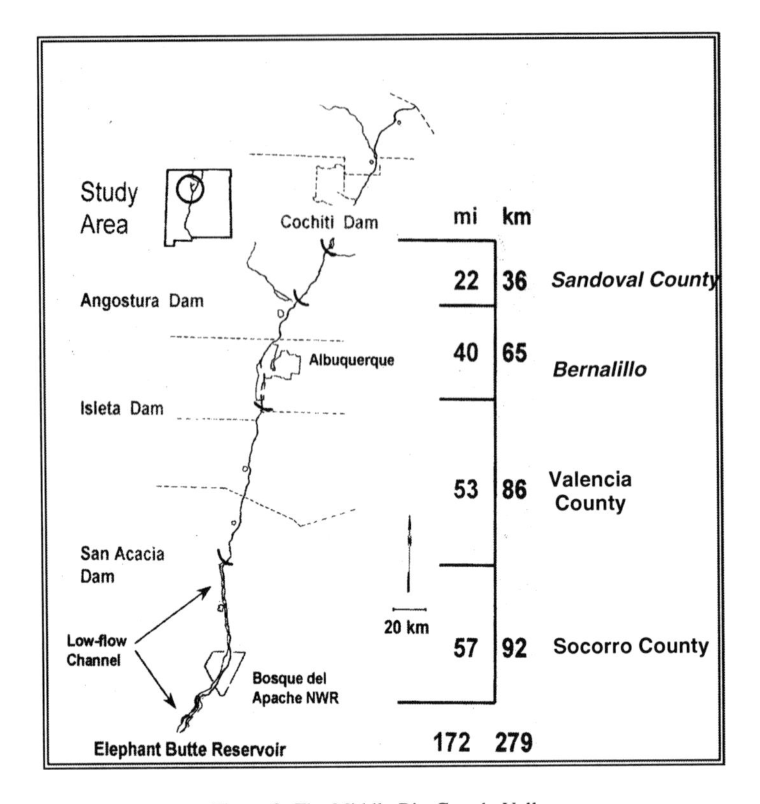

Figure 2. The Middle Rio Grande Valley

- Endangered Species Requirement: The SMS designated for the RGSM and habitat restoration: In-stream river flow in excess of the SMS used for habitat restoration and species preservation.

Each is discussed in further detail below.

2.1. Urban Water Demand

The only stretch of the river that has substantial urban consumption is the stretch that includes the city of Albuquerque. Albuquerque is the largest urban area in New Mexico with a metro population in excess of 800,000. Water usage is divided between residential (48%), commercial (27%), industrial (3%), and institutional (9%). Approximately 12% of consumption is attributable to system losses.[7] Currently, the urban demand of 85.15 million cubic meters per year (m^3/yr) (69,000 acre-feet per year (acf)) is being met by groundwater from the Albuquerque aquifer. However, this exceeds the annual steady-state withdrawal by as much as 23.45 million cubic meters (19,000 acf) per year.[8] The city of Albuquerque currently holds water rights for 59.23 million cubic meters (48,000 acf) from the San Juan Chama diversion project, which it is not using.[9] While the Albuquerque metropolitan area

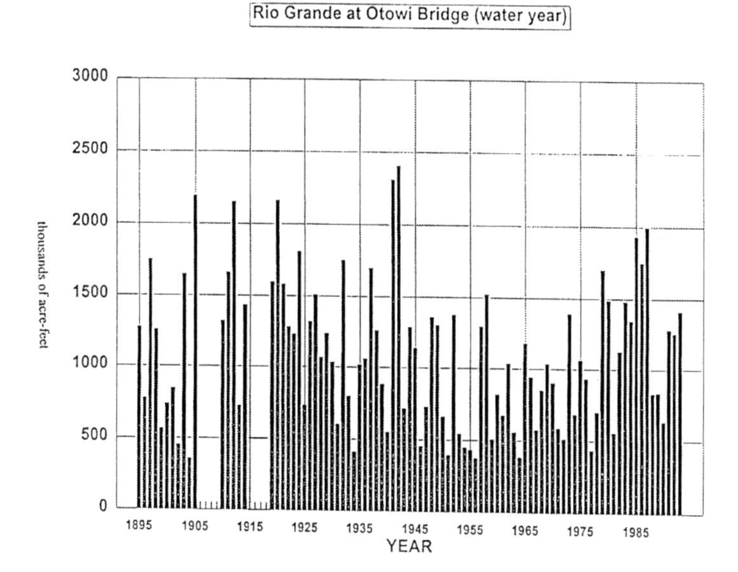

Figure 3. Rio Grande Hydrograph 1895 to 1992 (1000's of acre-feet). From Middle Rio Grande Council of Governments (1999)

is not currently using surface water to meet demands, the rate of mining from the aquifer and the growth in population (14.5% increase since 1990) suggest the current situation is untenable.

Albuquerque will have to rely increasingly, in the future, on its San Juan Chama surface water rights to meet demand. The city is implementing a management strategy that will combine groundwater withdrawals with these rights. A filtration plant, which is necessary for urban consumption of the surface water, is slated for completion in 2005, which will allow a reduction in groundwater use. However, depending on growth in the Albuquerque area, increased urban demands will compete with the more traditional agricultural use, as well as the SMS and the habitat restoration use. In the modeling and simulations, we assume the sustainable groundwater levels are used prior to surface water.

2.2. *Agricultural Water Demand*

Most of the southwest portion of the United States is desert. This is certainly true of New Mexico and the majority of the MRG valley. Despite this, agriculture plays an important economic and cultural role in the MRG valley. Currently, approximately 194 square kilometers (48,000 acres) of the more than 13,760 square kilometers (3.4 million acres) of total farmland (in the four counties) are irrigated.[10] While much of the area is not amenable to irrigation, more could be irrigated if water was available and the value added from irrigation was adequate. The importance of specific crops varies from area to area, but in general, corn, hay, wheat, and chilies are some of the more important crops. Much of the corn, hay and wheat is grown as feed for livestock, while the chile is grown solely for human consumption.

2.3. Safe Minimum Standard to Maintain Rio Grande Silvery Minnow

The RGSM historically was one of the most abundant and widespread fish in the Rio Grande, Pecos and Santa Rosa rivers in New Mexico. Today it is found only in the reach of the MRG from Cochiti Dam to Elephant Butte Reservoir (US FWS, 1999). On July 6, 1999 critical habitat was designated for the silvery minnow to ensure an SMS in-stream flow to help preserve the silvery minnow.[11]

2.4. Habitat Protection

The final demand component is in-stream flow above the SMS in order to help restore the ecological health of the river. In addition to the minnow, the Bosque is home to many species of wildlife that, through shrinking habitat have declining populations. There are eleven species of fauna in the MRG that are listed as endangered.

3. THE PLANNER'S PROBLEM

The planner, who wishes to maximize social welfare, must consider all of the competing uses, described above, as well as any applicable constraints. These constraints are imposed not only by the physical sciences, but also by the institutions. Thus, the planner's problem can be stated as:

$$
\begin{aligned}
&\max_{q} \mathrm{NB} \\
&\mathrm{s.t.}\, Q_1 = Q(0) \\
&\quad Q_2 \geqslant Q_1 - q_1, \\
&\quad Q_3 \geqslant Q_2 - q_2, \\
&\quad Q_4 \geqslant Q_3 - q_3, \\
&\quad Q_5 \geqslant Q_4 - q_4, \text{ and} \\
&\quad Q_j \geqslant q_{m_j} \geqslant \underline{q_j}.
\end{aligned}
\tag{1}
$$

where $q = \Sigma q_i = \Sigma q_j$, $i = 1, \ldots, 4$ competing uses of the water and $j = 1, \ldots, 4$ reaches of the MRG. Water available at the beginning of the first reach of the river, Q_1, is constrained by the maximum amount of water flowing into the reach from the previous reach of the river, $Q(0)$. The system is also constrained by the minimum amounts of water required to flow into each of the subsequent reaches (Q_2, Q_3, and Q_4) and into the lower Rio Grande, Q_5, due to water compacts with either Texas or Mexico.[12] In addition, there is a minimum water flow, q_{m_j}, that must be maintained through each of the j stretches of the river in order to meet the minimum SMS requirement, $\underline{q_j}$. Thus, the water flowing into each reach must also equal or exceed this minimum requirement.

The net benefits are estimated via traditional economic models. The constraints imposed by the physical science are incorporated via appropriate physical models.[13]

4. EMPIRICAL MODELING COMPONENTS AND DATA

Optimal water allocations across the competing uses along the four reaches of the river are estimated with a constrained optimization model that determines the optimal allocations that maximize net benefit.[14]

4.1. Agricultural Use

Water is an input into crop production. We calculate the net benefit of water as a function of the value of the final product, harvested crops. In this case the demand function relies on the form of the production function for the final product and relates net income per acre to water applied per acre.[15]

We assume the crops grown in the MRG are sold in a competitive market (i.e., the production from the MRG will not alter market price). Thus, the net benefit of MRG production is the producer's benefit, or simply total net income.

MRG agricultural water demand, q_A, is therefore a function of the net income of crop i per acre in reach j, NI_{ij}, and the number of acres of crop i harvested in reach j, A_{ij}, as well as a crop yield factor, a_i, which is simply a factor of the amount of water applied. That is; $q_A = q(NI_{ij}, A_{ij}, a_i)$.

ECONorthwest (1997) shows that under limited water conditions maximum crop yields per acre are obtained by maintaining the optimal water content per acre, q_i^*, and reducing the number of acres. Thus, we assume a constant cropping pattern (and thus the same optimal crop water requirement per acre) is maintained as the acreage in production varies. The total net benefits from agriculture, NB_A, are described by;

$$(2) \qquad NB_A = \sum_{j=1}^{4} \left(\frac{NI_j q_j}{q_i^*} \right)$$

The optimal cropping pattern water requirements and the average value per acre for each reach (or county) are reported in Table 1.

4.2. Urban Demand

Net benefits from urban use are estimated as the area under the demand curve, constrained by a minimum use (necessity) and a maximum use (year 2002 projected per capita use) consumption bounds. There is not an empirical demand estimate available for Albuquerque. Therefore, we constructed a Cobb-Douglas demand function using the current price/quantity combination as an anchor point. Urban demand for the representative consumer, q_U, is described by;

$$(3) \qquad q_U = \beta P_U^{\varepsilon},$$

Table 1. Agricultural Water Requirements

Reach (county)	Optimal water requirement[a] q_i^* (m^3 per m^2 per year)	Average value per hectare[b]
1 Sandoval	0.56	\$344
2 Bernalillo	0.61	\$385
3 Valencia	0.76	\$526
4 Socorro	0.72	\$378

[a] From Wilson and Lucero (1997)

[b] Based on prices taken from the New Mexico Department of Agricultural Census averaged over the years 1995 through 1997 and inflated to 1998 prices using the consumer price index and an expense to revenue ratio of 36% derived from US Dept of Agriculture Statistics for New Mexico Counties 1997.

where, P_U is the per unit price of water, β is a constant determined from current conditions,[16] and ε is the own price elasticity for urban water demand. Total urban demand is simple individual demand multiplied by the total number of consumers.

Own price elasticity is set equal to -0.50.[17] The minimum constraint is imposed to assure an allocation level to meet necessities. The minimum value is set at 371 liters per person per day (98 gallons per person per day.) This is the winter level consumption for Tucson, Arizona (assumed representative of Albuquerque).[18] The maximum consumption constraint is set at the maximum target level chosen by the City of Albuquerque, 662.4 liters per person per day (175 gallons per person per day).[19] Assuming a representative consumer, total net benefits for urban use are estimated by;

$$(4) \qquad \mathrm{NB_U} = \mathrm{N}\mu\left(\frac{\mathrm{q_U}}{(\mathrm{q_U})^{1/\varepsilon}} - \frac{\underline{\mathrm{q}}}{\underline{\mathrm{q}}^{1/\varepsilon}}\right),$$

where $\mu = \dfrac{\partial\beta^{-\varepsilon}\varepsilon}{(\varepsilon-1)}$. ∂ is a conversion factor to convert dollars per cfs to dollars per cubic feet per year (cfy). $\underline{q}$ is our minimum consumption level constraint, and N is the population in the metropolitan area.

4.3. *Ecological/Habitat Net Benefit*

Ecological or habitat benefits, NB_H, in the MRG are hypothesized to be derived from preservation of the endangered RGSM, NB_{SMS}, as well as from water in excess of the required SMS for the RGSM. The excess water results in benefits, NB_E, from restoration of the Bosque to a healthier state. Thus;

$$(5) \qquad NB_H = NB_{SMS} + NB_E.$$

Thus benefits are derived from the in-stream flow from the minimum quantity of water required to be in the stream at all point to meet the SMS for the RGSM, q_m and from in-stream flow, q_H, in excess of q_m, that provides additional ecological benefits.

Net benefits from the SMS are estimated via;

$$\text{(6)} \qquad NB_{SMS} = M \times WTP = V \times q_m$$

Where M is the number of households in New Mexico, V is the dollar benefit per cubic foot of the required in-stream flow. More specifically, $V = \frac{WTP}{q_m}$, where *WTP* is an individual household's willingness to pay to ensure the *SMS*. The *WTP* is taken from the CVM estimates of Berrens, Ganderton, and Silva (1996).

We extrapolate from the estimated *WTP* in order estimate benefits from in-stream flows in excess of $q_m (q_H)$. Specifically;

$$\text{(7)} \qquad NB_H = M \sum_{j=1}^{4} \frac{V}{q_m} l_j \left(q_{H_j} - \frac{q_{H_j}^2}{2\overline{q}_j} \right),$$

where $\overline{q}_j$ is the maximum amount of water which could be made available in reach j of the river and l_j is the proportionate distance of reach j.

FWS estimates q_m. Baseline q_m is 2.63 m^3/sec (93 cfs) and q_{SMSj} is 5.66 m^3/sec (200 cfs) in the stream at the San Acacia gauge. The WTP equals $30 per New Mexico household (1998$).[20]

5. SIMULATIONS AND RESULTS

The model detailed above is employed initially to estimate the base case using local data. We then estimate the upper and lower bounds on allocation changes and benefits that could be expected through normal in-stream flow variations by performing two simulations, one standard deviation in either direction of the base stream flow. This allows us to establish the range and bounds on allocations and benefits that we could expect from year to year. We then perform a series of simulations to assess the impact of other changes.

In the first set of simulations we substitute minimum and maximum WTP values found in the extant literature for the primary data. The next set of simulations incorporates the minimum and maximum estimates for climate change impacts on the in-stream flow. The third set of simulations substitutes the minimum and maximum SMS for endangered species preservation for the current prescribed level. We then perform a simulation that substitutes minimum and maximum groundwater exploitation levels that have been suggested as sustainable for the Albuquerque aquifer. The final simulation incorporates the projected population growth for the Albuquerque metro area and the state of New Mexico. The first set of simulations allows us to assess the impact on optimal allocation and the resulting net benefits to society when a benefit transfer is employed. That is, when transferred value data is substituted for primary data, how are allocations and benefits impacted. The second, third, and fourth set of simulations allows us to assess the impact of forecasted conditions, or future states of nature, based on models from the physical science realm. The final simulation allows us to assess the impact of forecast demographic conditions in the year 2030. Since no population models forecast declining population for the areas, we establish only an upper bound on the impact of population growth in the area.[21]

5.1. Base Case

The first optimization is the base case that incorporates Albuquerque's 2002 water use plan.[22] All base case parameters are presented in Table 2. The maximum flow at the beginning of the first reach of the MRG is equated to the average flow for August (the driest month of the year). In addition to those mentioned above, the base case parameters include the SMS estimated by the FWS for the silvery minnow, WTP for the SMS from Berrens *et al.*, and the elasticity estimate from Brown *et al.* Current population estimates for Albuquerque and the state are used.

Employing these parameter values the optimal allocation of water through the four stretches of the river, across competing uses is estimated, and subsequently, the net benefits associated with that allocation. Table 3 presents the optimal allocations while Table 4 presents the resulting net benefits.[23] In Table 3, note that the habitat allocations are the allocations in excess of the minimum required to meet q_m to account for natural losses in the reach. For example, in reach 4, the minimum required is 5.67 m^3/sec (200 cfs.)

Agricultural allocations occur in reaches 1 and 3 (R1 and R3) of the river, while in reaches two and four (R2 and R4) they are zero. Urban usage (which occurs only in R2) has a substantial allocation. H1 and H2 (habitat allocations in reaches 1 and 2) also have large allocations.[24] R4 does not receive any water under the optimal allocation. Net benefits accrued from the base case are presented in Table 4. The majority of benefits, 64%, are derived from urban welfare benefits.[25] SMS and habitat are second with 30% and agriculture accounts for 6%.

Table 2. Base Case Parameters

Parameter	Value
Water flow into Reach 1 of MRG (Q(0))	26.72 m^3/sec (943 cfs)
SMS through all reaches (q_m)	2.63 m^3/sec (93 cfs)
Own Price Elasticity (ε)	−0.5
WTP for SMS and Habitat	\$30 per NM Household
Population of Albuquerque Metro Area (N)	513,430
NM Population (M)	619,000
Groundwater Withdrawal (q_{GW})	1.95 m^3/sec (69 cfs)

Table 3. Base Case Allocations

Use	Quantity (m^3/sec)	Quantity (cfs)
q_{A1}	2	70
q_{A2}	0	0
q_{A3}	1	36
q_{A4}	0	0
q_{U2}	4	139
q_{H1}	13	471
q_{H2}	13	452
q_{H3}	0	0
q_{H4}	0	0

5.2. *Simulations*

As discussed above, the simulations performed include minimum and maximum variations in base stream flow (to bound the impact of the natural variations in the river), the projected population growth, minimum and maximum WTP values from BT studies, minimum and maximum variation in base stream flow associated with climate change projections, and minimum and maximum variations in sustainable ground water usage levels to represent the changes in the hydrologic model forecasts made since 1990. Table 5 presents the change in the relevant parameter value for each of these simulations as well as the cite or rationale for the employed value.[26]

Table 6 presents the percentage change in the optimal allocations across sectors and reaches and Table 7 reports the percentage changes in net benefits. Negative changes are indicated by parentheses. In cases where the base allocation was zero and the new allocation was some number greater than zero the percentage change is designated as 100%.

Table 6 reports several interesting results. First, regardless of the scenario, agricultural use of water in R4 is never optimal. The second note of interest is the change in the allocation to urban. The only case in which the upper bound on aggregate urban consumption is not binding is in the $\overline{q}_m$ simulation. In this case water is allocated to in-stream flow from q_{U2}. We also observe q_{A1} and q_{H3} allocations increasing in the $\overline{WTP}$, while the allocations to q_{A3}, q_{H1} and q_{H2} decline.[27] Finally, it is of interest to note that in all but the $\underline{WTP}$ simulation, q_{A1} and q_{H1} are affected

Table 4. Net Benefits from Base Case

Allocation	Net benefit ($ millions)	Percent of total allocation
Agriculture	$3.6	6%
Urban	$40.8	64%
SMS and Habitat	$18.9	30%
TOTAL	$63.3	

Table 5. Simulation Values

Simulation	Value	Cite
$\underline{Q}(0)$	18.41 m^3/sec (650 cfs)	CH2MHILL (1999)
$\overline{Q}(0)$	35.41 m^3/sec (1250 cfs)	CH2MHILL (1999)
$\underline{WTP}$	$20 per household	Berrens, Ganderton, and Silva (1996)
$\overline{WTP}$	$140 per household	Sanders, Walsh and McKean (1990)
$\underline{q}_{CLIM}$	25.1 m^3/sec (886 cfs)	Southwest Assessment (1999)
$\overline{q}_{CLIM}$	28.33 m^3/sec (1000 cfs)	Southwest Assessment (1999)
$\underline{q}_m$	1.27 m^3/sec (45 cfs)	US FWS (1999)
$\overline{q}_m$	5.67 m^3/sec (200 cfs)	US FWS (1999)
$\underline{q}_{GW}$	1 m^3/sec (35.3 cfs)	Kernodle (1995)
$\overline{q}_{GW}$	3.82 m^3/sec (135 cfs)	Kernodle (1995)
N (Metro Population)[a]	794,660	Multiple Sources
M (NM Population)[b]	1,116,837	

[a] Projection estimates based on annual perentage increases forecasted for state planning and development district 3 (Middle Rio Grande Council of Governments, 1999). Assumes the population served by the water system will grow at the same rate as the district.

[b] Calculated from US Census Bureau data (Adelamar and Alcantara, 1997) and projected household size data from Middle Rio Grande Council of Governments (1997).

Table 6. Percentage Change from Base Allocation[29]

USE	$\overline{Q}(0)$	$\underline{WTP}$	$\overline{WTP}$	$\underline{q}_{CLIM}$	$\overline{q}_{CLIM}$	$\underline{q}_m$	$\overline{q}_m$	$\underline{q}_{GW}$	$\overline{q}_{GW}$	N&M
q_{A1}	241	0	35	(56)	56	82	(100)	(20)	115	(97)
q_{A2}	100	0	0	0	0	0	0	0	0	0
q_{A3}	0	0	(87)	0	0	(41)	(100)	0	0	0
q_{A4}	0	0	0	0	0	0	0	0	0	0
q_{U2}	0	0	0	0	0	0	(4)	0	0	55
q_{H1}	(100)	0	(24)	26	(26)	(46)	46	14	(79)	66
q_{H2}	(28)	0	(8)	0	0	(8)	0	0	0	5
q_{H3}	0	0	100	0	0	100	100	0	0	0
q_{H4}	0	0	0	0	0	0	0	0	0	0

Table 7. Percentage Change from Base Case Net Benefits

USE	$\overline{Q}(0)$	$\underline{WTP}$	$\overline{WTP}$	$\underline{q}_{CLIM}$	$\overline{q}_{CLIM}$	$\underline{q}_m$	$\overline{q}_m$	$\underline{q}_{GW}$	$\overline{q}_{GW}$	N&M
Agriculture	199	0	(3)	(39)	39	43	(100)	(14)	79	(66)
Urban	0	0	0	0	0	0	(5)	0	0	55
Habitat	(1)	(33)	367	0	0	4	(1)	0	0	81
TOTAL	11	(10)	109	(2)	2	4	(9)	(1)	4	56

and they move in the opposite directions. This is because of the trade-offs we make between use within a reach of the river and the constraints that must be observed between reaches.

We observe spatial re-allocations in several of the simulations, for example, the reallocation from q_{A1} to q_{H1} in the $\underline{q}_{CLIM}$ simulation. We also observe sectoral reallocations, with the allocations to agriculture being the most volatile.

In order to assess the societal impact of these simulations, we analyze the percentage change from base case net benefits. These results are presented in Table 7. Overall changes, for the various scenarios range from an almost 10% decline in benefits to a 109% increase. In many cases, changes within the sectors are substantial. As might be expected, the agricultural sector experiences fluctuations in net benefit levels in all but one scenario. Habitat net benefits exhibit the largest fluctuations in percentage change. The population and $\overline{q}_m$ simulations are the only ones which impact net benefits in all sectors.

The information presented in Table 7 allows us to (in a naïve fashion) place bounds around natural variations, population growth (a one-sided bound), and uncertainty bounds around both the benefit transfer and the physical data. As stated previously, the minimum and maximum parameter values employed in the different simulations are based on the minimum and maximum estimates found either in the extant literature or forecast from the physical science models. Graphing the percentage changes in net benefits (from the base case) allows us a visual representation of the uncertainty associated with the various components of the optimization model. Graph 1 presents the results for total net benefits percentage changes and Graph 2 focuses on agriculture. When there is no change from the base case, the simulation results are not plotted. In some cases, the percent change exceeds the bounds on the graphs (for example, population on Graph 1). In these cases, the actual percent change is presented numerically next to the appropriate bar.

Comparing the bounds on percentage changes in net benefits allows us to qualitatively compare the impacts of uncertainty for the different scenarios discussed. The easiest judgments to make from the graphs is where there is the greatest uncertainty in our current data and models. From Graph 1, we note that while there is some variation in the net benefits attributable to the current climate models, this uncertainty falls within the range of net benefits we find in the natural in-stream flow variation. In fact, the only variations that fall outside the range of the natural in-stream flow are those associated with either the WTP benefit transfer or the population growth forecast. This suggests the impacts of employing the WTP BT or the current population projections will result in far greater impacts on net benefits than the variations associated with the climate forecast.

This is further illustrated by Graph 2 that depicts the impacts on agriculture. In this case, the climate scenarios result in a substantially larger variation than does the WTP benefit transfer, but both are dwarfed by the negative impact on agriculture brought about by the projected population growth. In the case of agriculture, the variation associated with the appropriate in-stream flow is also substantial as is the upper bound variation for groundwater.

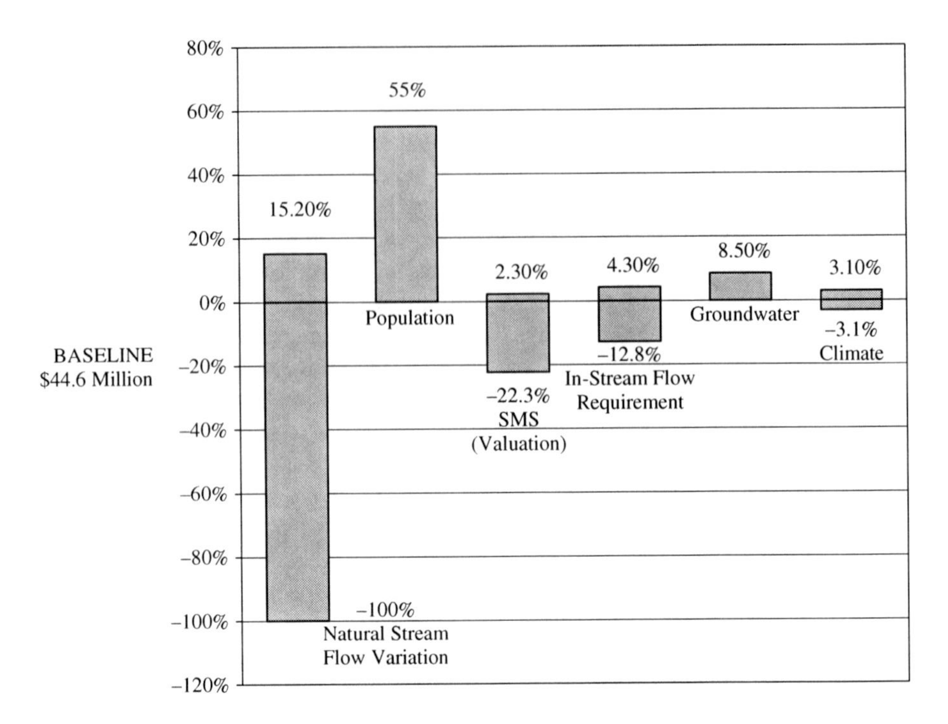

Graph 1. Changes in Total Net Benefits

6. CONCLUSIONS

Our optimization model for the Middle Rio Grande valley allows us to compare the relative uncertainties associated with using benefit transfers, and physical science models (that are usually described as too large scale to be accurate). The first question we sought to answer is:

"How much does the surrounding uncertainty of the benefit transfer, climate information, and other forecast information impact policy decisions in reallocation issues?"

Our analysis suggests, that in the case of water allocations in the Middle Rio Grande Valley, while the uncertainty associated with climate information impacts policy decisions, these bounds on changes in net benefits are substantially smaller than the bounds associated with normal in-stream variations, population growth projections, or the WTP benefit transfer. Within the agricultural sector, the ranges associated with climate or the BT are substantially smaller than other ranges we find (again, for example, see the population results).

The second question we sought to answer is;

"Where should research efforts be focused in order to improve analyses on which policy decisions are based?"

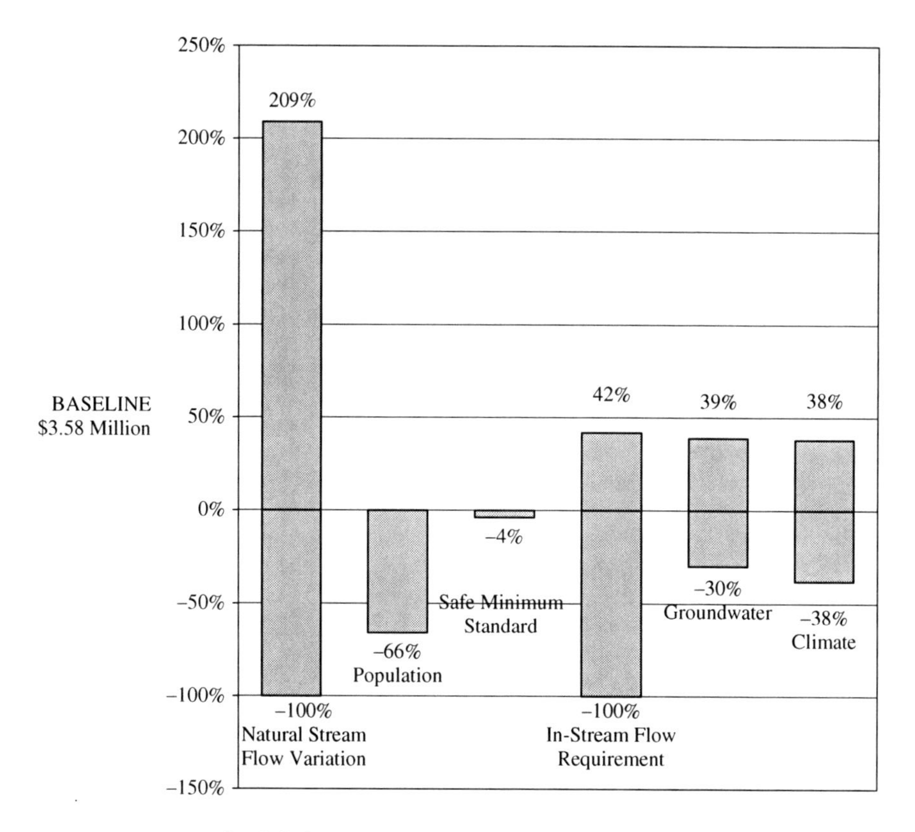

Graph 2. Changes in Net Benefits Attributed to Agriculture

This is the "value of information" question. Where do we make the largest gains in precision (per unit of effort) from improved information? The results from this case study suggest quite strongly that both population forecasts and the benefit transfer (in this case, the WTP for habitat) are critical factors. Thus, efforts focused on improved population growth models as well as improved demand models may prove more beneficial to policy decisions than further improvements on climatological models.

David Brookshire is professor, Janie Chermak is associate professor, and Richard DeSimone is research assistant, Department of Economics, University of New Mexico, Albequerque, NM, USA. *This paper was presented at the EVE Workshop, Lillehammer, Norway, October, 1999. The paper has since been revised and has benefited from comments by Olvar Bergland, Ståle Navrud, Richard Ready, Eirik Romstad, Ian Bateman, Clive Spash, Susana Mourato and other participants. We would like to thank the Southwest Regional Assessment Program for partial financial support, which was received from the US Global Change Research Program in conjunction with the White House Office of Science and Technology Policy.*

7. NOTES

[1] Of course, any uncertainty in the original study will be carried forward.

[2] For additional discussion of the considerations and potential problems of the benefit transfer method see, for example Desvouges *et al.* (1998).

[3] For a detailed history of water in New Mexico see Clark (1987).

[4] On average the MRG receives about 1.36 cubic km (1.1 million acre-feet) of water per year with a standard deviation of 0.65-cubic km (525,000 acre-feet). About two-thirds of the water, 0.9 cubic km (730,000 acre-feet), reaches Elephant Butte Reservoir and the remainder, 0.46 cubic km (370,000 acre-feet), is consumed in the MRG Valley. Agriculture accounts for about 30% of the consumption, municipal users 10% and natural losses account for about 60%. The interstate compact with Texas and Colorado requires New Mexico to deliver water to Elephant Butte Reservoir on a sliding scale based on the annual flow measured at Otowi Gauge in Northern New Mexico near the town of Los Alamos. The average annual required delivery is 66% of the Otowi flow.

[5] The proposed SMS derived from an FWS test in which water flow was released at Cochiti and allowed to flow directly to San Marcial without diversion. The test was conducted at a time when the stream bed was initially dry in the stretch between San Acacia and San Marcial. The results indicated a flow of 5.67 cubic meters per second measured at the San Acacia Gauge was sufficient to keep the stream bed wet between San Acacia and San Marcial. The FWS does not claim this to be the definitive SMS, the RGSM recovery plan establishes a three year time goal (beginning in July 1999) to determine the appropriate SMS and associated flow conditions.

[6] For a more thorough discussion of in-stream flow, economic impacts, endangered species, and habitat restoration see Watts et al. (2000).

[7] City of Albuquerque (1997).

[8] City of Albuquerque Public Works Department (1999).

[9] The San Juan Chama diversion redirects water from the Colorado River Basin through a series of tunnels and pipe systems in Northern New Mexico to the Rio Grande Basin.

[10] Adelamar and Alcantara (1997).

[11] The primary cause of species reduction in the MRG was determined to be fragmentation and loss of habitat caused by river management (US FWS, 1999). Agricultural diversion dams fragment the river creating barriers impacting the fish's ability to swim back upstream to spawn. River channelization for flood control has narrowed and straightened the river eliminating much of the backwater areas where the fish thrive. Releases from Cochiti timed to benefit agriculture have interfered with the spawning period and caused large un-natural flows washing the fish from their protective backwaters. River water diversions for agriculture have resulted in large stretches of dry river bed or river bed with small isolated pools, particularly in late summer. Agriculture accounts for 90% of all river diversions in the MRG with the average annual diversion being 48% of annual flow and at times, particularly during low flow summer months, 100% of flow in the lower reaches. On this basis the FWS established three elements of critical habitat required to sustain the RGSM (US FWS 1999); stream morphology, water quantity, and water quality.

[12] Our model is run for the average low flow (i.e., worst case scenario) month of August while compact and treaty obligations are met over a water year. Given that August flows most likely contribute little to the overall obligations, we do not impose the downstream obligations in our model.

[13] A complete discussion and derivation of the model and the data are presented in the technical appendices which are available from the authors.

[14] The optimization model was solved employing a conjugate gradient algorithm.

[15] Water is provided to the farmer by the Middle Rio Grande Conservancy District (MRGCD) at a nominal fee to cover operating costs, there is no actual charge for water. Consequently, we included the cost of water delivery in the overall net income per acre figure.

[16] From City of Albuquerque Public Works Department (1998).

[17] From Brown, *et al.* (1996).

[18] It could be argued that the water requirement to sustain life is substantially lower than 371 liters per person per day. Given this, our net benefit estimates for urban demand can, most likely, be considered conservative.
[19] City of Albuquerque (1995).
[20] From Berrens, Ganderton, and Silva, (1996). The CVM specifically asked for the contribution level an individual would be willing to make to a trust fund in order to provide a minimum in-stream flow in the MRG to protect the silvery minnow. In our analysis we assume the trust fund to be in place at $t = 0$ and to be perpetual.
[21] This is certainly an area for further research. Given the variations of population growth for different areas, population could be incorporated and a BT conducted.
[22] Urban water usage is constrained at both the upper and lower bound. At the lower bound usage is constrained to 371 liters per capita per day which is assumed to be the minimum subsistence level. The upper bound on usage is set at 662.4 liters per capita per day (175 gpd per capita), which is the target set by the city of Albuquerque in its 2002 plan.
[23] In Table 3, the first subscript (e.g., "A") refers to the use (A = agriculture, U = urban and H = habitat). Similarly, the second subscript (e.g., 1) refers to the reach of the river where the allocation occurs.
[24] Habitat allocation can occur either from a minimum flow constraints or from a net benefit allocation.
[25] The upper bound constraint on urban water demand was binding in this base case optimization.
[26] In this table, as well as all subsequent tables, a line under the symbol is used to denote the minimum, while a line over the symbol is used to denote the maximum (e.g., $\underline{q}_{GW}$ and $\overline{q}_{GW}$ represent minimum and maximum groundwater withdrawals, respectively).
[27] While this may not appear readily intuitive, this reallocation occurs due to an increase in the willingness to pay for habitat, the length of each reach, lj, and the constraints on the model (i.e., the quantity of water that is required in-stream and the natural loss). These factors result in both sectoral and spatial reallocations.

8. REFERENCES

Adelamar N, Alcantara (1997) Population Projections for The State of New Mexico by County, Volume B, New Mexico Population outlook in the 21st Century, US Census Bureau, Washington, D.C., US Printing Office.

Aigner DJ, Leamer E.E (1984) Estimation of Time-of-Use Pricing Response to the Absence of Experimental Data: An Application of the Methodology of Data Transferability, Journal of Econometrics, 26:204–28.

Berrens, RP, Ganderton, P, Silva CL (1996) Valuing the Protection of Minimum In-stream Flows in New Mexico, Journal of Agricultural and Resource Economics 21 (2): 294–309.

Brown, FL, Nunn SC, Shomaker J.W, Woodard G (1996) The Value of Water, A Report to The City of Albuquerque in Response to REF-95-010-SV.

Caves, DW, Christensen LR, Herriges JA (1984), Consistency of Residential Customer Response to Time-of-Use Electricity Pricing Experiments, Journal of Econometrics, 26:179–204.

CH2MHILL (1999) Low Flow Hydrology of The Rio Grande and Relation to Water Availability for The Rio Grande Silvery Minnow. A Draft Report Prepared for The Albuquerque, New Mexico Public Works Department.

City of Albuquerque Public Works Department (1998) Water & Sewer Rate Ordinance Bill Number F/S 0-7.

City of Albuquerque (1997) Water Resources Management Strategy, The City of Albuquerque Public Works Department, Twentieth Council Bill No. R-176. Enactment No 40-1997.

City of Albuquerque (1995) Resolution adopted a long range-range water conservation strategy for the City of Albuquerque and the properties served by the city's water utility.

City of Albuquerque Public Works Department (1999) Water Resources Planning, A New Resources Strategy Pamphlet.

Clark IG (1987) Water in New Mexico: A History of Its Management and Use, University of New Mexico Press, Albuquerque, NM.

Desvousges WH, Johnson FR, Banzhaf H S (1998) Environmental Policy Analysis with Limited Information: Principles and Applications of the Transfer Method, Edward Elgar Publishing.

Douglas AJ, Johnson R (1993) In-stream Flow Assessment and Economic Valuation: A Survey of Nonmarket Benefits Research, International Journal of Environmental Studies, 43: 89–104.

ECONorthwest 1997 Water Management Study: Upper Rio Grande Basin (Draft Report). Prepared for the Western Water Policy Review Advisory Commission.

Feather P, Hellerstein D (1997) Calibrating Benefit Function Transfer to Assess the Conservation Reserve Program, American Journal of Agricultural Economics, 79: 151–162.

Harrison D, Nicholds AL, Bittenbender SL, Berkman ML (1993) External Costs of Electricity Utility Resource Selection in Nevada, Report to Nevada Power Company, March, Cambridge, MA: National Economic Research Associates.

Kernodle JM, McAda DP, Thorn CR (1995) Simulation of Groundwater Flow in The Albuquerque Basin, Central New Mexico 1901–1994, US Geological Survey Water Resources Investigations Report 94–4251.

Kirchhoff S, Colby B, LaFrance J.T (1997) Evaluation the Performance of Benefit Transfer: An Empirical Inquiry, Journal of Environmental Economics and Management, 33: 75–93.

Kirchhoff S (1998) Benefit Function Transfer vs Meta-Analysis as Policy Making Tools: A Comparison Workshop on Meta-Analysis and Benefit Transfer: State of the Art and Prospects: Tinbergen Institute, Amsterdam.

Kohler DF, Mitchell BM (1984) Response to Residential Time-of-Use Electricity Rates: How Transferable are the Findings? Journal of Econometrics, 26:141–178.

Lee R, Krupnick AJ, Burtraw D (1995) Estimating Externalities of Electric Fuel Cycles: Analytical Methods and Issues, Washington, DC: McGraw-Hill/Utility Data Institute.

Middle Rio Grande Council of Governments (1999) Draft Surface Water Budget.

Middle Rio Grande Council of Governments of New Mexico (1997) Socioeconomic Estimates and Forecasts to 2050 For State Planning and Development District 3 and Southern Santa Fe County (MRGCOG R-97-3).

Patrick RH (1990) Rate Structure Effects and Regression Parameter Instability Across Time-of-Use Electricity Pricing Experiments, Resources and Energy 12 (2):179–195.

Rowe RD, Benow SS, Bird LA, Callaway JM, Chestnut LG, Eldridge MM, Lang CM, Latimer DA, Murdock JC, Ostro BD, Patterson AK, Rae DA, Whilte DE (1994) New York State Environmental Externalities Cost Study, Report Prepared for Empire State Electric Energy Research Corporation and New York State Energy Research and Development Authority, Boulder, CO. RCG/Hagler, Bailey, Inc.

Smith VK, Kaoru Y (1990) Signals or Noise? Explaining the Variation in Recreation Benefit Estimates, American Journal of Agricultural Economics, May:421–433.

Southwest Assessment (1999) Draft Southwest Assessment Summary Report for the US Global Change Research Program, January 1, 2000.

State of New Mexico Department of Agriculture (1995) New Mexico Agricultural Statistics.

State of New Mexico Department of Agriculture (1996) New Mexico Agricultural Statistics.

U.S. Fish and Wildlife Service (1999) Draft Recovery Plan for The Rio Grande Silvery Minnow.

U.S. Water Resource Council (1983) Economic and Environmental Principles and Guidelines for water and related Land Resources Implementation Studies.

Water Resources Research Special Issue, 28:651–752.

Watts G, Noonan W, Maddux H, Brookshire D (2001) The Endangered Species and Critical Habitat Designation: Economic Consequences for the Colorado River Basin, Endangered Species Protection in the United States: Biological Needs, Political Realities, Economic Consequences, eds, T. Shogren and J. Tschirhart, Cambridge University Press.

Williams M (1985) Estimating Urban Residential Demand for Water Under Alternative Price Measures, Urban Economics, 18:213–225.

Wilson BC, Lucero A.A. (1997) Water Use By Categories in New Mexico Counties and River Basins, and Irrigated Acreage in 1995 New Mexico, State Engineer Office Technical Report 49.

Workshop of the Association of Environmental Economists (1992) Utah, U.S.

NICK HANLEY, ROBERT E. WRIGHT, BEGONA ALVAREZ-FARIZO

ESTIMATING THE ECONOMIC VALUE OF IMPROVEMENTS IN RIVER ECOLOGY USING CHOICE EXPERIMENTS: AN APPLICATION TO THE WATER FRAMEWORK DIRECTIVE*

1. INTRODUCTION

The Water Framework Directive (2000/60) will bring about major changes in the regulation and management of Europe's water resources. Major changes include:

- a requirement for the preparation of integrated catchment management plans, with remits extending over point and non-point pollution, water abstraction and land use;
- the introduction of an EU-wide target of 'good ecological status' for all surface water and groundwater, except where exemptions for 'heavily-modified' water bodies are granted;
- the introduction of full social cost pricing for water use; and
- the incorporation of estimates of economic costs and benefits in catchment management plans.

How exactly regulators will interpret 'good ecological status' is at present not finalised. However, it is clear that it represents a wider set of parameters than the chemical and biological measures of water quality that have previously dominated EU water quality regulation, such as Biological Oxygen Demand or Ammonia (NH_3) levels. In this paper, we use three indicators of ecological status which ordinary people see as important, but which are also consistent with regulator's expectations about the scientific interpretation of this concept. We take ecological status to be determined by three broad factors: healthy wildlife and plant populations; absence of litter/debris in the river; and river banks in good condition with only natural levels of erosion. Recent assessments for UK waterbodies indicate that a significant fraction of rivers, lochs (lakes), estuaries and coastal waters will require improvements if they are to meet 'good ecological status' (DETR, 1999; Scottish Executive, 2002).

One main focus in this paper is therefore on the values people place on improvements in these three indicators, and thus on the non-market economic benefits of moves towards good ecological status. Whilst benefit estimates do exist for implementation of the Water Framework Directive (WFD), these are at present highly

*Reprinted from Journal of Environmental Management 78(2006) 183–193 with permission from Elsevier.

S. Navrud and R. Ready (eds.), Environmental Value Transfer: Issues and Methods, 111–130.

incomplete (WRc, 1999; Scottish Executive, 2002). However, we are also interested in the practicalities of environmental management using environmental valuation. Valuation exercises are expensive and time consuming, and regulators are very unlikely to have the time or money to commission original valuation studies for every catchment. Benefits transfer, the process of taking estimates from one context and adjusting and then applying them to another, is therefore likely to be important. Accordingly, we conduct a benefits transfer test across two similar rivers, to see what errors are likely to be experienced if benefits transfer procedures are used as part of implementing the WFD.

In what follows, Section 2 briefly describes the Choice Experiment method of environmental valuation and outlines some current issues in benefits transfer. Section 3 describes the case study rivers and survey design. Section 4 presents results, whilst Section 5 concludes.

2. METHODOLOGICAL APPROACH

2.1. *Choice Experiments*

The methodology we use to estimate the value of improvements in river ecology is Choice Experiments. Choice experiments (CE) are becoming a popular means of environmental valuation (Hanley, Mourato and Wright, 2001; Bennett and Blamey, 2001). Choice experiments are one example of the stated preference approach to environmental valuation, since they involve eliciting responses from individuals in constructed, hypothetical markets, rather than the study of actual behavior. The Choice Experiment technique is based on random utility theory and the characteristics theory of value: environmental goods are valued in terms of their attributes, by applying probabilistic models to choices between different bundles of attributes. By making one of these attributes a price or cost term, marginal utility estimates can be converted into willingness-to-pay estimates for changes in attribute levels, and welfare estimates obtained for combinations of attribute changes. The decision to use a CE approach here was driven by the desire to estimate values for different component parts, or aspects, of water quality, as interpreted by the WFD. These component parts constitute the attributes in the CE design detailed below.

2.2. *Previous Studies of River Ecology Changes using Choice Experiments*

Several authors have previously used CE to estimate the value of improvements in river quality. Adamowicz, Louviere and Williams (1994) studied people involved in water-based recreation in Alberta. They recruited a sample of 1,232 members of the general public, from which a 45% response rate was achieved. The attributes used were landscape terrain, fish size, catch rate, water quality, facilities (e.g. campsite), distance from home and fish species present. The authors found significant effects on utility from changes in fish size, catch rate, water quality and distance from home.

Burton et al. (2000) studied public preferences for catchment management plans in the Moore Catchment, Australia. This area is subject to problems of salinity, eutrophication and flooding, which are all linked to farming activities. Two populations were surveyed, one in the city of Perth and one in rural towns. The attributes used were area of farmland affected by salinity, area of farmland planted with trees, ecological impacts on off-farm wetlands, risk of major flood, changes in farm income, and annual contribution to management plan. The main findings were that the importance people placed on the cost attribute depended on their attitudes to environmental responsibility. Adverse impacts on wetlands, and losses (but not gains) in farm incomes also had significant impacts on utility.

Heberling et al. (2000) studied the benefits of reducing pollution from acid mine drainage in western and central Pennsylvania. Focus groups helped identify the attributes used in the questionnaire: water quality, miles of river restored, travel time from home to site, easy access points, and household costs. Water quality was measured according to what uses could be made of the stream, and took the levels 'drinkable' and 'fishable' and 'swimmable'. The water quality variables were statistically significant determinants of choice in the majority of models, with costs always being significant.

A closely-related technique to choice experiments is contingent ranking. Georgiou et al. (2000) used contingent ranking to estimate the benefits of water quality improvements in the River Tame in Birmingham. People were asked to rank three combinations of four attributes. These were:

- type of fishing (trout/salmon and good game; some game fish species return; a few game fish species return; fish stocks extinct)
- plants and wildlife (otters survive; increase in number and types of insects and greater numbers of bids; more plants and waterfowl; very limited wildlife)
- boating and swimming (both, boating only, swimming only, neither); and
- cost (extra council taxes): £2.50/month, £1.25/month, £0.42/month, zero)

Responses were used to estimate Willingness to Pay for marginal reductions in Biological Oxygen Demand and total ammonia.

Studies using other stated preference techniques to estimate values of ecological changes in river quality may also be found in the literature, notably contingent valuation studies of improvements in low-flow conditions (Hanley et al., 2003; Garrod and Willis, 1996). However, we do not review them here.

A conclusion from this brief review of existing literature is that no study exists which uses choice experiments to estimate the value of improvements in the concept of ecological status as embodied within the Water Framework Directive. Our study is a first step in this policy-relevant direction. Before detailing our study design, however, it is important to review another aspect of environmental valuation relevant to the Directive: Benefits Transfer.

2.3. Benefits Transfer

Valuation studies are extensive and time-consuming. For this reason, the policy community has become increasingly interested in benefits transfer techniques (Bateman et al., 2002). Benefits transfer (BT) is a method for taking value estimates from original studies, and adjusting them for use in some new context. The two main approaches to BT are:

- the transfer of adjusted mean values. Mean Willingness to Pay (WTP) estimates taken from the original study or studies are adjusted to account for differences in the environmental characteristics of the new site/context, and/or for differences in the socio-economic characteristics of the affected population at the new site.
- the transfer of benefit functions. Benefit functions are regression equations which explain variations in WTP and/or preferences across individuals according to variations in socio-economic factors and, in some cases, environmental characteristics. A benefits function can be used to produce estimates of WTP.

In both cases, meta analysis (that is, the quantitative analysis of a collection of past studies) can be used to inform the BT process.

Much academic work has taken place in the past 10 years, testing alternative BT methods, and assessing their accuracy. The academic jury is still 'out' on the validity of BT. Studies by Bergland et al. (1995); Barton (2002) and Rozan (2004) largely reject the validity of benefits transfer, both in terms of the transfer of adjusted mean values and the transfer of benefit functions. Brouwer (2000) surveys seven recent benefits transfer studies and finds that the average transfer error is around 20–40% for means and as high as 225% for benefit function transfers, whilst Ready et al. (2001) find a transfer error of around 40% in a multi-country study on the health benefits of reduced air pollution. Shrestha and Loomis (2001) find an average transfer error of 28% in a meta-analysis model of 131 US recreation studies. As Barton points out, though, even fairly small transfer errors (11–26% in his case) can be rejected using the statistical tests favoured by economists. However, this has not stopped the development of large BT software packages, such as the EVRI package, developed by Environment Canada, for use in policy analysis.

One debate on-going at present is whether more complex BT approaches necessarily do better than simple ones. Barton (op cit) finds a simple adjusted means transfer gets closer to original site values than the transfer of benefit functions. The opposite finding, however, is reported in Desvouges et al. (1998). Finding acceptable benefits transfer methods is essential to the wider use of environmental valuation in policy. However, the standards of accuracy required in academic work may exceed those viewed as tolerable by policy-makers, especially in prioritising or filtering alternative investments in water quality.

3. STUDY DESIGN

3.1. Physical Context

We located our choice experiment in the context of improvements to the ecology of the River Wear, in County Durham, England; and the River Clyde, in Central Scotland. These were chosen as broadly representative of the kind of waterbodies in the UK where moderate improvements in water quality are likely to be needed in order to meet Good Ecological Status.

The River Wear catchment extends from Burnhope Moor in the Pennines to the North Sea. Population is concentrated in the eastern half of the area, which includes Durham and Sunderland. Throughout much of the 20th century the lower sections of the river were heavily polluted by industry and mining, but have now recovered and support a migratory fishery. The focus for this study is that part of the River Wear which flows through the city of Durham, and which is graded as 'C' on the Environment Agency's General Quality Assessment scheme (interpreted as 'fair' quality). Existing problems include litter, algal growth and acidity problems due to mine drainage. Problems also exist with loss of bankside vegetation, increased erosion, and a decline in habitat and associated fish and wildlife populations. Within the Wear are many man-made structures built in and across river channels. These have important impacts on the way the river functions, altering flows and gravel movements and hindering migration of fish upstream. In terms of recreational uses the River Wear is important as a coarse and game fishery and also as a centre for other water-based recreation (formal and informal). The river also plays an important role in recreation and tourism.

The River Clyde is approximately 121 km long. During its journey from its source in the Beattock Hills to its tidal estuarine limits in Glasgow, its quality varies greatly. Discussion with regulators (the Scottish Environmental Protection Agency) led to the selection of the Clyde from Lanark to Cambuslang Bridge as the area for study. This mainly urbanised stretch has recreational and tourist attractions, and encompasses areas of great beauty like the Falls of Clyde, but also has some of the most problematic stretches in terms of water quality. Most of this section was graded 'B' using the Scottish river classification system, which is equivalent to the 'C' grade for the Wear under the General Quality Assessment classification system (i.e. fair quality, but in need of improvement to reach 'good ecological status').

3.2. Steps in Choice Experiment Design

Focus groups were recruited from local residents living around the two rivers in both case study areas in order to (i) gauge local attitudes to the rivers and to their problems (ii) investigate current uses of the two rivers and (iii) identify the attributes by which the rivers could best be characterised. We also gauged reaction to the idea of the need to pay for improvements in river ecology. As a result of group discussions, backed up by discussion with officers from both the Environment Agency (the regulator in England) and the Scottish Environmental Protection Agency, three river quality attributes were chosen for the CE. These were in-stream ecology, aesthetics/appearance, and bankside conditions, and are shown in detail in Figure 1. Each attribute was set at one of two levels. The 'fair' level was described in such a way as to be consistent with current conditions on the Rivers Wear and Clyde. The 'good' level was consistent with regulators' expectations as to what will likely constitute good ecological quality status under the Water Framework Directive. Note that none of these attributes are necessarily consistent with what an ecologist would choose in terms of either indicators of the ecological

Environmental attributes and levels used in the choice experiment		
	"GOOD" LEVEL	"FAIR" LEVEL
Ecology	• Salmon, trout and course fish (e.g. pike) • A wide range of water plants, insects and birds	• Only course fish (e.g. pike) • A poor range of water plants, insects and birds
Aesthetics / Appearance	• No sewage or litter	• Some sewage or litter
River Banks	• Banks with plenty of trees and plants • Only natural erosion	• Banks with few trees and plants • Evidence of accelerated erosion

Figure 1. Environmental attributes and levels used in the choice experiment

health of a waterbody, or underlying factors driving changes in ecological status: they merely represent the characteristics of 'water quality' as perceived by the general public.

A cost or price attribute was established as higher water rates payments by households to the local sewerage operator, Northumbria Water, for the R Wear sample; and to the local authority (Lanarkshire Council) for the R. Clyde sample[1]. Focus groups generally accepted the idea that improvements had to be paid for, and water rates were viewed as a realistic payment mechanism. The price vector used in the design was {£2, £5, £11, £15, £24}, and was chosen based on previous contingent valuation studies in the UK of river improvements.

Attributes and levels were then assigned into choice sets using a fractional factorial design. Due to the simple nature of the design, blocking of the choice sets (that is, introducing an additional attribute to dis-aggregate choice sets into manageable groups) was not necessary. Each respondent answered 8 choice questions. Each question consisted of a three-way choice: option A and option B, which gave an improvement in at least one attribute for a positive cost; and the zero-cost, zero-improvement status quo. Each choice card showed the attribute levels pictorially; a preceding section of the questionnaire explained the importance of each attribute to overall ecological quality [2]. Options A and B can be thought of as representing the outcomes of alternative catchment management plans for each river, with their associated costs.

Sampling was undertaken with a randomised quota-sampling approach, using in-house surveys by trained market research personnel in the autumn of 2001. We collected 210 responses for each river. Whilst this sample is rather small, it is comparable to others reported in the CE literature (e.g. Hanley et al., 2002; Bergmann et al., 2005). A larger sample size would, however, lead to lower standard errors and greater confidence in interpretation of our results.

4. METHODOLOGY

4.1. Statistical Model

The method of Choice Experiments is an application of Lancaster's 'characteristics theory of value' combined with 'random utility theory', and is therefore firmly based in economic theory. Individuals are asked to choose between alternative goods, which are described in terms of their attributes, one of which is price (or some proxy for price). Consider the two alternatives case ($C = 2$). The underlying utility function of individual 'i' is of the form:

$$U_{ij} = U(X_j, P_j) \tag{1a}$$

$$U_{ik} = U(X_k, P_k) \tag{1b}$$

where: 'X_j' and 'X_k' are vectors of attributes describing alternatives 'j' and 'k' and 'P_j' and 'P_k' are the prices or costs associated with each of the alternatives. Individual 'i' will choose alternative 'j' over alternative 'k', if and only if:

$$U_{ij} > U_{ik}. \tag{2}$$

That is, the total satisfaction received from 'consuming' alternative 'j' exceeds that received from alternative 'k'. If $U_{ij} = U_{ik}$ then the individual is indifferent between the two alternatives. If $U_{ij} = 0$ and $U_{ik} = 0$, then the individual receives no satisfaction from either alternative.

The utility functions associated with the comparison in Eq. (2) may be partitioned into two components:

$$U_{ij} = V(X_j, P_j) + \varepsilon(X_j, P_j), \tag{3a}$$

$$U_{ik} = V(X_k, Pk)) + \varepsilon(X_k, P_k), \tag{3b}$$

where the first term on the right-hand side of each of these expressions is deterministic and observable (sometimes referred to as an indirect utility function) while the second term is random and unobservable. ? (?) attribute this randomness to a variety of factors, including unobserved attributes, unobserved taste variations and measurement errors. Therefore, the probability that individual 'i' will choose alternative 'i' over alternative 'j' is:

$$\text{Prob}_i(j|C) = \text{Prob}(V_{ij} + \varepsilon_{ij} > V_{ik} + \varepsilon_{ik}), \tag{4}$$

where 'C' is the complete set of alternatives (in this case two alternatives, 'j' and 'k') and 'ε_{ij}' and 'ε_{ik}' are error terms.

In order to make Eq. (4) empirically tractable, assumptions must be made regarding the structure of the error terms. The usual assumption is that the errors are Gumbel-distributed and independently and identically distributed. This implies that:

$$\text{(5)} \qquad \text{Prob}_i(j|C) = \exp(\mu V_{ij})/\Sigma_C \exp(\mu V_{iC}),$$

where 'μ' is a scale parameter which is inversely proportional to the standard deviation of the error distribution. This parameter cannot be separately identified and is therefore typically assumed to be one. This assumption implies a constant error variance and also implies that as $\mu \rightarrow \infty$ the model becomes deterministic.

In order to derive an explicit expression for this probability, it is necessary to make an assumption regarding the distribution of the error terms (discussed below). As mentioned above, $V(\cdot)$ is composed of attributes describing each alternative. If $V(\cdot)$ is linear in its arguments and additive with a constant term (θ) then the indirect utility functions are:

$$\text{(6a)} \qquad V_{ij} = \theta_0 + \alpha P_j + \beta' X_{ij},$$

$$\text{(6b)} \qquad V_{ik} = \theta_0 + a P_k \beta' X_{ik},$$

and Eq. (5) becomes:

$$\text{(7)} \qquad \text{Prob}_i(j|C) = \exp[\mu(\theta_0 + \alpha P_j + \beta' X_{ij})]/\Sigma_C \exp[\mu(\theta_0 + \alpha P_j + \beta' X_{ij})],$$

In order to derive an explicit expression for this probability, it is necessary to know the distribution of the error terms. A typical assumption is that they are independently and identically distributed with an extreme-value (Weibull) distribution. This distribution for the error term implies that the probability of any particular alternative being chosen as the most preferred can be expressed in terms of the logistic distribution, which results in a specification known as the 'conditional logit model' or (less correctly) the 'multinomial logit model' (McFadden, 1974):

$$\text{(8)} \qquad \text{Prob}(j|C) = \exp(\theta_0 + \alpha P_j + \beta' X_j)/\Sigma_C \exp(\theta_0 + \alpha P_c + \beta' X_C).$$

Eq. (8) can be estimated by conventional maximum likelihood procedures (see Greene, 1997). Therefore, standard Likelihood ratio-based tests can be used to test restrictions on the parameters (or group of parameters), to test for differences in parameters across sub-groups (e.g. men versus women), and to evaluate goodness-of-fit (e.g. calculate pseudo R^2-values)

Individual-specific characteristics (shifters) that affect utility, such as income, education, marital status etc., can also be included in this specification. However, since these characteristics do not vary across the alternatives, such variables cannot be entered into Eq. (8) as linear arguments (e.g. an individual's education is the same regardless of whether he/she chooses alternative 'j' or 'k'). Such variables can only be included by interacting them multiplicatively with the attributes or the constant.

Adamowicz, Louviere and Williams (1994) show that estimates of consumers surplus associated with changes in the level of attributes can be easily derived from the estimates of this multi-nomial logit model. This calculation is based on an interpretation of the parameter of the price attribute being equal to the marginal utility of income. For the case of two alternatives, this involves summing the marginal values for each attribute when moving from a lower level of the attribute to some higher level of the attribute (for the case of linear demand). More formally, if 'X' is composed of '$X_1, X_2, \ldots, X_a$' attributes the implicit price (or willingness-to-pay) associated with any individual attribute, 'a' is:

$$p_a = -\beta_a/\alpha, \tag{9}$$

where from 'α' is the parameter estimate of the price variable 'P' and 'β_a' is the parameter estimate of the specific attribute 'X_a'. Standard errors and confidence intervals can also be calculated for these implicit prices, although there is still considerable discussion relating to what is the most appropriate method to use (Poe et al., 1994, 1997).

4.2. *IIA*

An important implication of this specification is that selections from the choice set must obey the 'independence from irrelevant alternatives' (IIA) property (or Luce's Choice Axiom; see Luce, 1959). This property states that the relative probabilities of two options being selected are unaffected by the introduction or removal of other alternatives. This property follows from the independence of the error terms across the different options contained in the choice set. If a violation of the IIA hypothesis is observed, then more complex statistical models are necessary that relax some of the assumptions used. These include the multinomial probit (Hausman and Wise, 1978), the nested logit (**?**, **?**), the random parameters logit model (Meijer and Rouwendal, 2000; Revelt and Train, 1998; Train, 1998; Train, 2003; Wedel and Kamakura, 2000) and the heterogeneous extreme value logit (Allenby and Ginter, 1995). There are numerous formal statistical tests than can be used to test for violations of the IIA assumption, with the test developed by Hausman and McFadden (1984) being the most widely used.

4.3. *Zero-bids and Status Quo Responses*

In most CVM studies a significant proportion of respondents usually report 'zero bids'. Likewise in CE studies it is often the case that a significant proportion of respondents select the 'status quo' option. In this sense, status quo responses are analogous to zero bids. In both cases, this implies that they are not willing-to-pay for the changes specified in the design. Zero bids and status quo responses may be categorised into three types. The first are 'genuine zero bids', where the respondent indicates that they not willing to pay anything because they do not value it in a utility sense. The second are 'protest bids', where the respondent reports a zero bid for reasons other than the respondent placing a zero value on the good in question. For example, the respondent disapproves of the principle of paying for environmental protection since they believe it should be required by law. The third are 'don't know' responses, where the respondent is simply uncertain about the amount they are willing-to-pay, noting that this amount could of course be zero.

Zero bids and status quo responses do not necessarily mean that an individual is unwilling to pay anything. It is likely the case that many of the respondents who report that at not willing-to-pay anything in hypothetical questioning, would actually pay something if they were required to do so 'in reality'. In CVM studies individuals who report zero bids are often excluded from the modelling. That is, the analysis is restricted to only individuals who report positive bids or only to only individuals who report positive bids and genuine zero bids, with individuals who report protest bids or who give 'don't know' responses being excluded (such information can be obtained with ancillary questions).

These empirical strategies are problematic since the samples used may be 'self-selected.' Individuals who report positive and genuine zero bids may be very different to those individuals who report protest bids or 'don't know'. If this is the case, they it is likely the case that any regression-based modelling aimed at evaluating the impact of factors thought to impact on willingness-to-pay may be biased. Alvarez-Farizo et al. (1999) have developed a statistical framework aimed at addressing this form of 'sample selection bias' in the estimation of CVM bid curves. However, the authors are unaware of any research that has tried to apply a similar framework in CE. Such an extension should be possible in principle, although the econometrics required to estimate a model along these lines are not straightforward. No attempt has been made to do this in this paper. Therefore, the reader should keep in mind this potential weakness in the modelling that follows.

4.4. Benefit Transfer and Pooling

An important concept in environmental economics is the notion of 'benefit transfer'. In a nutshell, benefit transfer is the ability to take the results from one 'study site' and apply them to other 'policy sites'. That is, being able to construct estimates of willingness-to-pay applicable based on set of parameters that applicable to a wide range of sites. In terms the substantive problem considered in this paper, it means being able to take the estimates of willingness-to-pay obtained from the River Wear and apply them to the River Clyde (or vice-versus). The time and cost advantages of being able to do this are obvious.

From a statistical point of view, the assessment of benefit transfer concerns testing for the equality of parameters and willingness-to-pay values 'across equations'.

Such tests are straightforward to carryout. In terms of the multi-nominal logit model, the test of the equality of parameters across models is the maximum likelihood extension of the 'Chow test for a structural break' (Chow, 1960). The test of equality of willingness-to-pay estimates across models is an application of the 'Wald test for non-linear restrictions' (Wald, 1939, 1943). These tests are described further below and an excellent discussion of the technical details can be found in Greene (2003). More generally, applicability of the principle of benefit transfer is based on the extent to which data from different samples can be pooled. These tests provide evidence about whether pooling is 'statistically acceptable' which therefore provides the researcher with 'hard evidence' on whether benefit transfer is advisable.

5. RESULTS[3]

Table 1 reports the results of the Hausman test for IIA. This test was carried out on a pooled sample of both survey sites ('Both Rivers') and individually for each survey site ('River Wear' and 'River Clyde'). In all three cases the acceptance of IIA was firmly rejected with the Hausman statistic being very large and statistically significant well below the one per cent level. This suggests that estimating the model as a multi-nomial logit could generate misleading results.

As mentioned above, there are various models that can be used to estimate a CE model in the presence of IIA. The approach that we follow here is the random parameters logit model, which is becoming increasingly popular in applied research. One limitation of the mult-nomial logit model is that it assumes that preferences are homogenous and only one parameter is estimated for each attribute. However, it is likely that preferences differ across individuals. In addition, an individual's preference should not vary across the eight choice set questions they were asked in the survey. The random parameters logit model allows for such variation in preferences across individuals and adjusts for error correlation across the choices made by each individual. Application requires assumptions being made about the distribution of preferences. It is assumed that preferences relating to the three attributes are heterogeneous and follow a normal distribution while preferences towards price are assumed to be homogenous. Therefore, separate parameters are estimated for each individual for each of the three attributes along with a single parameter for all for price.

Table 1. Hausman test for IIA

(1)	(2)	(3)
Sample:	Statistic	Significance level
Both rivers	124.7	$P < 0.01$
River wear	104.6	$P < 0.01$
River clyde	35.5	$P < 0.01$

Before turning to a discussion of the results, it is worth noting that in the River Wear sample, 23.8 per cent of respondents selected the 'status quo' while in the River Clyde sample 27.1 per cent did so. However, only a small number of individuals selected the 'status quo' for all eight of the choice set questions they were asked. These percentages are lower that what is often the case in CVM studies but are not small, so it is unclear whether the results may be biased because of protest bidding, 'don't know' responding, etc.

The estimates presented below are based on models that do not include any covariates. In a set of models not reported here, a series of respondent-specific control variables were included in the specification. These variables were: Whether the respondent has ever visited the site; household income; amount of water bill, the respondent's age; and whether the respondent has children. Inclusion of these variables did not have much impact on the estimates so the discussion below is based on the 'simplest' model that includes the three attributes (a_1, a_2 and a_3) price (p) and alternative specific-constants [$\alpha(A)$ and $\alpha(B)$]. These constants can be thought of as representing all other determinants of utility for each option not captured by the attributes.

Table 2 presents the estimates of the model. Turning first to the multi-nomial logit estimates [Columns (3), (5) and (7)], in all three samples the three attributes have the expected positive signs and all are statistically significant below the one percent level. Likewise, in all three samples, price has the expected negative sign. However, price is not statistically significant at even the generous ten percent level in the River Clyde sample [Column (7)]. Turning next to the random parameters logit estimates [Columns (4), (6) and (8)], in all three samples the three attributes have the expected positive signs and all are statistically significant below the one percent level.Therefore,with respect to the attributes both estimators are generating similar results. It is important to note that in the River Clyde sample price is now statistically significant at the five per cent level. These estimates confirm that people 'value' and are prepared to pay for water quality improvements and such improvements are valued 'even more' the lower the cost associated with obtaining them.

Also shown in Table 2 are the standard deviations and standard errors for the parameters of the random parameters logit estimates. It is interesting to note that the standard deviation for the 'river ecology' attribute is statistically significant at the five per cent level or lower in all three samples. The standard deviation for the 'aesthetics' attribute is only statistically significant (below the one percent level) in the River Clyde sample. The standard deviation of the 'banksides' attributes is not statistically significant in any of the samples. These results suggest two things relating to preferences. The first is that the major component of preference heterogeneity is preferences towards 'river ecology'. The second is that preference heterogeneity in the River Clyde sample compared to the River Wear sample is 'larger'. Put slightly differently, preferences appear to be more homogenous amongst River Wear respondents. Of course the key question that needs to be answered is why do preferences differ so much across these sites?

Table 2. Model estimates

(1)	(2)	(3)	(4)	(5)	(6)	(7)	(8)
	Sample:	Both Rivers		River Wear		River Clyde	
Estimator		Logit	RP Logit	Logit	RP Logit	Logit	RP Logit
a_1	River Ecology	0.281*** [0.034]	0.336***[0.048]	0.298*** [0.048]	0.346*** [0.066]	0.267*** [0.048]	0.333*** [0.069]
a_2	Aesthetics	0.236*** [0.36]	0.290*** [0.048]	0.294*** [0.053]	0.343*** [0.071]	0.188*** [0.050]	0.246*** [0.067]
a_3	Banksides	0.300*** [0.025]	0.362*** [0.040]	0.308*** [0.036]	0.340*** [0.062]	0.298*** [0.036]	0.370*** [0.056]
p	Price	–0.028*** [0.005]	–0.037*** [0.007]	–0.048*** [0.007]	–0.057*** [0.011]	–0.089 [0.065]	–0.172** [0.009]
$\alpha(A)$	Constant for Option A	0.719*** [0.063]	0.806*** [0.089]	1.022*** [0.093]	1.108*** [0.138]	0.430*** [0.08]	0.499*** [0.118]
$\alpha(B)$	Constant for Option B	0.509*** [0.074]	0.581*** [0.093]	0.0822*** [0.106]	0.884*** [0.144]	0.203*** [0.104]	0.271*** [0.131]
Standard deviations of parameters							
$\sigma(a_1)$	River Ecology	–	0.501*** [0.173]	–	0.512** [0.241]	–	0.494** [0.249]
$\sigma(a_2)$	Aesthetics	–	0.682 [0.191]	–	0.533 [0.328]	–	0.793*** [0.252]
$\sigma(a_3)$	Banksides	–	0.008 [0.34]	–	0.075 [0.96]	–	0.028 [0.356]
–2lnL		7,021	7,014	3,472	3,470	3,523	3,517
N		420	420	210	210	210	210

(1) Standard error in parentheses;
(2) '*' = statistically significant at the 10 per cent level; '**'= 5 per cent level; and '***'= 1 per cent level.

It is clearly the case that if the parameters were identical in numeric values for the River Wear and River Clyde samples then it would be acceptable to estimate the model on the two samples pooled together and apply the results to both sites. That is benefit transfer would be applicable. As mentioned above the degree of similarity between the parameters can be formally tested by the maximum likelihood analogue of the Chow test. More formally, this test is a test of the difference in the parameters across the two samples:

$$(10) \qquad \beta'\text{s(Wear)} - \beta'\text{s(Clyde)} = 0$$

The results of this test for both the multi-nomial logit and random parameters logit models are shown in Table 4. In both cases, the equality of parameters is firmly rejected. The Chi-square values are large and are statistically significant well below the one per cent level. This suggests that the structures of the choice models—which are representative of underlying indirect utility functions—are significantly different from each other for the two samples. More generally it suggests that benefit transfer is not advisable.

Table 4 reports the implicit prices, along with their standard errors, obtained by applying Eq. (9). These values are the amount of money individuals are willing-to-pay for the specified improvement given in the table. Most of these prices are statistically significant below the one percent. It is important to note that the multinomial logit and random parameter logit models generate a set of implicit prices that are very similar for the River Wear sample [Columns (5) and (6)]. This suggests that preference heterogeneity is likely not a factor of much importance and the prices are robust. For the River Clyde sample, the multi-nomial logit model generates prices that are not statistically significant to zero [Column (7)]. However, the random parameters logit model gives prices for the River Clyde sample that are statistically significant at the five per cent level [Column (8)]. The fact that these prices are not significant in the multi-nomial logit model but are significant in the random parameters logit model demonstrates the potential importance of controlling for preference heterogeneity in choice experiments.

Ignoring statistical significance for the moment, the point estimates in Table 3 suggest that willingness-to-pay is higher in the River Clyde sample compared to

Table 3. Likelihood ratio test of parameter equality

(1)	(2)	(3)	(4)
Logit		Random parameters logit	
χ^2 value	Significance level	χ^2 value	Significance level
27.1	$P \leq 0.01$	27.8	$P \leq 0.01$

(1) Test is: $\beta'\text{s(Wear)} - \beta'\text{s(Clyde)} = 0$.

Table 4. Willingness-to-pay (implicit prices) estimates for improvements in water quality

(1)	(2)	(3)	(4)	(5)	(6)	(7)	(8)
Sample:		Both Rivers		River Wear		River Clyde	
Attribute	Improvement	Logit	RP Logit	Logit	RP Logit	Logit	RP Logit
River Ecology	From 'fair' to 'good'	£20.17*** [3.03]	£18.19** [2.75]	£12.54*** [1.92]	£12.19** [1.99]	£60.08 [40.13]	£38.70** [16.95]
Aesthetics	From 'fair' to 'good'	£16.91*** [3.00]	£15.68*** [2.62]	£12.35*** [2.15]	£12.07*** [2.09]	£42.38 [29.29]	£28.57** [13.05]
Banksides	From 'fair' to 'good'	£21.53*** [3.48]	£19.57*** [2.83]	£12.92*** [1.92]	£12.67*** [1.85]	£67.08 [47.83]	£42.99** [19.49]

(1) Standard error in parentheses;
(2) '*' = statistically significant at the 10 per cent level; '**' = 5 per cent level; and '***' = 1 per cent level.

Table 5. Wald test for willingness-to-pay (implicit price) equality

(1)	(2)	(3)	(5)	(6)
Estimator	Logit		Random parameters logit	
Price	Wald χ^2 statistic	Significance level	Wald χ^2 statistic	Significance level
River Ecology (p_1)	616.6	$P \leq 0.01$	172.2	$P \leq 0.01$
Aesthetics (p_2)	194.4	$P \leq 0.01$	61.4	$P \leq 0.01$
Banksides (p_3)	795.9	$P \leq 0.01$	234.1	$P \leq 0.01$

Test is: $p_j(\text{Wear}) - p_j(\text{Clyde}) = 0$.

the River Wear sample. This is the case for both estimators. It is also interesting to note that for the River Clyde sample the random parameters logit model gives prices that are about two-thirds of those given by the multi-nomial logit model. In the River Clyde sample, controlling for preference heterogeneity 'pushes' the prices closer to those found in the River Wear sample. In other words, controlling for preference heterogeneity appears to increase the probability that benefit transfer is advisable (Table 4).

In order to explore the notion of benefit transfer more formally, Table 5 shows the results of the Wald test aimed at evaluating whether the implicit prices are the same across the two models. More formally, this test is concerned with the difference in the estimated prices:

$$p_j(\text{Wear}) - p_j(\text{Clyde}) = 0 \tag{11}$$

This test is carried out for the prices generated by both the multi-nomial logit and random parameters logit models. In all cases, equality of prices is firmly rejected. The Chi-square values are all large and all are statistically significant well below the one per cent level.

6. CONCLUSIONS

In this paper, we were interested in seeing (1) what values people place on improvements to watercourses such as are envisaged under the Water Framework Directive and (2) whether choice experiments provide encouraging evidence for benefits transfer in this context. With regard to the former point, three attributes were selected to represent the concept of 'good ecological status' under the Directive: river ecology, which represents aquatic life including fish, plants and invertebrates; aesthetics, which represent the amount of litter in the river; and the quality of banksides both in terms of vegetation and in terms of erosion. For the River Wear, we found that people place insignificantly different values on these three aspects of the quality of rivers. One possible interpretation of this is that all three are seen as equally valid indicators of a 'healthy river', which is all people really care about. Another is that the amount of information provided to respondents was insufficient for them to distinguish between the three attributes. Given this finding, it would have in fact been more straightforward to use Contingent Valuation to value the change from fair to good water quality: however, this was not something the researchers could know prior to undertaking the CE. For the River Clyde, larger differences were found in attribute values, with aesthetic improvements being valued appreciably lower than either river ecology or bankside conditions.

The second purpose of this paper was to carry out tests of benefits transfer. This was thought to be important, since the Water Framework Directive will impose a considerable burden on regulators to compare the costs and benefits of river basin management plans. Finding acceptably-accurate means of benefits transfer will be a vital component of this task. We used an identical survey instrument to value identical improvements on two rivers which are both classified as being of 'fair' quality currently. However, both benefits transfer tests were rejected here: preferences and values differ significantly across the two samples. This is a similar finding to that reached by Morrison et al. (2002), who largely reject transferability of values and preferences in a choice experiment study of wetlands conservation in Australia. We found that people living near the Clyde valued improvements to their local river more highly than people in Durham valued identical improvements to their local river; despite the fact that the former sample was lower income than the latter. This is surprising to the extent that the demand for environmental quality is typically assumed to increase with income—although the elasticity of WTP with respect to income is less than one, implying that poorer groups are willing to give up higher fractions of their income for environmental improvements than richer groups (Kristrom and Reira 1996; Hokby and Soderqvist 2003). This is what we find: people living near the Clyde appear willing to exchange a larger fraction of their income for local environmental improvements than better-off people living near the Wear. Other possible reasons why those living near the Clyde might value improvements more highly than those living near the Wear include differences in quality of nearby rivers (substitute sites), differences between the two rivers in terms of their natural characteristics (e.g. hydrology, scale), differences in cultural attitudes to the two rivers, and different uses to which the two rivers are currently put.

Finally, work clearly needs to progress on finding acceptable methods of benefits transfer for water quality improvements under the Water Framework Directive, since it is hard to see how it can be fully implemented in Europe without such a benefits transfer system being set up. Choice experiments do seem promising in this regard, since they can incorporate variations in both environmental quality and socio-economic characteristics across sites, which would seem a priori to be the biggest drivers of differences in value. The present study shows that simple choice experiments may not be capable of delivering such benefit transfers within conventional limits of statistical significance. However, it may well be that policy-makers will view much lower levels of accuracy as acceptable in practice. The question is: how close is close enough?

ACKNOWLEDGEMENTS

This paper is based on data collected during a study for the UK Department of the Environment, Food and Rural Affairs. We thank Maggie Dewar for research assistance.

7. NOTES

[1] Sewage treatment is privatised in England but remains a public service in Scotland.
[2] For a copy of the questionnaire, please contact the corresponding author.
[3] The statistical package LIMDEP with NLOGIT (Version 3) was used to estimate the models and perform the statistical tests (Greene, 2003).

REFERENCES

Adamowicz W, Louviere J, Williams M (1994) Combining stated and revealed preference methods for valuing environmental amenities. Journal of Environmental Economics and Management 26: 271–292

Allenby G, Ginter J, (1995) The effects of in-store displays and feature advertising on consideration sets. International Journal of Research in Marketing 12: 67–80

Alvarez-Farizo B, Hanley N, Wright RE, Macmillan D (1999) Estimating the benefits of agri-environmental policy: econometric issues in open-ended contingent valuation studies. Journal of Environmental Planning and Management 42: 23–43

Barton D, (2002) The transferability of benefits transfer. Ecological Economics 42: 147–164

Bateman I, Carson R, Day B, Hanemann M, Hanley N, Hett T, Jones-Lee M, Loomes G, Mourato S, Ozdemiroglu E, Pearce DW, Sugden R, Swanson J (2002) Economic Valuation with Stated Preference Techniques: a Manual. Edward Elgar, Cheltenham

Bennett J, Blamey R (2001) The Choice Modelling Approach to Non-Market Valuation, Cheltenham UK and Northampton, MA, USA: Edward Elgar

Bergland O, Magnussen K, Navrud S (1995) Benefit Transfer: Testing for Accuracy and Reliability.DiscussionPaper#D-03/1995.Department of Economics and Social Sciences, The Agricultural University of Norway

Bergmann A, Hanley N, Wright R (2005) Valuing the attributes of renewable energy investments. Energy Policy forthcoming

Brouwer R (2000) Environmental value transfer: state of the art and future prospects. Ecological Economics 32 (1): 137–152

Burton M, Marsh S, Patterson J (2000) Community Attitudes towards Water Management in the Moore Catchment. Paper to Agricultural Economics Society conference, Manchester

Chow G (1960) Tests of equality between sets of coefficients in two tests of equality between sets of coefficients in two linear regressions. Econometrica 28: 591–605

Desvouges WH, Johnson FR, Banzhaf HS (1998) Environmental Policy Analysis with Limited Information. Edward Elgar, Cheltenham

DETR (1999) Prospective costs and benefits of implementing the proposed EU water resources framework directive. Report to DETR, London, prepared by WRc

Garrod G, Willis K (1996) Estimating the benefits of environmental enhancement: a case study of the River Darrent. Journal of Environmental Planning and Management 39 (2): 189–203

Georgiou S, Bateman I, Cole M, Hadley D (2000) Contingent ranking and valuation of water quality improvements. CSERGE discussion paper 2000–18, University of East Anglia

Greene W (1997) Econometric Analysis, New York: Macmillan

Greene WH (2002) NLOGIT (Version 3), Econometric Software Inc. Plainview, New York

Greene WH (2003) Econometric Analysis, fifth ed. Prentice Hall, New York

Hanley N, Mourato S, Wright R (2001) Choice modelling approaches: A superior alternative for environmental valuation? Journal of Economic Surveys 15: 3

Hanley N, Wright RE, Koop G (2002) Modelling recreation demand using choice experiments: Rock climbing in Scotland. Environmental and Resource Economics 22: 449–466

Hanley N, Schlapfer F, Spurgeon J (2003) Aggregating the benefits of environmental improvements: Distance-decay functions for use and non-use values. Journal of Environmental Management 68: 297–304

Hausman J, Wise D (1978) A conditional probit model for qualitative choice: Discrete decisions recognizing interdependence and heterogeneous preferences. Econometrica 42: 403–426

Hausman J, McFadden D (1984) Specification tests for the multinomial logit model. Econometrica 52: 1219–1240

Heberling M, Fisher A, Shortle J (2000) How the Number of Choice Sets Affects Responses in Stated Choice Surveys.Cincinnati, USEPA, Cincinnati

Hokby S, Soderqvist T (2003) Elasticities of demand and willingness to pay for environmental services in Sweden. Environmental and Resource Economics 26 (3): 361–383

Kristrom B, Reira P (1996) Is the income elasticity of environmental improvements less than one? Environmental and Resource Economics 7: 45–55

Louviere J, Henscher D, Swait J (2000) Stated Choice Methods: Analysis and application. Cambridge: Cambridge University Press

Louviere J. (2001) Choice experiments: an overview of concepts and issues. In: Bennett, J., Blamey, R. (eds), The Choice Modelling Approach to Environmental Valuation. Edward Elgar, Cheltenham

Luce RD (1959) Individual Choice Behaviour: A Theoretical Analysis. Wiley, New York

McFadden D (1974) Conditional logit analysis of qualitative choice behaviour. In: Zarembka, P. (ed), Frontiers in Econometrics. Academic Press, New York

McFadden D, Train K (2000) Mixed MNL models for discrete response. Journal of Applied Econometrics 15: 447–470

Meijer E. Rouwendal J (2000) Measuring welfare effects in models with random coefficients, SOM-theme F: Interactions between Consumers and Firms. AKF, Copenhagen

Morrison M, Bennett J, Blamey R (2002) Choice modelling and tests of benefits transfer. American Journal of Agricultural Economics 84 (1): 161–170

Poe GL, Severance-Lossin EK, Welsh MP (1994) Measuring the difference (X - Y) of simulated distributions: A convolutions approach. American Journal of Agricultural Economics 76: 904–915

Poe GL, Welsh MP, Champ PA (1997) Measuring the difference in mean willingness to pay when dichotomous choice contingent valuation responses are not independent. Land Economics 73: 255–267

Ready R, Navrud S, Day B, Doubourg R, Machado F, Mourato S, Spanninks F, Rodriquez M (2001) Benefits Transfer: Are Values Consistent across Countries?. Department of Agricultural Economics, Penn State University

Revelt D, Train K (1998) Mixed logit with repeated choices: Households choices of appliance efficiency level. Review of Economics and Statistics 80: 647–657

Rozan A (2004) Benefit transfer: A comparison of WTP for air quality between France and Germany. Environmental and Resource Economics 29 (3): 295–306

Scottish Executive (2002) Costs and Benefits of Implementation of the EU Water Framework Directive (2000/60) in Scotland. Report to the Scottish executive, Edinburgh by WRc

Shrestha R, Loomis J (2001) Testing a meta analysis model for benefit transfer in international outdoor recreation. Ecological Economics 39 (1): 67–84

Train K (1998) Recreation demand models with taste differences over people. Land Economics 74: 230–239

Train K (2003) Discrete Choice Models with Simulation. Cambridge University Press, Cambridge

Wald A (1939) Contributions to the theory of statistical estimation and testing hypothesis. Annals of Mathematical Statistics 10: 299–326

Wald A (1943) Tests of statistical hypotheses concerning several parameters when the number of observations is large. Transactions of the American Mathematical Society 54: 426–482

Wedel M, Kamakura W (2000) Market Segmentation: Conceptual and Methodological Foundations. Kluwer, Boston

WRc (1999) Potential Costs and Benefits of Implementing the Proposed Water Resources Framework Directive. Final report to DETR, London

IAN J. BATEMAN AND ANDREW P. JONES

CONTRASTING CONVENTIONAL WITH MULTI-LEVEL MODELING APPROACHES TO META-ANALYSIS: EXPECTATION CONSISTENCY IN UK WOODLAND RECREATION VALUES*

1. INTRODUCTION

The past two decades have witnessed an increasing reliance upon benefit-cost analysis (BCA) as a tool for project appraisal and to inform decision making. In the UK, a typical example of this trend is provided by the 1995 Environment Act which brought into being the Environment Agency (EA) and imposed 'general duties' upon the Agency to take account of the costs and benefits arising from its policies (H.M. Government, 1995). For many agencies, particularly those which have explicitly environmental or public good responsibilities, the assessment of benefits necessitated by adopting BCA approaches has led to a growing interest in tools for the monetary valuation of preferences for environmental goods and services. Consequently, expressed preference methods such as contingent valuation (CV) and conjoint analysis (CA) together with revealed preference techniques such as hedonic pricing (HP) and individual and zonal travel cost (TC) have enjoyed an unprecedented increase in application. However, use of such methods raise theoretical, empirical and practical issues. At a theoretical level certain of these various techniques yield different measures of value. Furthermore, the validity of certain modes of application and analysis have been subject to criticism and are associated with recognized biases, exhibited as empirical regularities within the published literature. These issues place an onus upon the analyst to explain to decision makers the consequences of adopting certain study designs. However, from a decision perspective a further and pressing practical issue concerns the fact that individual applications incur both direct and time related costs. Consequently the proliferation of valuation studies has coincided with increased interest in the potential for benefit transfer.

Rosenberger and Loomis (2000) define benefit transfer as "the application of values and other information from a 'study' site with data to a 'policy' site with

*This paper originally appeared in the journal Land Economics. The authors are grateful to the journal and its Editor, Professor Daniel Bromley, for permission to reproduce this paper within the current volume. CSERGE is a designated research center of the UK Economic and Social Research Council (ESRC) who also support the PEDM. The authors are grateful to Simon Gillam, Alastair Johnson and the UK Forestry Commission who provided further funding for this research.

S. Navrud and R. Ready (eds.), Environmental Value Transfer: Issues and Methods, 131–160.

little or no data" (p1097). A number of approaches to undertaking transfers are available[1] including simple transfer of unadjusted point estimates, transfer of benefit demand functions and meta-analysis. As the simplest approaches cannot incorporate the characteristics of a given site within the transfer exercise, considerable attention is being given to the development of methods for transferring benefit demand functions (Loomis, 1992; Bergland et al., 1995; Loomis et al., 1995; Downing and Ozuna, 1996; Kirchhoff et al., 1997; Brouwer and Spaninks, 1999; Brouwer and Bateman, 2000). However, results are mixed with some studies reporting considerable success while others indicate abject failure. Given this and the empirical difficulties of such studies, a substantial literature has developed regarding the applications of meta-analysis techniques as a basis for benefit transfer.

Meta-analysis is the statistical analysis of the summary findings of prior empirical studies for the purpose of their integration (Glass, 1976; Wolf, 1986). Developed over the last thirty years, it has most commonly been applied in the fields of experimental medical treatment, psychotherapy, and education. Typically, these experiments took place in well-controlled circumstances with standard designs. Deviation from such specifications increases the problems with any cross-analysis (Glass et al., 1981)[2].

Despite problems, meta-analysis offers a transparent structure with which to understand underlying patterns of assumptions, relations and causalities, so permitting the derivation of useful generalizations (Hunter et al., 1982). It permits the extraction of general trend information from large datasets gleaned from numerous studies which would otherwise be difficult to summarize. In comparison with other benefit transfer techniques, Rosenberger and Loomis (2000) identify three advantages of adopting a meta-analysis approach: (i) it typically collates information from a greater number of studies, (ii) it is relatively straightforward to control for methodological differences between valuation source studies, (iii) benefit transfer is readily affected by setting explanatory variable values to those at the desired target site be it a previously surveyed, unsurveyed or just proposed (i.e. currently non-existent) site.

Table 1 extends reviews by Van den Bergh et al., (1997) and Smith and Pattanayak (2002) to provide a brief summary of meta-analysis studies in this area. As can be seen, while analyses have addressed a number of issues, the bulk of applications have been within the field of recreation benefits valuation.

The empirical applicability of meta analysis to any given context is determined by the number, quality and comparability of studies available to the researcher (Desvousges, et al., 1998). Here there is a difficult trade-off between the desire to expand analyses so as to enhance the applicability of results to different goods, provision changes, locations, contexts, etc., and the consequent increase in data demands which such expansions entail. For example, Rosenberger and Loomis (2000) consider a wide range of outdoor recreation activities (10 separate categories ranging from fishing to rock climbing to snowmobiling) across a very extensive area, the US and Canada. This analysis requires a large valuation dataset

Table 1. Meta-analysis studies in environmental and resource economics

Subject area	Study authors
Recreation benefits	Bateman et al., (1999a, 2000), Markowski et al., (2001), Rosenberger and Loomis (2000), Shrestha and Loomis (2001), Smith and Kaoru (1990a), Sturtevant et al. (1995), Van Houtven et al., (2001), Walsh et al. (1990, 1992),
Price elasticity in TC studies	Smith and Kaoru (1990b)
CV versus revealed preference	Carson et al. (1996)
Multiplier effects of tourism	Baaijens et al. (1998), Van den Bergh et al., (1997, Ch9)
Wetland functions	Brouwer et al., (1999), Woodward and Wui (2001)
Groundwater quality	Boyle et al., (1994), Poe et al., (2001)
Price elasticity for water	Espey et al., (1997)
Urban pollution valuation	Smith (1989), Smith and Huang (1993), Smith and Huang (1995), Schwartz (1994), Van den Bergh et al., (1997, Ch10)
Noise nuisance	Button (1995), Nelson (1980), Van den Bergh et al., (1997, Ch4)
Congestion and transport	Button and Kerr (1996), Van den Bergh et al., (1997, Ch13 and 14), Waters (1993)
Visibility and air quality	Desvousges et al., (1998), Smith and Osborne (1996)
Endangered species	Loomis and White (1996)
Valuation of life estimates	Mrozek and Taylor (2002), Van den Bergh et al., (1997, Ch11)

and their study utilizes 682 value estimates from 131 separate studies. By contrast the meta-analysis presented in this paper considers just one type of activity, recreation in open-access woodlands, and just one geographical area, Great Britain, a land area just over 1% the size of that considered by Rosenberger and Loomis. Our analysis is initially restricted just to measures obtained by application of the CV method yielding a dataset of 44 value estimates from 11 studies. A second analysis supplements these data with results obtained from 6 TC studies, bringing the total number of value estimates to 77[3]. While this is less than the size of the Rosenberger and Loomis dataset (reflecting the fewer number of studies conducted in Great Britain) the much smaller geographical boundaries of our study, and its focus upon just one activity, mean that data are placed under considerably less stress, enhancing the reliability of resultant benefit transfer estimates. The disadvantage of this focus is that our results are not readily applicable to other activities or to areas outside Great Britain.

The study described here embraces two objectives. The minor of these concerns the extent to which meta-analysis confirms expectations, derived from theory and empirical regularities, regarding the relationship of values derived from the various permutations of study design represented in our assembled dataset. In so doing we seek to highlight to decision makers (and researchers) the influence upon value estimates of adopting different methods or analytical techniques and so directly address concerns regarding the variability of valuation estimates for apparently similar goods. As a direct extension of this investigation we address the issue of whether, after allowing for design choice, different authors are associated with significantly different valuation estimates. Evidence for such effects would constitute a substantial criticism of valuation studies raising the charge that authors tailor findings to the desires of those commissioning research. However, the principal objective of this study is analytical as we detail alternate approaches to the construction of meta-analysis models.

The first and second meta-analyses presented here are conducted by applying conventional regression techniques to, initially, the subset of CV estimates and subsequently to the full set of CV and TC estimates. These analyses provide a basis for illustrating the limitations of such conventional regression techniques in comparison with a third analysis obtained through application of multilevel modeling (MLM) methods (Goldstein, 1995) to the full dataset of CV and TC observations. As discussed in detail subsequently, the MLM approach allows the researcher to explicitly incorporate potential nested structures within the data, permitting examination of a number of key issues and criticisms of both meta-analysis and valuation studies. Crucially, the MLM approach allows the researcher to relax strong and commonly adopted assumptions regarding the independence of estimates with respect to the numerous natural hierarchies within which they reside. For example, we might expect estimates derived for a given forest to be more similar than those obtained from different forests. Furthermore, whilst not in accord with any theoretical expectation, it might be observed that estimates produced within the same study (or, as highlighted above, by the same author) were more similar than other estimates[4]. While many previous meta-analyses have failed to acknowledge this issue by implicitly assuming independence between estimates (e.g. see some of the studies reported in Van den Bergh et al., 1997), others have adopted weighting approaches, typically by dividing the data associated with each estimate by the number of estimates within the study concerned (e.g. Markowski et al., 2001; Mrozek and Taylor, 2002)[5]. However, both approaches are flawed; independence ignores the real possibility of similarity between nested estimates, while weighting schemes such as those described here result in all studies receiving equal weight irrespective of the fact that we have more information about those containing higher numbers of valuation estimates. Furthermore, such studies typically only address the nesting of estimates within studies and ignore other equally plausible hierarchies such as the nesting of estimates within sites or within authors.

By explicit incorporation of data hierarchies within the analysis, the MLM approach both provides insight into areas in which the independence assumption fails to hold and, through improved modeling of such nesting, ensures that standard errors on parameter estimates are correctly estimated and the significance of explanatory variables accurately assessed. Such a meta-analysis can then defensibly be used to investigate the extent to which valuation estimates conform to expectations. This then links together our analytical objectives with the validity aims of the paper. We can used our refined MLM model to examine both expected differences, such as those associated with different methods and analytical techniques, and unexpected differences, such as the clustering of estimates within authors as described above.

The remainder of the paper is organized as follows. In Section 2 we provide some background to the case study and detail the theoretical and empirical expectations embraced by this application. Section 3 sets out and reports conventional meta-analyses of our data. Section 4 repeats this process for our MLM based model discussing in detail the nature of his approach and how it differs from the conventional approach. Section 5 concludes by highlighting advantages and limitations of the MLM approach, examining the implications of our findings for the validity of valuation exercises and distilling messages for policy makers within this area.

2. THE RECREATIONAL VALUE OF FORESTS

2.1. Background and Data

In terms of land use, British forestry has always been the poor cousin of agriculture. A history of deforestation meant that, by 1900, only 4% of England and Wales and 2% of Scotland and Ireland was forested, by far the lowest level in Europe (Rackham, 1976). The establishment of the UK Forestry Commission in 1919 has done much to reverse this trend and over 10%[6] of the land area of Great Britain is under woodland today. This constitutes the largest single source of open-access land, generates approximately 24–32 million recreational visits per annum (NAO, 1986; Benson and Willis, 1990; 1992), and produces a national aggregate consumer surplus value estimated at between £40 million (Bateman, 1996) and well over £50 million (Benson and Willis, 1992) at current prices. From an economic perspective, the recreational value of forestry is therefore one of its most important benefit streams.

The initial stage of any meta-analysis involves a survey of the relevant literature to identify potential base data studies. Table 2 presents summary details from some 30 studies of UK woodland recreation value yielding over 100 benefit estimates. As can be seen these studies embrace a diversity of recreation value units including per annum, capitalized and per forest values. This variety is not readily incorporated within a meta-analysis and so our study concentrates upon the largest single group of estimates; the per person per visit values.

As outlined above, an initial analysis focused solely upon those estimates obtained from applications of the CV method. Here survey respondents were asked to state their willingness to pay (WTP) for the recreational value of the forests concerned[7]. Table 2 indicates that there are 8 studies yielding 28 estimates of the direct 'use value' of the recreational services provided by forests. Three studies also asked respondents about their WTP for both the present and possible future use (or 'option value'; Weisbrod, 1964; Pearce and Turner, 1990) of forests providing a further

Table 2. Studies of open-access woodland recreation value in Great Britain

Value type	Recreation value unit	Valuation method	No. of studies	Date conducted[1]	No. of value estimates	Value range (£, 1990) (m = million)
Use	Per person per visit.	CV	8[a]	1987–1993	28	£ 0.28 – £ 1.55
Use + option	Per person per visit.	CV	3[b]	1988–1992	16	£ 0.51 – £ 1.46
Use	Per person per visit.	ZTC	3[c]	1976–1988	17	£ 1.30 – £ 3.91
Use	Per person per visit.	ITC	3[d]	1988–1993	16	£ 0.07 – £ 2.74
Use	Per person per year	CV	3[e]	1989–1992	7	£ 5.14 – £ 29.59
Use	Per household capital[2]	CV	3[f]	1990	3	£ 3.27[3] – £ 12.89
Use	FC forests/ conservancy[4]	TC	1[g]	1970	13	£0.1 m – £1.1 m
Use	Total UK value	TC	6[h]	1970–1998	6	£6.5 m – £62.5 m
–	All studies	–	30	1970-1998	106	–

Notes:
1 = Dates refer to the year of study survey rather than publication date.
2 = These studies use a once-and-for-all willingness to pay per household question.
3 = We have recalculated this figure by including those who refused to pay as zero bids.
4 = The FC at the time divided the area of Great Britain into a number of Forest Conservancies and large forests to which these estimates relate.

Study references:
a = Whiteman and Sinclair (1994); Hanley and Ruffell (1991); Bishop (1992); Willis and Benson (1989); Hanley (1989); Willis et al (1988); Bateman and Langford (1997); Bateman, Lovett and Brainard (2003)
b = Bishop (1992); Willis and Benson (1989); Willis et al (1988)
c = Benson and Willis (1992); Hanley (1989); Everett (1979)
d = Willis and Garrod (1991); Bateman (1996); Bateman et al., (1996)
e = Whiteman and Sinclair (1994); Bishop (1992); Bateman (1996)
f = Hanley and Munro (1991); Hanley and Ecotec (1991); Hanley and Craig (1991).
g = H.M. Treasury (1972)
h = H.M. Treasury (1972); Grayson et al (1975); NAO (1986); Willis and Garrod (1991); Benson and Willis (1992); Bateman (1996).

16 estimates of this wider recreational value. In total therefore, these studies yield 44 value estimates[8].

A second analysis was conducted by expanding the dataset to include a further 23 per person per visit value estimates obtained from TC studies. These estimates can be further subdivided. There are 16 individual TC estimates of which 9 use ordinary least squares (OLS) estimators. A further 7 use maximum likelihood (ML) estimators[9]. There are also 17 zonal TC (*ZTC*) estimates all of which use OLS estimators.

Taken together, these CV and TC studies yield 77 value estimates across 21 forests (methods were well represented across these forests[10]). The following list of variables which might potentially influence value estimates were identified:

Method: A set of four binary variables indicating the method/estimation technique adopted to produce the value estimate:

CV = 1 for contingent valuation method used; 0 otherwise

ITCols = 1 for individual travel cost method with ordinary least squares estimators used; 0 otherwise

ITCml = 1 for individual travel cost method with maximum likelihood estimators used; 0 otherwise

ZTC = 1 for zonal travel cost method with ordinary least squares estimators used; 0 otherwise

Option (CV studies only): 1 = use value plus option value requested in WTP question, 0 = use value alone

Elicit (CV studies only): A set of five binary variables identifying the WTP elicitation method employed (variable names as follows: *OE* = open ended, *IB* = iterative bidding, *PC* = payment card, *PCH* = high range payment card, *DC* = dichotomous choice).

Forest: A set of 21 binary variables identifying each of the forests included in at least one of the studies (variable/forest names as follows: *Mercia, Thames Chase, Gt. Northern Forest, Aberfoyle, Derwent Walk, Whippendell Wood, New Forest, Cheshire, Loch Awe, Brecon, Buchan, Newton Stewart, Lorne, Ruthin, Castle Douglas, South Lakes, North York Moors, Durham, Thetford, Dean, Dalby*)

Author: A set of six binary variable identifying authors common to a set of studies (studies can be identified via notes to Table 2; variable names as follows: *Bateman, Bishop, Everett, Hanley, Whiteman, Willis*)

Year: Continuous variable; the number of years before (negative) or after (positive) the base year (1990)

Table 3 reports summary descriptive statistics for the value estimates disaggregated by the various *Method* and *Author* variables. All values were adjusted to a common base year (1990) set roughly in the middle of the density of collated estimates. The table highlights two important features of the dataset that are the subject of subsequent investigation. First, the data is dominated by estimates derived from studies conducted by Willis et al., reflecting their leading role in this field. Second, while the number of estimates is too small to permit calculation of confidence intervals, values do appear to vary by *Method* (e.g. the *ZTC* estimates appear to be substantially higher than those from other approaches) and possibly by Author (although it is clearly important to control for the effect of *Method* here). These initial observations provide focal points for the analyses described subsequently.

2.2. Theoretical and Empirically Derived Expectations

Taken together, theory and empirical regularities reported in the valuation literature provide a rich set of expectations regarding how our valuation estimates may vary according to the differing combinations of valuation methods and analytical

Table 3. Per person per visit woodland recreation value estimates (£, 1990) disaggregated by study author and valuation/estimation method

Method	Whiteman & Sinclair	Hanley et al.	Bishop	Willis et al.	Bateman et al.	Everett	All
CV	3 0.78 (0.66– 0.93) [0.14]	6 1.30 (0.85– 1.55) [0.27]	4 0.89 (0.46– 1.46) [0.46]	28 0.71 (0.28– 1.29) [0.27]	3 1.08 (0.47– 1.55) [0.55]	0 – – –	44 0.84 (0.28– 1.55) [0.36]
ITCols	0 – –	0 – –	0 – –	6 1.46 (0.47– 2.74) [0.84]	3 1.35 (1.07– 1.58) [0.26]	0 – –	9 1.42 (0.47– 2.74) [0.68]
ITCml	0 – – –	0 – – –	0 – – –	6 057 (0.07– 1.13) [0.47]	1 1.20 (1.20– 1.20) [–]	0 – – –	7 0.66 (0.07– 1.20) [0.49]
ZTC	0 – –	1 2.14 (2.14– 2.14 [–]	0 – –	15 2.53 (1.58– 3.91 [0.66]	0 – –	1 1.30 (1.30– 1.30) [–]	17 2.43 (1.30– 3.91) [0.71]
All	3 0.78 (0.66– 0.93) [0.14]	7 1.41 (0.845– 2.14) [0.40]	4 0.89 (0.46– 1.46) [0.46]	55 1.27 (0.07– 3.91) [0.95]	7 1.21 (0.47– 1.58) [0.38]	1 1.30 (1.30– 1.30) [–]	77 1.24 (0.07– 3.91) [0.83]

Cell contents are as follows:
Number of estimates
Mean value (£/person/visit)
(Range: minimum to maximum value)
[StDev of values]

techniques from which they were obtained. This means that we can use our various meta-analyses to examine the extent to which value estimates conform to these expectations. If we were to assume that all our meta-analyses were equally robust we could use them to provide a commentary upon the validity of our valuation estimates. However, as highlighted previously, we have good reason to suspect that our MLM meta-analyses provide a superior alternative to conventionally estimated models. Therefore we can reverse the direction of our test by examining the differing extents to which our various meta-analyses provide results which conform to expectations. Here improved conformity with expectations may be taken as indicating superior performance of a given meta-analysis technique.

What then are the relationships which we might expect to observe within our valuation estimates? Considering the subset of CV studies first, an initial expectation is that questions seeking to elicit the sum of option plus use value should yield higher values than those addressing use values alone (Pearce and Turner, 1990).

Staying within the CV studies, theory also provides clear guidance regarding the impact of changing WTP elicitation method across the various permutations identified in our list of variables. Carson et al., (1999) extend earlier work by Hoehn and Randall (1987) to provide a comprehensive critique of the incentive compatibility of differing WTP elicitation approaches. They note that a simple open ended (OE) WTP question, such as 'What are you willing to pay?', is liable to free-riding behavior, typically leading to understatement of WTP. Conversely, following the work of Farquharson (1969), Gibbard (1973) and Satterthwaite (1975), Carson et al., show that "no response format with greater than a binary response can be incentive compatible without restrictions on preferences" (p.11)[11]. This provides a powerful argument in favor of CV studies adopting the single bound dichotomous choice (DC) format wherein respondents may only choose to accept or reject an interviewer specified discrete WTP sum. For our purposes the DC approach also provides a useful benchmark for testing the theoretical compatibility of our various meta-analyses. For example, we can expect that estimates of WTP derived from OE elicitation techniques should be below those provided by the DC format. Similarly, the iterative bidding (IB) approach, in which respondents can bid up or down from a given starting point, opens the possibility of free riding again resulting in values which are lower than those derived from DC designs. However, these theoretical expectations can be modified in the light of empirical regularities, repeatedly observed in the literature. So, for example, IB studies have been shown to exhibit significant starting point biases (Roberts et al., 1985; Boyle et al., 1985) and in comparative tests have provided value estimates which lie below those given by DC methods but above those derived from OE formats (Bateman et al., 1995). The situation with payment card (PC) approaches, in which respondents choose values from a range presented to them, is equally complex. While recent years have seen a renaissance in the use of PC approaches (Rowe et al., 1996), they fail an incentive compatibility test and in the face of free-riding are again likely to yield under-estimates of true WTP. Furthermore, changes in the PC range given to respondents may induce psychological effects resulting in further biases. It is an examination of one such possible bias which yields our high range payment card (PCH) study (Bateman 1996; Bateman, Lovett and Brainard, 2003) which compares various payment cards including those deliberately designed to stretch well beyond the distribution of woodland WTP measures as obtained by a more typical PC range. This test found that measures derived from the PCH were significantly higher than those obtained from other, conventionally designed, PC approaches.

In summary, we have a variety of theoretically and empirically derived expectations regarding elicitation effects in CV studies. If we rely solely upon theoretical expectations then, in the presence of strategic free-riding within non-incentive compatible formats, we might expect DC derived measures to exceed those obtained from other formats. However, if we temper these theoretical expectations with observed empirical regularities then, while we would still expect OE estimates to be below those from DC studies we would expect IB values to lie between these values. PC measures also suffer incentive incompatibility although we expect those obtained from the PCH format to exceed those from other PC analyses.

Widening our analysis to include the TC estimates again both theory and practice provide some guidance regarding expectations. Comparing these with CV estimates, while the latter yield direct Hicksian welfare measures of WTP, TC methods provide Marshallian consumer surplus estimates. The relationship of these measures depends upon the relative shape of the underlying compensated and uncompensated demand curves for the goods and provision changes concerned (Just et al., 1982; Boadway and Bruce, 1984). Carson et al., (1996) review 83 studies from which 616 comparisons of CV to revealed preference (RP; including TC) estimates are drawn, yielding a whole sample mean CV:RP ratio of 0.89 (95% CI = 0.81 to 0.96), i.e. CV estimates were found to be significantly lower than TC values.

As noted in the preceding section, we can identify a number of distinct types of TC analysis. Certain of the TC based estimates of woodland recreation value rely upon theoretically inappropriate OLS estimation techniques (labeled above as *ITCols* measures). Such techniques are liable to lead to over-estimates of benefits due to an inability to reflect the truncation of non-visitors within an on-site TC survey sample. In contrast other estimates (labeled as *ITCml* measures) have been derived using appropriate maximum likelihood estimators which explicitly model the truncation of non-visitors and are not upwardly biased in this respect. There are also a number of zonal TC (*ZTC*) estimates. These are also likely to yield over-estimates of values both because, in this instance, all used OLS estimators and because of a systematic upward bias in most zonal estimates of travel time and distance (and hence consumer surplus) recently identified by Bateman et al., (1999b).

Taken together these theoretical and empirical factors lead to clear expectations regarding the relationships which should hold in our meta-analyses. In summary these are that, within CV estimates those derived from OE methods should yield the lowest values and that IB estimates should lie above these but below those from DC formats. The relation with PC estimates is less clear other then PCH estimates should exceed those from other PC designs. A general expectation is that TC studies should produce higher values than CV analyses and that within TC estimates those from ZTC and *ITCols* designs should be higher than *ITCml* measures.

Considering the remaining variables identified from our set of estimates, the *Forest* variables are included to identify any influences that variations in the nature of individual sites (e.g. facilities) may have upon stated WTP. We have no theoretical expectations regarding these variables, however empirical work by Brainard et al., (1999, 2001) examining the drivers of demand for forest recreation found that site facilities had very little discernable impact upon observed demand for woodland recreation which was instead driven primarily by locational factors (a result which supports the use of TC methods). This would suggest that the *Forest* variables are likely to prove relatively weak predictors of variation between value estimate. The *Year* variable seems most likely to reflect perceived changes in the availability or desirability of open-access woodland recreation over time and therefore as no prior expectation (although its observed sign is clearly of policy interest). Finally, while we have no theoretical expectation that the *Author* variables should impact significantly upon values, if this did prove to be the case it would constitute a problem for valuation research, giving support to the criticism that some authors yield unusually high or low value estimates.

Together these expectations provide a basis for validating and comparing our various meta-analyses. As outlined above these open, in Section 3, with conventionally modeled analyses initially for just our CV estimates after which we expand to include the TC estimates. Section 4 then re-estimates the latter model using MLM techniques.

3. CONVENTIONALLY ESTIMATED META-ANALYSES

3.1. *Conventional Meta-Analysis of the CV Per Person Per Day Values*

Our initial meta-analysis applied conventional regression methods to our set of CV derived value estimates. This restriction removed the *Method* variables defined previously. However, all other variables were considered within this analysis. Within the *Elicit* variables the *DC* dummy was omitted as an incentive compatible base case against which all other elicitation effects could be observed. Collinearity between the *Author* and *Forest* variables was too high to permit their simultaneous inclusion within a single model (e.g. all studies by Hanley et al., were conducted in Aberfoyle forest although others also undertook studies in this forest). When tested separately the *Forest* variables proved, as per expectations outlined above, to be almost always insignificant predictors of WTP. Given this, the first model reported in this paper concentrates instead upon the *Author* variables. Here we hold the Bateman et al. studies as the base case (as these fall roughly in the middle of values reported by other researchers) and include all other *Author* dummies. Inspection of the *Year* variable indicated little variation across CV estimates relative to our wider dataset and this variable was reserved for subsequent analysis. Tests indicated that a linear model performed better than other functional forms yielding the model given in Table 4[12].

The model detailed in Table 4 fits the data well and conforms well with our theoretical and empirical expectations. The *Option* variable provides a strong, positive and highly significant influence upon stated WTP; as expected respondents facing a 'use plus option value' question stated higher WTP sums than those facing 'use value alone' questions. The *Elicit* variables also conform well with prior expectations. Compared to the incentive compatible *DC* base case all methods yield negative departures (suggesting the anticipated presence of free riding strategies) except for the *PCH* method (where the psychological pressure exerted by the high range payment card seems to have raised stated WTP above that predicted via the *DC* approach; a result which is just significant at the 10% level). The size

Table 4. Conventionally modeled meta-analysis of CV estimates of per person per visit recreation values (£, 1990) for open-access woodland in Great Britain

Variable	Coefficient	95% CI	p
Intercept	1.061	0.999–1.684	< 0.001
Option	0.419	0.290–0.549	< 0.001
OE	−0.443	−0.784– −0.102	0.032
IB	−0.419	−0.901– −0.064	0.144
PC	−0.129	−0.53–0.276	0.589
PCH	0.489	−0.052–0.971	0.090
Bishop	0.065	−0.303–0.434	0.764
Hanley	0.497	0.156–0.838	0.017
Whiteman	0.161	−0.215–0.537	0.466
Willis	−0.118	−0.443–0.208	0.538

$R^2 = 0.718$; $R^2(\text{adj.}) = 0.643$; $n = 44$

and significance of estimated coefficients also conforms well with expectations with the *OE* method exerting the largest downward pressure upon estimates (this being the only effect which is clearly significant at the 5% level), while the *IB* approach results in a lesser negative effect followed by the *PC* results, with both of these proving insignificant. Overall this ordering conforms in all aspects with our prior expectations providing some considerable support for this model. However, this is not the case for our set of *Author* variables. Here the expectation is of no significant effect and while this is generally the case, this is not true of the *Hanley* variable which yields a clearly significant positive effect. This latter result is somewhat worrying as it appears to suggest that reported valuation estimates are partly dependent upon the researcher carrying out the study. We therefore move to our wider dataset, boosted by the TC estimates and re-examine this and the other issues raised above.

3.2. *Conventionally Estimated Meta-Analysis of the CV and TC Per Person Per Day Values*

The analysis was subsequently expanded by the addition of the 23 estimates of per person per visit woodland recreation values obtained using TC methods. In addition to increasing the total observations to 77, this also adds the set of *Method* variables which defines the four method/estimation combinations used (*CV, ITCols, ITCml* and *ZTC*, of which the *CV* studies are held as the base case in subsequent analyses)[13]. The *Elicit* variables were omitted from this analysis as they did not apply to the TC studies, however the expanded period covered by the wider dataset permitted inclusion of the *Year* variable. Models were estimated using conventional regression technique[14]. Table 5 details results for a number of model specifications. In each case, tests of functional form indicate that the linear specification performs roughly as well as other standard forms and is retained for comparability and ease of interpretation.

In Table 5, Model A restricts investigation to the 21 *Forest* variables referring to study site effects, reporting only the three most significant of these dummies. Even these prove highly insignificant, a result which conforms to our expectations as set out previously. In model B these variables are removed in favor of the *Method* dummies which yields a dramatic increase in explanatory power. Perhaps more importantly the sign and significance of these variables conforms well with our prior expectations. Remembering that CV studies form our base case, we find no significant effect from the *ITCml* variable (a result which persisted throughout our analysis such that we omit this variable from subsequent analyses in Table 5 for

Table 5. Conventional meta-analyses of CV and TC estimates of per person per visit recreation values (£, 1990) for open-access woodland in Great Britain

		Models				
		A	B	C	D	E
Intercept		1.1980	0.8368	0.6687	0.6796	0.7697
		(0.1057)	(0.0764)	(0.0862)	(0.0886)	(0.0910)
		[11.34]	[10.95]	[7.75]	[7.67]	[8.46]
		{0.000}	{0.000}	{0.000}	{0.000}	{0.000}
Option				0.2717	0.2626	0.3414
				(0.1436)	(0.1469)	(0.1434)
				[1.89]	[1.79]	[2.38]
				{0.063}	{0.078}	{0.020}
Forest:	*Cheshire*	−0.3780		−0.4029	−0.4153	−0.3962
		(0.3839)		(0.2163)	(0.2203)	(0.2109)
		[−0.98]		[−1.86]	[−1.88]	[−1.88]
		{0.328}		{0.067}	{0.064}	{0.065}
	Loch Awe	0.5653		0.4379	0.4212	0.4154
		(0.4881)		(0.2760)	(0.2812)	(0.2690)
		[1.16]		[1.59]	[1.50]	[1.54]
		{0.251}		{0.117}	{0.139}	{0.127}
	Aberfoyle	0.4445		0.5491		
		(0.3104)		(0.1799)		
		[1.43]		[3.05]		
		{0.156}		{0.003}		
Method:	*ZTC*		1.5973	1.6988	1.7253	1.8461
			(0.1447)	(0.1378)	(0.1418)	(0.1427)
			[11.04]	[12.33]	[12.17]	[12.94]
			{0.000}	{0.000}	{0.000}	{0.000}
	ITCols		0.5876	0.8005	0.7910	0.7994
			(0.1854)	(0.1767)	(0.1805)	(0.1727)
			[3.17]	[4.53]	[4.38]	[4.63]
			{0.002}	{0.000}	{0.000}	{0.000}
	ITCml		−0.1811			
			(0.2062)			
			[−0.88]			
			{0.383}			

(Continued)

Table 5. *(Continued)*

		Models				
Author:	*Hanley*				0.4926 (0.1955) [2.52] {0.014}	0.4390 (0.1881) [2.33] {0.023}
Year						0.0755 (0.0276) [2.74] {0.008}
	R^2 (adj.)	0.020	0.531	0.690	0.678	0.705
	n	77	77	77	77	77

Cell contents are: Estimated coefficient
(s.e.)
[t-value]
/p-value/

where:
Dependent variable = recreational value (£) per person per visit;
Option = 1 where the value estimate relates to the sum of use plus option value and 0 where the value estimated is use value alone[21];
Cheshire = 1 for studies conducted at Cheshire forest and 0 otherwise;
Loch Awe = 1 for studies conducted at Loch Awe forest and 0 otherwise;
Aberfoyle = 1 for studies conducted at Aberfoyle forest and 0 otherwise;
ITCols = 1 if study uses the individual travel cost method with an OLS estimator and 0 otherwise;
ITCml = 1 if study uses the individual travel cost method with a ML estimator and 0 otherwise;
ZTC = 1 if study uses the zonal travel cost method (all employ OLS estimators) and 0 otherwise;
Hanley = 1 if study conducted by Hanley et al., and 0 otherwise.
Year = Continuous variable; the number of years before (negative) or after (positive) the base year (1990);

which the base case now becomes *CV* and *ITCml* estimates) but strongly significant and positive effects associated with the *ITCols* and *ZTC* variables. This result again confirms our prior expectations, suggesting that these latter estimates are upwardly biased.

Model C adds the Option and previously considered *Forest* variables producing a further substantial improvement to model fit which does not change substantially in remaining models. As expected the *Option* variable yields a positive and significant ($\alpha = 10\%$) effect upon values. Interestingly, two of the *Forest* variables also prove significant. However, as mentioned previously, one of these, the site variable for *Aberfoyle* forest, is strongly correlated with the author variable *Hanley* (all of the Hanley et al., studies were conducted at Aberfoyle although other authors also provide estimates for this forest). Given the insignificance of all but one other of the *Forest* variables and our results of Table 4 it seems reasonable to investigate the possibility that it is this *Author* variable which is the root of this effect. Accordingly the move from Model C to Model D exchanges the *Aberfoyle* variable with that for *Hanley*, the latter also proving significant.

Taken together, the results of Model D and that reported in Table 4 could be seen as supporting the argument that valuation estimates may be subject to authorship effects. An alternative explanation is that the Hanley et al., estimates are elevated because of some characteristics of the Aberfoyle site for which they were estimated. Yet a further explanation might be that this result is in some way a product of the conventional modeling approach adopted in this meta-analysis. All of these possibilities are explored subsequently.

Model E adds the final variable *Year* into the analysis. This yields a small, positive and significant coefficient. The result is not particularly robust, becoming insignificant ($p = 0.181$) when the oldest estimate (that provided by Everett (1976)) is omitted, yet even then the sign and size of the coefficient remain similar ($\beta = 0.0526$). This suggests that, given a longer data period, a positive trend in valuations might become more clearly established. While emphasizing statistical uncertainties regarding this result, its general message seems plausible, suggesting an increasing relative interest in outdoor, environmentally based recreation over the last three decades and echoing the seminal work of Krutilla and Fisher (1975).

In summary, with the exception of the *Hanley* variable, the relationships detailed in Model E conform well to expectations. Values are positively related to the *Option* variable which in this best fit model is now significant at the 5% level. Similarly the *Method* variables *ITCols* and *ZTC* both have significant and positive coefficients reflecting their expected relationship with the CV and *ITCml* values which form the base case of this analysis. Here the only *Forest* variable to prove significant ($\alpha = 10\%$) is that for *Cheshire*. The negative coefficient on this variable may reflect the high visitor congestion observed in studies of this forest (Willis and Benson, 1989)[15]. As noted, the positive and significant *Year* variable also seems highly plausible. Model E also provides the best fit to our data and, given the generally desirable characteristics noted above, provides a typical example of a meta-analysis estimated using conventional statistical modeling approaches. We now consider an alternative to this approach and examine the extent to which this may provide superior insight into the nature and robustness of these postulated relationships.

4. AN MLM APPROACH TO META-ANALYSIS

The various models reported in Tables 4 and 5 all assume independence between estimates. However, in recent years a suite of 'Multilevel Modeling' (MLM) techniques have been developed within the fields of epidemiology and education research to allow the researcher to relax this assumption and develop models which explicitly incorporate natural hierarchies or 'levels' within which data is clustered (Goldstein, 1995). This is achieved by modeling the residual variance of estimates in two parts; that due to the effect of given levels upon estimates, and that remaining due to true unexplained error. In effect, this approach allows for the possibility that variation within value estimates may differ between levels thus violating the independence assumption. In order to relax this assumption and examine the advantages of an MLM approach, this technique is now applied to our meta-analysis of woodland recreation estimates[16].

A potential limitation of the application of conventional regression techniques in meta-analysis occurs if the observations being modeled possess an inherent hierarchy. Within conventional estimation strategies, some of the variables used to predict recreation may be specific to each individual study (examples being the study design and elicitation method used). However, others, such as the author, study or forest to which a given estimate pertains may be constant across a set of such estimates. These former categorizations can be conceptualized as higher level variables, and in this sense the data may be viewed as possessing a hierarchical structure. The data structure from the above examples can be seen as actually corresponding to a range of hierarchical levels; of value estimates (level 1) within studies (level 2), of value estimates (level 1) within forests (level 2), or alternatively of value estimates (level 1) within authors (level 2)[17]. Given sufficient data, this hierarchy could be extended with further levels representing, for example regions or even nations.

Hierarchical data structures cannot be easily accommodated within the traditional regression framework. Here, the values of author or study location related variables must be collapsed to the level of the individual value estimate and simply replicated across all observations sharing those characteristics. This procedure is problematic in that it provides no information on, for example, the probability of estimates made in the same forests, or by the same authors producing similar value estimates. This limitation may be circumvented, as employed in the examples given in Tables 4 and 5, by the use of dummy variables to indicate forest location or authorship. However, this solution can present difficulties. With the present data, there are only a limited number of authors and forest sites, and hence the number of dummy variables that need to be added to the models are manageable. However, it is readily apparent that any model estimated using dummies will quickly become extremely large and complex if the dataset contains numerous observations at each level of the hierarchy.[18]

An alternative to the use of dummy variables to model hierarchical data structures is to fit a series of separate regression models. For example, separate models could be fitted for each forest or author. However, this approach defeats the objective of meta-analyses when the variables found to be significant may differ between models. Furthermore, unreliable results may be produced due to small sample sizes when there are relatively few estimates for each forest, as in the present case.

Aside from methodological considerations, a further limitation of traditional analyses stems from the fact that they may contain poorly estimated parameters and standard errors (Skinner et al., 1989). Problems with standard error estimation arise due to the presence of intra-unit correlation; the fact that recreation value estimates from studies within the same forest, or by the same author, may be expected to be more similar than those drawn from a random sample. If intra-unit correlation is small, then reasonably good estimates of standard errors may be expected (Goldstein, 1995). However, where intra-unit correlation is significant then conventional regression strategies will tend to under-estimate standard errors, meaning that confidence intervals will be too short and significance tests will too often reject the null hypothesis.

For simplicity, a two level hierarchy of i value estimates (at level 1) within j authors (at level 2) is considered in the examples below. As with a traditional generalized linear model, the observed responses y_{ij} are the published mean per person per visit recreation value estimates in 1990 pounds sterling. Considering a situation with just one explanatory variable, *OPTION* (defined as before) being tested, a simple model may be written as:

$$(1) \qquad y_{ij} = \beta_{0j} + \beta_1 OPTION_j + \in_{ij}$$

Here the subscript i takes the value from 1 to the number of value estimates in the model, and the subscript j takes the value from 1 to the number of authors in the sample. Using this notation, items with two subscripts ij vary from estimate to estimate. However, an item that has a j subscript only varies across authors but is constant for all the estimates made by each author. If an item has neither subscript it is constant across all studies and authors.

As the authors included in the analysis are treated as a random sample from a population, Equation (1) may be re-expressed as:

$$\begin{aligned} & \beta_{0j} = \beta_0 + \mu_j \\ (2) \qquad & \hat{y}_{ij} = \beta_0 + \beta_1 OPTION_{ij} + \mu_j \end{aligned}$$

Where β_0 is a constant and μ_j is the departure of the j-th author's intercept from the overall value. This means that it is an author level (level 2) residual that is the same for all estimates nested within an author. In other words this term describes, after holding constant the effect of the explanatory variables within the model, the residual influence of the author in determining the outcome for each individual mean WTP estimate they published.

The notations expressed in Equation (2) can be combined. Introducing an explanatory variable *cons*, which takes the value 1 for all estimates (and hence forms a constant or intercept term), and associating every term with an explanatory variable, the model becomes as shown in Equation (3):

$$(3) \qquad y_{ij} = \beta_0 cons + \beta_1 OPTION_{ij} + \mu_{0j} cons + \in_{0ij} c$$

Finally, the coefficients can be collected together and written as:

$$(4) \qquad \begin{aligned} y_{ij} &= \beta_{0ij} cons + \beta_1 OPTION_{ij} \\ \beta_{0ij} &= \beta_0 + \mu_{0j} + \in_{0ij} \end{aligned}$$

In Equation (4), both μ_j (the level 2 or author level residuals) and $\in_{ij}$ (the level 1 or estimate level residuals) are random quantities whose means are estimated to be equal to zero. A comparison between the multilevel model expressed in Equations

(3) and (4) and the original non-hierarchical structure depicted in Table 4 illustrates the tenet of multilevel models. Traditionally the residual error term of a model, $\in$, is seen as an annoyance and the aim of the modeling process is to minimize it's size. With multilevel models the error term is of pivotal importance in model estimation. Rather than a single error term being estimated, it is stratified into a range of terms, each representing the residual variance present at each level of the hierarchy. Viewed in this sense, μ_j represents author level effects, whilst $\in_{ij}$ represents those operating at the level of the value estimate.

If, after holding constant the influence of the x_{ij} explanatory variables in the model, $\mu_j > \in_{ij}$, then this would suggest that some factors associated with the authors themselves are of greatest importance in explaining the residual variation in WTP estimates. If instead $\mu_j < \in_{ij}$ then some un-modeled factor associated with the elicitation of each estimate (which, for example, could be associated with the characteristics of each specific study, or might simply be random variation in each elicited WTP value) is more important. A common scenario is that, whilst both μ_j and $\in_{ij}$ are large in a model containing few x_{ij} explanatory variables, both will decrease as further explanatory variables are added and the residual variance in the model is explained.

The structure presented in Equation (4) is known as a variance components model (Lin, 1997). For ease of interpretation the estimated parameters may be classified as either being of a *fixed* or *random* nature. The fixed parameters are those for which a just single coefficient is estimated, and hence correspond to those that would be found in a conventional analysis. In this example both *CONS* and *OPTION* are fixed. In contrast, the random parameters are those where individual estimates are made for every unit at each level of the hierarchy. Here both μ_j and $\in_{ij}$ are random, as a value of $\in_{ij}$ is estimated for each value estimate (at level 1 of the model) and a value of μ_j is estimated for each author (at level 2 of the model). Hence, in terms of model interpretation, it is the stratification of the error term to form these random parameters that differentiates a multilevel model from more traditional regression analysis techniques. Remembering that $OPTION_{ij}$ is a dummy variable that represents whether the elicited WTP requested use plus option value ($OPTION = 1$) or use value alone ($OPTION = 0$), the variance components model depicts the relationship between *OPTION* and the value estimate as being constant, but (provided $\mu_j > 0$) recreation values are modeled as being higher for some authors than others.

Whilst there are various methods available for parameter estimation in multilevel models, an approach known as Iterative Generalized Least Squares (IGLS) was adopted in our subsequent analysis. The statistical theory underpinning IGLS is described in detail by Goldstein (1995). Briefly, initial estimates of the fixed parameters are derived by traditional regression methodologies ignoring the higher-level random terms. The squared residuals from this initial fit are then regressed on a set of variables defining the stricture of the random part to provide initial estimates of the variances/covariances. These estimates are then used to provide revised estimates of the fixed part, which is in turn employed to revise the estimates of the random part, and so on until convergence. Crucially, a difficult estimation problem is decomposed into a sequence of linear regressions that can be solved efficiently and effectively, providing maximum-likelihood estimates[19].

It is important to note that the slopes and intercepts which are estimated for units within level 2 and above of the hierarchy will not be the same as those that would be obtained from a traditional generalized linear solution. They are in-fact residuals which have, to a greater or lesser extent, been shrunken towards the average regression line giving the predicted relationship between mean WTP and the explanatory variables across all authors. Taking our example of a 2-level model, at the author level, if $\sigma^2_{e0} = \text{var}(\in_{0ij})$ and $\sigma^2_{u0} = \text{var}(u_{0j})$ then each author level residual is estimated using Equation (5):

$$(5) \qquad \hat{u}_j = \frac{n_j \sigma^2_{u0}}{n_j \sigma^2_{u0} + \sigma^2_{e0}} \tilde{y}_j$$

Here, n_j is the total number of estimates produced by author j, $\tilde{y}_j$ is the raw residual associated with the author (the mean estimate level residual for all estimates made by author j) and $\hat{u}_j$ is the shrunken residual. From this, it can be seen that if n_j is large and there are many value estimates made by an author, then the predicted level-2 residuals will be closer to the raw residual than when n_j is small. If n_j is small, then the residual will be shrunken towards the mean. Similarly if σ^2_{e0} is large and there is high variability in the recreation value estimates produced by an author, then the predicted residual will also be shrunken. In this sense, the MLM approach provides conservative estimates of variability at different levels of the hierarchy where units based on a small sample or a very variable outcome are considered to provide little information. This is particularly pertinent here because, as has already been considered, the statistically significant positive coefficient observed for *Hanley* in Table 5 was based on studies that were all conducted at a single forest (Aberfoyle).

A multilevel re-analysis of the meta-analysis data was undertaken using the MLwiN package (Rasbash et al., 2000) developed by the Multilevel Models Project at the Institute of Education, London. Three sets of model were produced; one with a hierarchy of WTP estimates nested within authors, one of estimates within study locations and finally one of estimates within published studies. The results of the model of estimates nested within authors are given in Table 6. Here those CV elicitation techniques which produced estimates which were insignificantly different from the incentive compatible DC approach have been merged with the latter to yield a base case set of estimates from which departures are estimated. This leaves two CV elicitation techniques; the variable *CVOE* identifying those CV estimates produced using the OE format, while *CVPCH* refers to CV studies using the PCH format.

Although technically different, the fixed parameters in the model in Table 6 can be interpreted in the same way as an ordinary regression. The results detailed in the fixed part of the model now conform entirely with our theoretically and empirically derived expectations. As before the Option variable yields a significant and positive effect. Considering the CV elicitation variables and remembering the DC results

Table 6. MLM model estimates

Variable	Coefficient	95% CI	p
	FIXED EFFECTS		
Constant	0.703	0.952 – 0.954	< 0.001
Option	0.391	0.083 – 0.699	0.013
CVOE	−0.593	−1.081 – −0.105	0.018
CVPCH	0.887	−0.089 – 1.863	0.075
ZTC	1.917	1.609 – 2.220	< 0.001
ITCols	0.823	0.452 – 1.193	< 0.001
ITCml	0.041	−0.355 – 0.436	0.841
Year	0.071	0.010 – 0.132	0.021
	RANDOM (HIERARCHICAL) EFFECTS		
	Variance	**95% CI**	**p**
Level 1 (Value estimate) *Variance* σ^2_{e0}	0.218	0.142 – 0.295	< 0.001
Level 2 (Author) *Variance* σ^2_{u0}	0.021	−0.021 – 0.121	0.673

−2*loglikelihood (IGLS) = 86.759 (n = 77)

form our base case, the CVOE variable is associated with a clear negative effect which the coefficient on CVPCH is strongly positive; both results conforming to expectations. Turning to the Method variables, as expected the ZTC and *ITCols* variables both produce strong positive effects while, as observed previously, the *ITCml* variable is statistically insignificantly different to the DC base case; given that these are the most theoretically and methodologically defensible of the TC and CV methods this seems a reassuring result. Finally, the positive and significant coefficient on the *Year* variable is reconfirmed.

In summary therefore, the fixed part of the multilevel model reported in Table 6 conforms entirely with our prior expectations. However, one of the prime objectives of fitting such a model was to determine if, after controlling for the variables in the fixed part, there was still statistically significant variation in WTP estimates between authors. These random effects are shown in the lower part of Table 6. This part of the model is relatively simple. Although the multilevel methodology involves estimating a separate intercept value for each author (μ_j) and a separate residual for each value estimate ($\in_{ij}$), the variance between the two levels of the model may be neatly summarized by the two parameters σ^2_{u0} and σ^2_{e0}. These are the same parameters used in the calculation of the shrinkage factor illustrated in Equation (5) and are known as variance parameters, as they indicate the variance in the μ_j and $\in_{ij}$ terms respectively. Hence a comparison of the values of σ^2_{u0} and σ^2_{e0} shows the relative importance of author (level 2) and estimate (level 1) effects in determining the variability of WTP values that is not explained by the fixed parameters in the model.

The parameter estimates for both σ^2_{u0} and $\sigma^2_{e0}$are greater than zero, suggesting that variability between estimates and between authors remains after controlling for the explanatory variables that were included in the fixed part of the model. Taking the ratio of these estimates suggests that approximately 9% of unexplained variation in elicited recreation value is associated with author effects. However, the calculation of t-statistics for each coefficient shows that whilst statistically significant residual variation remains between estimates at level 1 ($t = 5.59, p < 0.001$) the effect of authorship at level 2 does not reach statistical significance ($t = 0.41, p > 0.05$). In other words, the multilevel analysis suggests that an author effect is present but is not statistically significant.

Although in conflict with the earlier findings from the conventional regression analysis, such a result accords with theoretical expectations that recreation values should not vary significantly according to study authorship. This provides substantial (if, on its own, insufficient) support for the practice of placing monetary values upon preferences for non-market environmental goods. The author specific results are illustrated in Figure 1 where the value of u_j estimated for each individual author is presented in rank order along with corresponding 95% confidence intervals. The figure shows that, in the multilevel analysis, studies by *Hanley* et al. are still predicted to give the highest recreation values and those by *Willis* et al. the lowest. However, the confidence intervals clearly overlap. This represents a reduction in variance from the situation observed in Tables 4 and 5 where estimates provided by *Hanley* et al. were found to significantly differ from that of other authors. The reduction of variance is due to the effects of the conservative estimation strategy

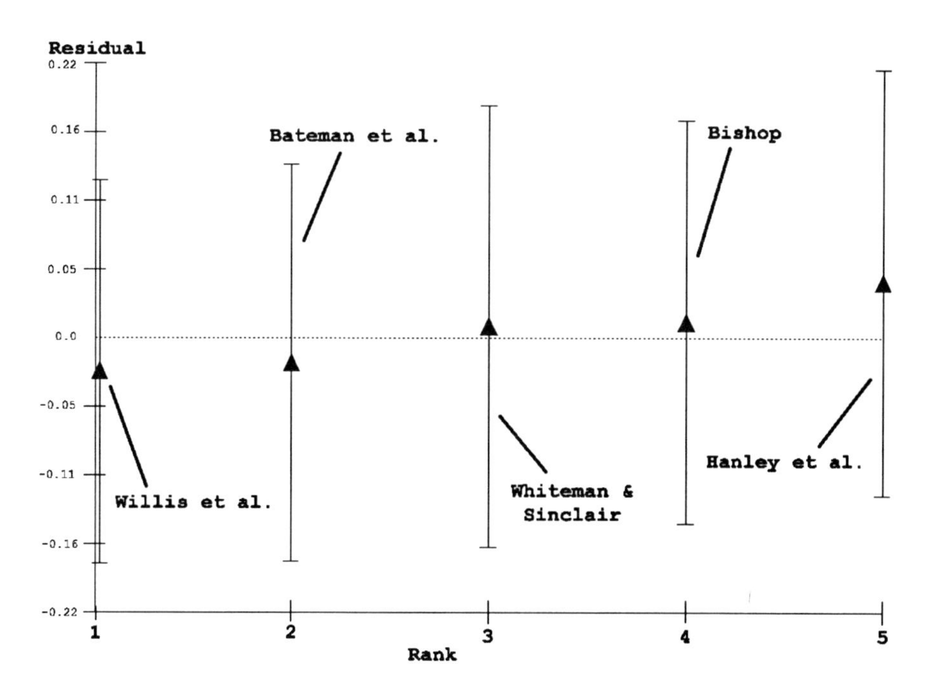

Figure 1. MLM author level residuals

implemented in Equation (5) where residuals are shrunken towards the mean value. The contrast between Tables 4 and 5 and the findings of Table 6 provides a clear justification for the application of MLM techniques to meta-analysis studies.

The shrinkage illustrated by Figure 1 has interesting implications for the comparison of results between multilevel and non-multilevel models. The message from the multilevel model is that variation is present between authors but, because of the magnitude of the variance and the size of the sample, it cannot be said to be statistically significant. Hence we are making a statement about the importance of context (in this case authorship) and composition (the remaining unexplained variation in between WTP estimates). The traditional regression approach used previously did the opposite; it told us little about the overall roles of context and composition, but it did highlight two authors with rather different patterns of responses from the rest of the sample. From this comparison, it is clear that, whilst the conclusions reached may be different from those of a conventional analysis, the multilevel approach is prudent if the intention of the analysis is to quantify whether there are overall contextual influences (in this case associated with different authors) on the measured outcome (recreation value).

The earlier conventional analyses also found evidence of a *Forest* (site) effect where recreation values for *Cheshire* were significantly lower than the rest of the sample, and those for *Loch Awe* and *Aberfoyle* were relatively higher (see Table 5). To test if any evidence of between-site heterogeneity remained after a multilevel approach was taken, the model presented in Table 6 was refitted, but this time authorship at level 2 was replaced by *Forest* identifiers. The fixed effect coefficient values and levels of significance were not found to differ greatly from the previous example and are hence not replicated here. However in this case the values of σ^2_{u0} (now for forests) and σ^2_{e0} (for value estimates) were estimated at 0.010 ($t = 0.73$, $p > 0.05$), and 0.212 ($t = 5.63$, $p < 0.01$) respectively. In similar fashion to the model for authors, these results show strong variation between estimates, but only a limited forest site effect (accounting for under 5% of the total residual variance). Figure 2 shows the forest level residuals ranked with 95% confidence intervals. In order to maintain legibility only those forests mentioned previously are identified. As with the original non-multilevel analysis, *Cheshire* shows the greatest negative residual (and hence correspondingly lower than predicted WTP values), whilst *Loch Awe* and *Aberfoyle* yield the highest positive residual values. However, again confidence intervals clearly overlap, thus conforming to our empirically derived prior expectation that forests do not exert significant impacts upon recreation values (although as noted before, their location may influence the quantity of visits).

Finally, we also considered the possibility of significant between-study heterogeneity (i.e. looking at the clustering of estimates within studies). As per our examination of author effects, we do not expect variance within estimates to differ significantly between studies. Analysis clearly confirmed this expectation with values of σ^2_{u0} (between study variance) and σ^2_{e0} (for value estimates) estimated at 0.017 ($t = 0.55$, $p > 0.05$), and 0.216 ($t = 5.68$, $p < 0.01$) respectively.

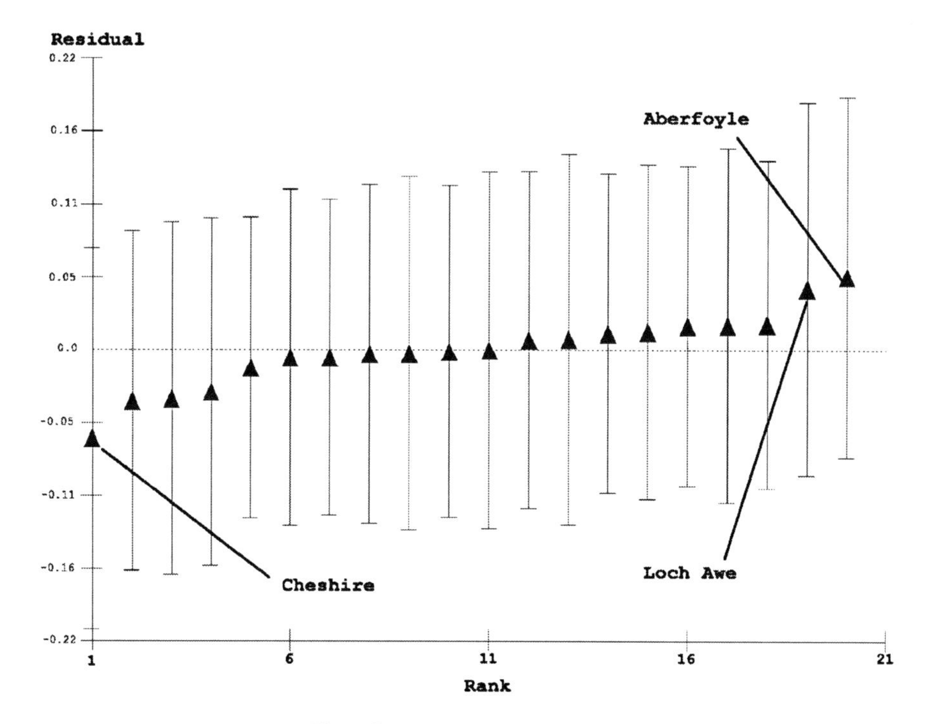

Figure 2. MLM forest level residuals

5. DISCUSSION AND CONCLUSIONS

There are a numerous routes through which benefit transfer and meta-analysis research may be taken forward. These include improvements in the conduct and reporting of new studies, the specific incorporation of benefit transfer and meta-analysis requirements within their design, and the re-analysis of past work. The present paper goes some way towards highlighting a novel way in which this latter aim might be best realized. We have compared the application of traditional regression and novel MLM methodologies to meta-analyses of British woodland recreation values. While both sets of results generally conform well to expectations derived from either theoretical considerations or empirical regularities, our conventional regression findings suggest that certain authors and forests are associated with large recreation value residuals. However, the more sophisticated and conservative MLM approach shows that these residuals are not large enough (or are not based on a large enough sample size) to be differentiated from variation that might be expected by chance. In so doing it is only these MLM based models which conform in all respects to prior expectations, a finding which underscores the importance of adopting such approaches which explicitly model the hierarchical nature of almost all meta-analysis datasets.

Here we have fitted only simple two level MLM models. More complex structures have not been implemented here for a number of reasons. No significant variation was observed between authors or survey site locations, and it is highly unlikely that a more detailed model hierarchy would have contradicted these findings. A second limitation to the use of more complex hierarchies concerns sample size; as models become more complicated there is an associated loss of degrees of freedom. In particular, the conservative estimation strategy used means that the presence of a small amount of level 2 variation in a simple two-level model may be shrunken to zero if a more complex structure is attempted. Whilst the dataset we have studied is comprehensive, it is based on a sample of just 77 observations, and hence has somewhat limited power. The increased number of observations that will result from more studies being undertaken would allow a greater complexity of models to be fitted.

Although the essential ideas of multilevel models were developed over 20 years ago, it is only recently that improvements in computing power and advances in our understanding of effective model implementation have meant that their execution has become a practical proposition (Bull et al., 1998). We are currently on a wave of innovation as use spreads from the original developers to the wider research community. Having said that, the multilevel approach retains some of the limitations of more traditional quantitative techniques, as well as introducing new ones.

In the MLM models presented here, influences on recreation values are modeled more powerfully than traditional techniques allow, yet the random parameters can ultimately offer only limited insight into the reasons behind between-author and between-forest variations in outcome. Preferences for complex, non-market environmental goods such as open-access recreation involve a detailed interplay between a wide range of factors that are difficult to quantify and may be subject to random variation. This unpredictability will undoubtedly introduce uncertainty into any model, multilevel or not, developed to identify and predict the important influences on such preferences. However, whilst multilevel models cannot remove this uncertainty, they can allow it to be more richly quantified and accounted for, and hence allow for systematic factors to be assessed.

Finally, our MLM estimated meta-analysis has some clear messages for both policy makers and economists who work within the applied policy arena. As noted, our results conform well with prior theoretically and empirically derived expectations. However, these expectations are not that value estimates will be invariant to choice of study methodology or analytical approach. Indeed the reverse is true. For example, as predicted by considerations of incentive compatibility, we show that CV estimates of recreation value derived from an OE elicitation technique will be significantly lower than those obtained by a DC approach. Similarly we show that TC values derived through inappropriate OLS estimators will be upwardly biased in comparison with those derived from maximum likelihood estimators or from CV studies using DC elicitation techniques. It is the responsibility of the economist to highlight these expected differences to the policy maker and to advise upon the most theoretically and methodologically appropriate approach to the issue at hand.

That said, the absence of significant author or study level impacts within our MLM meta-analyses is encouraging providing an argument against criticisms that, for example, certain authors produce unusually high or low valuation estimates. This analysis also has some specific messages for policy makers within the UK Forestry Commission. In particular, while some evidence for site effects was found in the conventionally estimated models reported in Table 5, these do not persist within the more sophisticated MLM analysis given in Table 6. As noted previously, this finding is in line with other research showing that, while visitor arrivals at UK woodlands are highly responsive to a variety of locational factors, they are somewhat less responsive to the facilities on offer at these sites (Brainard et al., 1999; 2001)[20]. Given this, the onus upon woodland policy makers within the UK context, appears to be upon using scarce resources to optimize site location rather than to extend the diversity of facilities within woodlands.[21]

6. NOTES

[1] For reviews of the issues raised by benefit transfer applications see Brookshire and Neill (1992), OECD (1994), Pearce and Moran (1994), Bergland et al., (1995), Van den Bergh et al., (1997) and Desvousges et al., (1998).

[2] Meta-analyses also face the problem that studies published in the available literature may over represent that subset of all studies which produce 'positive' or significant results if studies yielding 'negative' or non-significant findings tend not to be published.

[3] Details of all of these estimates are given in Table A1 in Bateman et al., (2000).

[4] As discussed subsequently, value estimates may also be 'cross classified', e.g. where different authors conduct studies at the same, as well as differing, forest sites.

[5] Markowski et al., (2001) clearly describe the weighting procedure used as follows: "For these models we weight the data to reflect the "oversampling" of estimates associated with studies with a large number of observations relative to others with just a few or one observation. To do this, we first determine how many estimates (k_j) in the sample are associated with each study (j) to define a study weight. We then divide the data associated with each observation (dependent variable and explanatory variables) by the weights (k_j) for each study. Thus, rather than all observations having equal weights in the estimation, which is the case for the basic model, each study has an equal weight in this estimation for models of per-day and per-trip welfare estimates." (p.12)

[6] This decomposes into 14.7% of Scotland, 12.0% of Wales and 7.4% of England. However, this is still well below an EU average of about 25% of land area under forestry (FICGB, 1992).

[7] Note that CV studies can be adapted to ask either WTP or willingness to accept compensation questions in respect of either gains or losses of the resource concerned (Mitchell and Carson 1989), although only the WTP format was used in the studies concerned.

[8] Further details of these studies are provided in Bateman et al., (2000).

[9] For a discussion of ML estimators see Maddala (1983).

[10] The distribution of estimates by forests and methods is as follows: 44 CV estimates across 20 forests; 9 ITCols estimates across 7 forests; 7 ITCml estimates across 7 forests; 17 ZTC estimates across 16 forests.

[11] While the DC method is incentive compatible, whether or not it is in practice also demand revealing (i.e. produces unbiased estimates of true WTP) is an ongoing source of debate (Green et al., 1998; Carson et al., 1999).

[12] Bateman et al., (1999a) use a reduced form of the model reported in Table 4 in their GIS based benefit transfer analysis of woodland recreation values.

[13] In addition we have one further *Author* category (*Everett*) and one extra *Forest* study site (*Dalby*).

[14] Equation (A1) in Bateman et al., (2000) details such a model showing effects for individual forests.

[15] By contrast the *Loch Awe* coefficient is positive (although not statistically significant) reflecting its somewhat remote and secluded location attracting a more 'dedicated' woodland user (as noted by Willis and Benson, 1989.

[16] We initially develop this approach in Brouwer et al., (1999). However, this earlier analysis is restricted to CV studies alone, considers only one form of potential data hierarchy and is complicated by the necessity of drawing upon studies of diverse resources taken from a number of countries; factors which make interpretation of findings problematic. The present study examines a single resource within a single country but considers three potential data hierarchies (while also providing a fuller account of the MLM modeling structure).

[17] If no two authors undertake a study in the same forest, then this may be extended to a three level hierarchy of WTP estimates (level 1) within forests (level 2) within authors (level 3). If multiple authors do study the same forests, then a more complex structure (known as cross classified) exists wherein estimates (level 1) are nested within a cross-classified level (2) of forests and authors. Such a case is not considered here (although it is the subject of ongoing research by the authors), but the theory of cross classified hierarchies is discussed in detail by Goldstein (1995).

[18] An example might be an international dataset of value estimates nested within hundreds of study locations.

[19] A limitation of IGLS for models with a binomial or Poisson distributed response variable (neither of which were used in the present application) is that is uses a method based on either marginal or penalized quasi-likelihood. This requires assumption of normally distributed variance above level one of the hierarchy.

[20] This is not to suggest that site facilities are irrelevant in attracting visitors. However, as shown by Brainard et al., locational factors provide much stronger predictors of demand. In part this may be because virtually all UK sites provide the basic walking and recreational amenities which characterize woodlands visits and are thus relatively little differentiated in terms of further cogent attributes.

[21] Note that all TC studies relate to use value alone.

7. REFERENCES

Baaijens SR, Nijkamp P, Montfort KV (1998) Explanatory meta-analysis for the comparison and transfer of regional tourism income multipliers, Regional Studies, 32 (9): 839–849

Bateman IJ (1996) An economic comparison of forest recreation, timber and carbon fixing values with agriculture in Wales: A geographical information systems approach, Ph.D. Thesis, Department of Economics, University of Nottingham

Bateman IJ, Langford IH, Turner RK, Willis KG, Garrod GD (1995) Elicitation and truncation effects in contingent valuation studies, Ecological Economics, 12 (2):161–179

Bateman IJ, Garrod GD, Brainard JS, Lovett AA (1996) Measurement, valuation and estimation issues in the travel cost method: A geographical information systems approach, Journal of Agricultural Economics, 47 (2): 191–205

Bateman IJ, Langford IH (1997) Budget constraint, temporal and ordering effects in contingent valuation studies, Environment and Planning A, 29 (7): 1215–1228

Bateman IJ, Lovett AA, Brainard JS (1999a) Developing a methodology for benefit transfers using geographical information systems: Modeling demand for woodland recreation, Regional Studies, 33 (3): 191–205

Bateman IJ, Brainard JS, Garrod GD, and Lovett AA (1999b) The impact of journey origin specification and other measurement assumptions upon individual travel cost estimates of consumer surplus: A geographical information systems analysis, Regional Environmental Change, 1 (1): 24–30

Bateman IJ, Jones AP, Nishikawa N, Brouwer R (2000) Benefits transfer in theory and practice: A review, CSERGE Global Environmental Change Working Paper, Centre for social and economic research on the global environment, University of East Anglia and University College London

Bateman IJ, Lovett AA, Brainard JS, (2003) Applied Environmental Economics: A GIS Approach to Cost Benefit Analysis, Cambridge University Press, Cambridge

Benson JF, Willis KG (1990) The aggregate value of the non-priced recreation benefits of the Forestry Commission estate. Report to the Forestry Commission, Department of Town and County Planning, University of Newcastle upon Tyne

Benson JF, Willis KG (1992) Valuing informal recreation on the Forestry Commission estate, Bulletin 104, Forestry Commission, Edinburgh

Bergland O, Magnussen K, and Navrud S (1995) Benefit transfer: Testing for accuracy and reliability, Discussion Paper #D-03/1995, Agricultural University of Norway

Bishop KD, (1992) Assessing the benefits of community forests: An evaluation of the Recreational use benefits of two urban fringe woodlands, Journal of Environmental Planning and Management, 35 (1):63–76

Boadway R, Bruce N (1984) Welfare Economics, Basil Blackwell, Oxford

Boyle KJ, Bishop RC, Welsh MP (1985) Starting point bias in contingent valuation bidding games, Land Economics, 61:188–194

Boyle KJ, Poe GL, Bergstrom J (1994) What do we know about groundwater values?: Preliminary implications from a meta-analysis of contingent valuation studies, American Journal of Agricultural Economics, 76 (5):1055–1061

Brainard JS, Lovett AA, Bateman IJ (1999) Integrating geographical information systems into travel cost analysis benefit transfer, International Journal of Geographical Information Systems, 13 (3): 227–246

Brainard JS, Bateman IJ., Lovett AA, (2001) Modelling demfor recreation in English woodlands, Forestry, 74 (5): 423–438

Brookshire DS, Neill H R (1992) Benefit transfers: Conceptual and empirical issues, Water Resources Research, 28 (3):651–655

Brouwer R, Spaninks FA (1999) The validity of environmental benefits transfer: Further empirical testing Environmental and Resource Economics, 14 (1): 95–117

Brouwer R, Langford IH, Bateman IJ, Turner RK (1999) A meta-analysis of wetlcontingent valuation and studies, Regional Environmental Change, 1 (1): 47–57

Bull JM, Riley GD, Rasbash J, Goldstein H (1998) Parallel implementation of a multilevel modelling package, University of Manchester, Manchester

Button KJ (1995) Evaluation of transport externalities: What can we learn using meta-analysis?, Regional Studies 29:507–517

Button KJ, Kerr J (1996) Effectiveness of traffic restraint policies: A simple meta-regression analysis, International Journal of Transport Economics 23:213–225

Carson, Flores NE, Martin KM, Wright JL (1996) Contingent valuation and revealed preference methodologies: Comparing the estimates for quasi-public goods, Land Economics, 72:80–99

Carson RT, Groves T, Machina MJ (1999) Incentive and informational properties of preference questions, Plenary Address, Ninth Annual Conference of the European Association of Environmental and Resource Economists (EAERE), Oslo, Norway, June 1999

Desvousges WH, Johnson FR, Banzaf HS (1998) Environmental Policy Analysis With Limited Information: Principles and Applications of the Transfer Method, Edward Elgar, Northampton, Mass

Downing M, Ozuna T (1996) Testing the reliability of the benefit function transfer approach, Journal of Environmental Economics and Management, 30:316–322

Espey M, Espey J, Shaw WD (1997) Price elasticity of residential demfor water: A meta-analysis, Water Resources Research, 33 (6):1369–1374

Everett RD (1979) The monetary value of the recreational benefits of wildlife, Journal of Environmental Management, 8:203–213

Farquharson R (1969) Theory of Voting, Yale University Press, New Haven

FICGB (Forestry Industry Committee of Great Britain) (1992) The Forestry Industry Year-Book 1991–92, FICGB, London

Gibbard A (1973) Manipulation of voting schemes: A general result, Econometrica 41:587–601

Glass GV (1976) Primary, secondary and meta-analysis of research, Educational Researcher, 5 (10):3–8

Glass GV, McGaw B, Smith ML (1981) Meta-analysis in social research, Sage Publications, Beverley Hills, CA

Goldstein H, (1995) Multilevel Statistical Models (2nd edition) Edward Arnold, London

Grayson AJ, Sidaway RM, Thompson FP (1975) Some aspects of recreation planning in the Forestry Commission in Searle, GAC (ed.) Recreational Economics and Analysis: Papers presented at the Symposium on Recreational Economics and Analysis, London Graduate School of Business Studies, January 1972, Longman, Essex

Green D, Jacowitz KE, Kahneman D, McFadden D (1998) Referendum contingent valuation, anchoring, and willingness to pay for public goods, Resource and Energy Economics, 20 (2):85–116

Hanley, ND (1989) Valuing rural recreation benefits: An empirical comparison of two approaches, Journal of Agricultural Economics, 40 (3):361–374

Hanley ND, Craig S (1991) Wilderness development decisions and the Krutilla-Fisher model: The case of Scotland's Flow Country Ecological Economics, 4:145–164

Hanley ND, Ecotec Ltd (1991) The Valuation of Environmental Effects: Stage Two Final Report. The Scottish Office Industry Department and Scottish Enterprise, Edinburgh

Hanley ND, Munro A (1991) Design bias in contingent valuation studies: The impact of information, Discussion Paper in Economics 91/13, Department of Economics, University of Stirling

Hanley ND, Ruffell RJ (1991) Recreational use values of woodl and features Report to the Forestry Commission, University of Stirling

HM Government (1995) Environment Act 1995, HMSO, London

HM Treasury (1972) Forestry in Great Britain: An Interdepartmental Cost/Benefit Study, HMSO, London

Hoehn JP, Randall A (1987) A satisfactory benefit cost indicator from contingent valuation, Journal of Environmental Economics and Management, 14 (3):226–247

Hunter JE, Schmidt FL, Jackson G (1982) Advanced Meta-analysis: Quantitative Methods for Cumulating Research Findings A Cross Studies, Sage, Beverly Hills

Just RE, Hueth DL, Schmitz A (1982) Applied Welfare Economics and Public Policy, Prentice Hall, Englewood Cliffs, NJ

Kirchhoff S, Colby BG, LaFrance JT (1997) Evaluating the performance of benefit transfer: An empirical inquiry, Journal of Environmental Economics and Management, 33:75–93

Krutilla JV, Fisher AC (1975) The Economics of Natural Environments: Studies in the Valuation of Commodity and Amenity Resources, Johns Hopkins University Press (for Resources for the Future), Baltimore, MD

Lin X (1997) Variance component testing in generalised linear models with random effects Biometrika 84:309–25

Loomis JB (1992) The evolution of a more rigorous approach to benefit transfer: Benefit function transfer, Water Resources Research, 28 (3):701–705

Loomis JB, Roach B, Ward F, Ready R (1995) Testing transferability of recreation demmodels across regions: A study of corps of engineers reservoirs Water Resources Research, 31:721–730

Loomis JB, White DS (1996) Economic benefits of rare and endangered species: Summary and meta-analysis, Ecological Economics, 18 (3): 197–206

Maddala GS (1983) Limited-Dependent and Qualitative Variables in Econometrics, Cambridge University Press, Cambridge

Markowski MA, Boyle KJ, Bishop RC, Larson DM, Paterson RW (2001) A cautionary note on interpreting meta analyses, unpublished paper, Industrial Economics Inc

McCullagh P, Nelder JA (1989) Generalized Linear Models, 2nd Edition, Monographs on Statistics and Applied Probability 37, Chapman Hall, London

Mitchell RC, Carson RT (1989) Using Surveys To Value Public Goods: The Contingent Valuation Method, Washington DC, Resources for the future

Mrozek JR, Taylor LO (2002) What determines the value of life? A meta-analysis, Journal of Policy Analysis and Management 21 (2): 253–70

NAO (National Audit Office) (1986) Review of Forestry Commission Objectives and Achievements Report by the Comptroller and Auditor General, National Audit Office, HMSO, London

Nelson JP (1980) Airports property values: A survey of recent evidence, Journal of Transport Economics and Policy 19:37–52

OECD (Organisation for Economic Co-operation Development) (1994) Project and policy appraisal: Integrating economics and environment, OECD, Paris

Pearce DW, Turner RK (1990) Economics of Natural Resources the Environment, Harvester Wheatsheaf, Hemel Hempstead

Pearce DW, Moran D (1994) The Economic Value of Biodiversity, Earthscan, London

Poe GL, Boyle KJ, Bergstrom JC (2001) A preliminary meta analysis of contingent values for ground water revisited, in Bergstrom JC, Boyle KJ., Poe GL (eds) The Economic Value of Water Quality, Edward Elgar, Northampton, MA

Rackham O (1976) Trees Woodlin the British Landscape, Dent, London

Rasbash J, Browne W, Goldstein H, Yang M, Plewis I, Healy M, Woodhouse G, Draper D, Langford I, Lewis T (2000) A Users Guide to MlwinN: Version 2.1, Institute of Education, London

Roberts KJ, Thompson ME, Pawlyk PW (1985) Contingent valuation of recreational diving at petroleum rigs, Gulf of Mexico, Transactions of the American Fisheries Society, 114:155–165

Rosenberger RS, Loomis JB (2000) Using meta-analysis for benefit transfer: In-sample convergent validity tests of an outdoor recreation database, Water Resources Research, 36 (4): 1097–1107

Rowe RD, Schulze WD, Breffle W (1996) A test for payment card biases, Journal of Environmental Economics Management, 31:178–185

Satterthwaite M (1975) "Strategy-Proofness Arrow Conditions: Existence Correspondence Theorems for Voting Procedures Welfare Functions," Journal of Economic Theory 10:187–217

Schwartz J (1994) Air pollution daily mortality: A review a meta analysis, Environmental Economics 64:36–52

Shrestha RK, Loomis JB (2001) Testing a meta-analysis model for benefit transfer in international outdoor recreation, Ecological Economics, 39:67–83

Skinner CJ, Holt D, Smith TMF (1989) Analysis of Complex Surveys Wiley, Chichester, UK

Smith VK (1989) Can we measure the economic value of environmental amenities?, Southern Economic Journal 56:865–878

Smith VK, Kaoru Y (1990a) Signals or noise? Explaining the variation in recreation benefit estimates, American Journal of Agricultural Economics, May 1990:419–433

Smith VK, Kaoru Y (1990b) What have we learned since Hotelling's letter? A meta-analysis, Economics Letters, 32:267–272

Smith VK Huang JC (1993) Hedonic models air pollution: Twenty five years counting, Environmental and Resource Economics 3:381–394

Smith VK, Huang JC (1995) Can markets value air quality? A meta-analysis of hedonic property values models, Journal of Political Economy, 103:209–227

Smith VK, Osborne L (1996) Do contingent valuation estimates pass a "Scope" test? A meta-analysis, Journal of Environmental Economics Management 31:287–301

Smith VK, Pattanayak SK (2002) Is meta-analysis a Noah's Ark for non-market valuation?, Environmental and Resource Economics, 22(1–2): 271–296

Sturtevant LA, Johnson FR, Desvousges WH (1995) A meta-analysis of recreational fishing Triangle Economic Research, Durham, North Carolina

Van den Bergh JCJM, Button KJ, Nijkamp P, Pepping GC (1997) Meta-Analysis in Environmental Economics, Kluwer Academic Publishers, AH Dordrecht, The Netherlands

Van Houtven GL, Pattanayak SK, Pringle C, Yang J-C (2001) Review meta-analysis of water quality valuation studies, Draft Report, prepared for US Environmental Protection Agency, Office of Water Office of Policy, Economics, Innovation, Washington, DC.

Walsh RG, Johnson DM, McKean JR (1990) Nonmarket values from two decades of research on recreation demand, in Link A, Smith VK (eds.) Advances in Applied Micro-Economics, Vol. 5:167–193 JAI Press, Greenwich, CT

Walsh RG, Johnson DM, McKean JR (1992) Benefit transfer of outdoor recreation demstudies, 1968–1988, Water Resources Research, 28(3):707–713

Waters WG (1993) Variations in the value of travel time savings: empirical studies the values for road project evaluation, *mimeo*, cited in Van den Bergh JCJM, Button KJ, Nijkamp P, Pepping GC (1997), Meta-Analysis in Environmental Economics, Kluwer Academic Publishers, AH Dordrecht, The Netherlands

Weisbrod BA (1964) Collective—consumption services of individual consumption goods, Quarterly Journal of Economics, 78:471–477

Whiteman A, Sinclair J (1994) The Costs Benefits of Planting Three Community Forests: Forest of Mercia, Thames Chase Great North Forest, Policy Studies Division, Forestry Commission, Edinburgh

Woodward RT, Wui Y-S (2001) The economic value of wetl and services: A meta-analysis, Ecological Economics, 37:257–270

Willis KG, Benson JF (1989) Values of user benefits of forest recreation: Some further site surveys Report to the Forestry Commission, Department of Town County Planning University of Newcastle upon Tyne

Willis KG, Benson JF, Whitby MC (1988) Values of user benefits of forest recreation wildlife. Report to the Forestry Commission, Department of Town County Planning, University of Newcastle upon Tyne

Willis KG, Garrod GD (1991) An individual travel cost method of evaluating forest recreation, Journal of Agricultural Economics, 42(1):33–41

Wolf F (1986) Meta-Analysis: Quantitative Methods for Research Synthesis, Sage, Beverly Hills

R. SHRESTHA, R. ROSENBERGER AND J. LOOMIS

BENEFIT TRANSFER USING META-ANALYSIS IN RECREATION ECONOMIC VALUATION

1. INTRODUCTION

Natural resource managers, regulatory agencies, and decision makers often encounter the situation where an economic analysis is mandatory, but time and funding for a new economic valuation study to conduct a full-scale assessment of benefits and costs are constrained. Benefit transfer is a practical tool commonly used to estimate values of non-market goods and services, and serves as an alternative to the entirely new primary valuation research (Luken et al., 1992; Desvousges et al., 1992; Morgan and Owens, 2001). The opportunity to use existing studies makes benefit transfer convenient to many analysts and decision makers. Some of the advantages of performing benefit transfers over conducting new economic valuation studies include lower cost, shorter estimation time, and relative independence from budgetary constraints. However, there are reasons to be concerned about the validity of the transferred values compared to the non-market values estimated from the 'first best' empirical studies. Thus, benefit transfer approaches require scrutiny before their application in a policy setting.

Various benefit transfer approaches have been used in natural resources and environmental valuation. First, a unit-day value transfer method was developed by the U.S. Water Resources Council and the U.S. Department of Agriculture (USDA) Forest Service (Loomis and Walsh, 1997; US Water Resources Council, 1973, 1979, 1983). Since the Resource Planning Act of 1974 (RPA), the USDA Forest Service has been periodically updating its unit-day value estimates summarizing existing recreation valuation studies. Application of RPA values is, however, seriously constrained by the rigidity of its point estimates. Second, a direct transfer approach, in which the benefit is transferred from the best-matched study, is frequently used in benefit transfer (Boyle and Bergstrom, 1992; Desvousges et al., 1992). Obviously, the difficulty in this approach is not only a matter of availability of the best-matched study, but also the correspondence of resource attributes and values between study and policy sites (Rosenberger and Phipps, this volume). Third, a benefit function transfer approach that accounts for the differences in site attributes is regarded as a potentially better approach to benefit transfer (Loomis et al., 1995; Loomis, 1992; Downing and Ozuna, 1996; Kirchhoff et al., 1997; Brouwer and Spaninks, 1999). The recreation valuation literature suggests benefit function transfer is a superior method in that it enables researchers to adjust the estimated demand function to calculate values of a resource at the new policy site. Fourth, benefit transfer with meta-analysis is a new approach, which extends the benefit function transfer by using a meta-analysis of existing studies to estimate the benefit function. This

S. Navrud and R. Ready (eds.), Environmental Value Transfer: Issues and Methods, 161–177.

approach accounts for the variability in benefit estimates with respect to the valuation methodologies and site attributes in benefit estimation, thus potentially better reflecting the values of resources (Rosenberger and Loomis, 2000b; Rosenberger and Phipps, 2002; Shrestha and Loomis, 2001).

Meta-analysis uses existing valuation studies to develop a benefit transfer function (BTF) that can be applied in estimating non-market values at unstudied policy sites. Meta-analysis is a statistical method that synthesizes existing research findings (Glass et al., 1981, 1976; Stanley, 2001). This method has been frequently used in numerous disciplines since the seminal work of Gene V. Glass in 1976. Meta-analysis in recreation valuation was introduced as meta-regression analysis (Walsh et al., 1989, 1992; Smith and Kaoru, 1990), particularly to explain variation in consumer surplus or willingness to pay (WTP) measures. Some recent applications of meta-analysis in economic valuation studies include groundwater quality (Boyle et al., 1994), air quality (Smith and Huang, 1995), endangered species (Loomis and White, 1996), air pollution and visibility (Smith and Osborne, 1996), health effects of air pollution (Desvousges et al., 1998), recreation (Rosenberger and Loomis, 2000a, 2000b; Shrestha and Loomis, 2001; Sturtevant et al., 1998), and wetlands (Brouwer et al., 1999, Woodward and Wui, 2001).

Most past recreation meta-analyses were focused on synthesizing valuation literature. However, meta-analysis could be an appropriate method for benefit transfer, which enables researchers to better account for the site/resource characteristics and methodological differences in their value estimates (Sturtevant et al., 1998). In recreation economic valuation, the empirical studies measure economic value of the resources in terms of WTP or consumer surplus for the study site or resource using non-market valuation methods, for example, contingent valuation (CVM) or travel cost methods (TCM). In meta-analysis these primary estimates of WTP enter as observations of the dependent variable in a regression model (Walsh et al., 1992; Smith and Kaoru, 1990; Sturtevant et al., 1998). Site attributes, valuation method, and user socioeconomic variables of the original research are incorporated as explanatory variables in the model. Once the meta-regression model is estimated, it may be used as a BTF to estimate WTP values of the policy site. Thus, the information and knowledge about the policy site that corresponds to the variables of the BTF would allow researchers to adjust the function to fit the characteristics of the policy site, providing more accurate benefit estimates for the policy site.

Benefit transfer using meta-analysis has several advantages. First, information is utilized from a number of studies providing a more robust measure of central tendency than can usually be obtained from any single study. Second, methodological differences in the original studies can be controlled for when estimating a value from the meta-regression equation. Third, by setting the explanatory variables specific to the policy site, the analyst can potentially account for the heterogeneity in the policy site attributes. For these reasons, the benefit estimate using meta-analysis is likely to be a better approximation of the value of the resource at the policy site.

However, the validity of meta-analytic benefit transfers is not necessarily evident. Despite its parsimony and convenience, there is concern among analysts and decision makers whether meta-analytic benefit transfers can accurately value non-market goods. A complete answer to this question is beyond the scope of a single study. However, our intention is to inform the debate through a discussion and illustration of meta-analytic benefit transfer procedures and potentials. We first develop a methodological framework and validity test statistics. Then, we estimate meta-regression models that serve as benefit transfer functions. Finally, we test the validity of our BTFs using out-of-sample studies from the U.S. and other countries around the world.

2. META-ANALYTIC BENEFIT TRANSFER METHODOLOGY

Benefit transfer using meta-analysis attempts to model an underlying valuation function that is composed of a broad range of original studies. This valuation function is more likely to account for the attributes and values of policy sites when used for benefit transfer than traditional benefit transfer methods. We adopt the framework proposed by Rosenberger and Phipps (2002, this volume). The valuation function (V) is proposed to be an envelope function of individual site specific valuation functions estimated from different studies, where the WTP value per unit of use for a site is a function of the attributes of the site.

2.1. Specification of the Meta-Analysis and Benefit Transfer Function

Meta-analysis uses multivariate regression techniques to estimate a benefit transfer function. In principal, if we can model the factors accounting for the variation in values measured from prior research, then we should be able to calibrate WTP values based on these factors. An explicit specification of the meta-regression based benefit transfer function can be written as

$$\text{(1)} \qquad WTP_{mn} = \alpha + \sum_{k=1}^{k} \beta_k x_{k,mn} + e_m + u_n$$

where, m denotes which candidate study the WTP estimate comes from ($m = 1, \ldots, M$), and n denotes the WTP estimate reported in the study ($n = 1, \ldots, N_m$). In the case where each study (m) provides a single estimate (n), then $N_m = 1$ and e_m collapses into u_n. In the case where each study provides one or more estimates, then we need to account for the common error across estimates (u_n) and the group-specific or panel error within a study (e_m). The total number of estimates $N = \sum_{m=1}^{M} N_m$. The variations in WTP_{mn} are to be explained by a vector of explanatory variables $k = 1, \ldots, K$, denoted by $x_{k,mn}$. Important explanatory variables may include resource attributes and uses, valuation methods, and socio economic characteristics. The estimates within the study may share, in part or whole, several explanatory variables, whereas the estimates across studies may differ in many of those exogenous variables. The fact that the estimates are not independent within a study leads to a nested error structure, that is, decomposed error variance at the study level, e_m and error at the estimation level, u_n, which are assumed to be normally distributed with zero mean and constant variances σ_e^2 and σ_u^2, respectively (Bijmolt and Pieters, 2001). In Equation (3), α is the intercept term and β_k is a vector of slope parameters of the benefit transfer function to be estimated.

The meta-regression specification (1) clearly shows the panel structure of the benefit transfer function. The panel data structure arises from the multiple WTP estimates reported in original studies. Furthermore, various studies report different numbers of estimates, thus constituting an unbalanced panel structure in the meta-regression model. Researchers have used various meta-regression specifications to account for the panel effects– including the specification of weighted least squares (WLS), generalized least squares (GLS), and explicit specification of panel models using fixed and random effects (Rosenberger and Loomis, 2000a; Sturtevant et al., 1998; Desvousges et al., 1998). But, it is also not uncommon to assume no panel effect and use ordinary least squares (OLS) estimator, where either all estimates are included or a single candidate estimate from each study is selected (Bijmolt and Pieters, 2001; Stanley, 2001).

2.2. *Validity Tests*

A fundamental issue in benefit transfer is its accuracy in providing estimates of values for different contexts, including unstudied sites, policies, and issues. One manner in which this accuracy is tested is in the form of consistency or convergent validity. The validity tests evaluate whether benefit transfer estimates converge or are consistent with known estimates for specific sites, policies, or issues. Thus, validity tests can only be conducted for sites with known values (for a review of past validity tests, see Rosenberger and Phipps, this volume). If the validity tests support the use of benefit transfer in cases where the true value is known, then we may be certain in its broader application in informing policy and conducting resource assessments.

The validity tests primarily examine the following hypothesis:

$$(2) \qquad \begin{aligned} H_0 &: WTP_{BTF} - WTP_{ACT} = 0 \\ H_a &: WTP_{BTF} - WTP_{ACT} \neq 0. \end{aligned}$$

The null hypothesis (H_0) is that the BTF predicted value (WTP_{BTF}) and the actual value (WTP_{ACT}) are the same. A rejection of the H_0 implies that the two WTP estimates are essentially not the same suggesting the rejection of the validity of the BTF estimates. Several parametric and non-parametric approaches are used to test for the validity of transfer values. Brouwer and Spaninks (1999) draw a list of such test-statistics from Bergland et al. (1995) to evaluate the validity of transfer values generated using various benefit transfer approaches. Table 1 lists three different tests of convergent validity between BTF predicted WTP and the actual value of the resource or site.

The value of λ provides the magnitude of convergence between BTF predicted and actual WTP estimates (Table: 1). Although it is not a formal hypothesis test, λ is often used to evaluate the validity of transfer values (Loomis, 1992; Rosenberger and Loomis, 2000b). A relatively smaller λ value suggests better performance of the BTFs in terms of convergent validity. Convergence of BTF predicted and actual WTP estimates can be tested using a paired t-test for equality of means (Sirkin 1995). In paired t-tests, statistical significance of the t-value indicates rejection of the null hypothesis $\mu_D = 0$, where μ_D is the mean difference between BTF predicted and actual WTP values (Table: 2). Thus, the convergent validity of the BTF requires that the null hypothesis is not rejected. Valdes (1995) used paired t-test for benefit transfer analysis of CVM studies in recreation sports fishing. Brouwer and Spaninks (1999) used similar statistics to test equality of average WTP estimates from different studies.

Another test for the validity is the correlation test that measures the direction and degree of association between the predicted and actual WTP estimates. The Pearson's correlation coefficients (r) can be evaluated using the hypotheses stated in Table: 3. The rejection of the $H_0 : r = 0$ implies significant correlation between the predicted and original WTP values, thus an indication of consistency between estimates. Carson et al. (1996) used a correlation test to evaluate the validity of non-market values elicited from revealed and stated preference methods. However, correlations do not inform us of the uni-directional causations. Rosenberger and Phipps (2002) use multivariate regression analysis to explore the directional causation of λ estimates in developing a meta-analytic benefit transfer model.

Table 1. Validity Tests For Benefit Transfer

	Objective	Hypothesis	Test procedure
1.	Analyze the similarity of BTF predicted and original WTP values	$\left[\frac{(WTP_{BTF} - WTP_{ACT})}{WTP_{ACT}}\right] * 100 = \lambda$	Percentage difference.
2.	Test for the equality of BTF predicted and original mean WTP values	H_o: $\mu_D = 0$ H_a: $\mu_D \neq 0$	Paired t-test.
3.	Test for the correlation between BTF predicted and original WTP values	H_o: $r = 0$ H_a: $r \neq 0$	Pearson's Correlation.

λ represents percentage difference between BTF predicted and actual WTP estimates. A relatively smaller λ (% difference) indicates convergence.

Table 2. Variables in Meta-Regression Benefit Transfer Functions

Variables	Description
Method variables:	
METHOD	1 if stated preference (SP) valuation, 0 if revealed preference (RP).
DCCVM	1 if SP and dichotomous choice elicitation technique
OE	1 if SP and open ended elicitation technique was used.
PAYCARD	1 if SP and payment card elicitation technique was used.
ITBID	1 if SP and iterative bidding elicitation technique was used.
RPSP	1 if SP and RP were used in combination.
INDIVID	1 if RP approach was an individual model.
ZONAL	1 if RP and zonal travel cost model was used.
RUM	1 if RP was a random utility model.
SUBS	1 if demand model incorporated substitute sites.
TTIME	1 if RP model included travel time variable.
MAIL	1 if survey type was mail.
PHONE	1 if survey type was phone.
INPERSON	1 if survey type was in person.
LOGLIN	1 if function has log dependent and linear independent variables.
LOGLOG	1 if function has log dependent and independent variables.
LINLIN	1 if function has linear dependent and independent variables.
VALUNIT	1 if consumer surplus was originally estimated as per day.
TREND	Year WTP estimate was recorded, coded as 1967 = 1, ..., 1996 = 30.
Site Variables:	
FSADMIN	1 if study sites were USDA Forest Service National Forests.
R1, R2, R3, R4, R5, R6, R8, R9, R10	Dummy for USFS Regions; R1=MT or ND, R2=CO, KS, NB, SD or WY, R3=AZ or NM, R4=ID, NV or UT, R5=CA or HI, R6=OR or WA, R8=Southeaststates, R9=Northeaststates, R10 = Alaska. There is no R7.
LAKE	1 if the recreation site was a lake.
RIVER	1 if the recreation site was river.
NATL	1 if the recreation site was the entire U.S..
FOREST	1 if the recreation site was a forest.
PUBLIC	1 if ownership of the recreation site was public.
DEVELOP	1 if a recreation site had developed facilities.
NUMACT	Number of different recreation activities the site offers.
Recreation activity variables:	
CAMP... GENREC	1 if the relevant recreation activity was studied: CAMP is camping, PICNIC is picnicking, SWIM is swimming, SISEE is sightseeing, NONMTRBT is non-motorized boating, MTRBOAT is motorized boating, OFFRD is off-road driving, HIKE is hiking, BIKE is biking, DHSKI is downhill skiing, XSKI is cross county skiing, SNOWMOB is snowmobiling, BGHUNT is big game hunting, SMHUNT is small game hunting, WATFOWL is waterfowl hunting, FISH is fishing, WLVIEW is wildlife viewing, ROCKCL is rock climbing, HORSE is horseback riding, and GENREC is general recreation.
Socioeconomic variables:	
INCOME... POPUL	State-level socioeconomic variables: INCOME is per capita income in \$1000, AGE is % of population over 65 years, EDU is % holding at least bachelor degree, BLACK is % African American, HISPAN is % Hispanic, POPUL is population size.

Table 3. Estimated Meta-Regression Benefit Transfer Functions

Variable	NATIONAL	NEAST	SEAST	WIMNT	PACAL
CONSTANT	19.159* (10.20)		26.541[a] (24.89)	66.234* (15.19)	10.044[a] (10.65)
METHOD	−17.598* (6.81)			−34.381* (12.17)	
DCCVM				27.066* (13.63)	
OE	−7.468* (3.82)				
RPSP	−38.170* (10.73)	na	na		na
PAYCARD	−28.333 (19.72)	na	na		na
INDIVID		38.927* (14.40)	52.653* (14.21)		
RUM		63.278* (29.500)		−28.392* (16.12)	30.445* (14.65)
SUBS	−20.277* (5.08)			−33.481* (17.75)	−18.851* (12.67)
MAIL			19.309 (12.01)		
PHONE	−18.626* (4.24)				
LOGLIN					22.642** (11.84)
LOGLOG		24.305* (12.11)			
VALUNIT	−5.820 (4.12)	−18.426* (9.00)		−10.023* (5.85)	
TREND	1.613* (0.38)				
FSADMIN	−20.056* (3.929)		−18.471 (11.36)	−14.595* (8.74)	
R2	−6.581 (5.077)	na	na	−8.791 (5.57)	na
R5	−10.448 (6.96)	na	na	na	
R6	−14.218* (4.55)	na	na	na	
R8	−8.756* (3.74)	na	pc	na	na
R9	−7.124* (3.83)	pc	na	na	na
R10	−14.980* (8.50)	na	na	na	
LAKE	−16.803* (6.61)	−30.097* (15.26)			
RIVER	17.747* (8.05)		−73.951* (26.40)	40.462* (26.84)	

(Continued)

Table 3. *(Continued)*

Variable	NATIONAL	NEAST	SEAST	WIMNT	PACAL
FOREST			17.792* (3.87)	−18.358* (10.29)	
PUBLIC	21.655* (5.66)	29.652* (6.18)	48.940* (27.78)		pc
DEVELOP			−65.216* (24.30)		
NUMACT				2.267* (0.72)	
CAMP					107.59*(33.30)
PICNIC			−25.683* (12.30)	−45.120 (15.82)	60.118* (33.37)
SISEE		78.925* (16.56)	−50.590* (27.36)		36.809* (10.05)
OFFRD	−7.898(5.08)	na	na	na	19.803 (10.05)
BIKE	−13.569 (7.63)	−58.772 (19.11)	−25.962* (15.92)		na
DHSKI		na	na	40.033 (17.10)	
XSKI		14.005 (8.20)	na		na
SNOWMOB	−20.299* (9.74)	na	na	20.026 (12.38)	na
BGHUNT	12.478* (3.42)	21.70* (13.94)	−48.391* (23.82)	19.070* (11.68)	34.536* (22.25)
WATFOWL	10.161* (4.25)	14.479* (8.24)	−57.781* (23.49)		17.827* (8.57)
FISH	9.057* (4.12)		−61.378 (25.07)		19.419* (7.81)
WLVIEW			−49.923 (23.42)		30.304* (15.94)
ROCKCL	39.738 (12.59)		na	28.222 (13.5557)	na
HORSE	−11.841 (5.11)	na	na	na	na
GENREC					24.721* (13.74)
Adjusted R^2	0.26	0.28	0.66	0.36	0.33
F-STAT	9.98* [25, 656]	8.89* [10, 195]	16.14* [14, 97]	10.08* [15, 222]	4.48* [12, 82]
N	682	206	112	238	95

* is $p < 0.10$ (all variables are $p <= 0.20$). Dependent variable is consumer surplus or WTP per person day. a is intercept term dropped in NEAST ($p > 0.9$), but SEAST ($p = 0.29$) and PACAL ($p = 0.34$) retained. na is no observation, pc is perfectly correlated. Figures in parentheses indicate standard errors (corrected for heteroscedasticity & serial correlation using Newey-West version of White correction). Figures in brackets are F-stat degrees of freedom.

2.3. *Benefit Transfer Tests Using Out-of-Sample Studies*

The predictive validity of meta-regression models could be analyzed in various ways. Rosenberger and Loomis (2000b) used in-sample convergent validity on average value transfer. In other cases, predictive validity is also analyzed in terms of a model's success in predicting the values of a holdout sample, much like a bootstrap process or Monte Carlo method (Loomis 1992). In real world benefit transfers, the analysts' main interest may be to ensure the consistency or convergence of BTF predicted value to the actual WTP of the site or resource. Thus, the most plausible way to test for the consistency or convergence between the BTF predicted and actual WTP values would be to test for the validity of the BTF predicted values against out-of-sample estimates.

An out-of-sample estimate of WTP refers to the value reported in the original study that is beyond the studies included in developing a meta-regression analysis BTF. These out-of-sample original values represent the unknown actual WTP value of the resource or site. Out-of-sample testing evaluates an existing meta-regression BTF's validity to estimate new WTP values. In our analysis of the BTF in subsequent sections, we test for consistency and convergence of the BTF predicted and original out-of-sample WTP values.

3. DATA SOURCES

3.1. *Data Collection for Meta-Analysis*

Meta-analytic benefit transfers use information derived from existing studies in their formulation. Thus, in our application, all past recreation use valuation studies in the U.S. were candidates for our meta-analysis. In the dataset, we updated the literature review by Walsh et al. (1989, 1992) with additional valuation studies available up to 1998. The data for years prior to 1993 were obtained from the MacNair (1993) database that also coded Walsh et al. (1989, 1992). We obtained a few additional details directly from the Walsh et al. (1989, 1992) study. Our database contains data from Walsh et al. (1989, 1992), MacNair (1993), and our current literature search. We performed meta-analysis on the full dataset that spans estimates reported from 1967 to 1998.

We searched recreation valuation studies using electronic databases including the American Economic Association's EconLit, First Search Databases, the University of Michigan-Dissertation and Thesis Abstracts, National Technical Information Service (NTIS) database and the Water Resources Abstract Index. We also used unpublished or gray literature including western regional research publication (W133) from 1987 to 1996. Inclusion of gray literature reduces a potential publication bias in meta-analysis (Stanley, 2001). Recreational fishing studies were not emphasized in our meta-analysis as they were assessed in two previous literature reviews—Sturtevant et al. (1998) and Markowski et al. (1997). However, fishing studies included in Walsh et al. (1989, 1992) and McNair (1993) were used.

A total of 131 recreation valuation studies were obtained, that provided 682 individual benefit estimates. Enough information was obtained to fully code for each of the variables listed in Table 2. Inclusion of candidate studies in the dataset was restricted to the studies that reported WTP value per unit of use such as recreation visitor day, trip, or year. All values were converted to WTP per person per activity day. An activity day corresponds with one day of recreation use regardless of amount of time spent in the activity.

3.2. Data Collection for Out-of-Sample Analysis

The out-of-sample WTP values of recreation use valuation studies in the U.S. include those that became available after the development of our BTFs. All WTP values are adjusted for inflation using the implicit price deflator. Our out-of-sample analysis includes 14 studies from the U.S. that are available from 1998 to 2001, which reported 44 estimates.

Similarly, the out-of-sample studies on international recreation use valuation were collected using databases, journals, and workshop proceedings. In addition to the inflationary adjustments in the WTP values reported in the international studies, purchasing power parity (PPP) indices were used to adjust the values to be comparable with U.S. estimates (World Bank, 1999). Thus, the WTP values obtained from all sources are adjusted to 1996 U.S. dollars. The international studies were retrieved from various recreation sites around the world (referred to as the rest of the world (ROW) from 1991 to 1999. There are 28 studies from 14 countries reporting 83 estimates.

4. META-ANALYSIS RESULTS

The meta-regression models to be used as the BTFs were estimated from the thirty-year recreation use valuation data in the U.S. The data used for our meta-analyses are panel data that contain multiple estimates from many studies. Multiple value estimates from a single source can result in systematic effects that are not accounted for in the specification of a classical regression model. A random or fixed effect specification has to be used to resolve such bias. However, previous testing for our dataset rejected panel effects in favor of equal effects (no panel effects) (Rosenberger and Loomis 2000a). Although panel effects are suspected in our data, the complexity of the data does not enable identification of the panels. Therefore, with the assumption of independent and identically distributed (iid) error terms, we use a classical ordinary least squares (OLS) technique to estimate our meta-regression models for benefit transfer.

Table 2 lists the variables coded and tested in specifying the BTFs. Optimized BTFs are estimated for national and regional scopes. Optimization was conducted by retaining only variables significant at $p \leq 0.20$ or better. While it is important to account for socioeconomic factors in a BTF, the information reported in the original studies on socioeconomic variables representing study specific measures was incomplete. We attempted to capture socioeconomic effects by including state-level economic and demographic variables. However, potentially due to aggregation effects of our state-level data, there are no statistically significant socioeconomic variables in the BTFs.

Five BTFs were estimated, one for each of the four geographic regions represented by U.S. census regions and one for the national model (Rosenberger and Loomis, 2000b). The regional BTFs are NEAST (northeastern states), SEAST (southeastern states), WIMNT (intermountain west), and PACAL (Pacific coast states and Alaska). PACAL is a combination of two census regions. Because of lack of degrees of freedom for Alaska, we tested coefficient equality using a Chow test (Greene, 1997) and found that we could combine Alaska with Pacific coast states. However, testing the four regions against one another yielded an F statistic of 3.66, suggesting at the 0.01 level of significance that at least one of the regional coefficients differs from one of the others. This means that there is lack of equality of regional coefficients and that separate models should be estimated for each census region. Thus, we estimated four regional BTFs including a combined model for PACAL (Table 3). Model specifications were based in part on Walsh et al. (1989, 1992) and Smith and Kaoru (1990). The explanatory power of the national BTF shown by adjusted R^2 is 0.27, slightly below that of Walsh et al.'s (1989, 1992) combined travel cost/contingent valuation model. The regional BTFs had greater explanatory power, ranging from 0.28 for NEAST, 0.33 for PACAL, 0.36 for WIMNT, to 0.66 for SEAST.

The signs and significance of the variables in the BTFs are consistent with past recreation valuation studies. For example, *METHOD* is negative indicating contingent valuation studies produce lower estimates of WTP than do travel cost studies, a result consistent with Carson et al. (1996), Walsh et al. (1989, 1992) and a recent review of the contingent valuation method literature by Carson et al. (2001). In our meta-analysis, only the big game hunting activity variable (*BGHUNT*) was consistently significant across national and all regional BTFs. Other activity variables have different influences across the BTFs (Table 3). The discussion on BTF results is not elaborated here since our intent is to use these results for benefit transfer analysis. For further elaboration, see Rosenberger and Loomis (2000b).

5. BENEFIT TRANSFER TEST RESULTS

To test the validity between BTF predicted and original out-of-sample WTP values, the predicted WTP values were calculated incorporating out-of-sample study characteristics. In doing so, relevant variables in the BTFs were set according to the original out-of-sample study characteristics (e.g., recreation activity, lake, stream or region). If unavailable from the out-of-sample study, they were set at zero or at the mean value for the BTF, as applicable. Our out-of-sample test approach has the advantage of including the effects of the variables on the out-of-sample original studies. This is usually not possible in real world benefit transfer when the transfer has to be performed without a pre-existing valuation study of a new policy site. In such cases, analysts will have to set explanatory variables of BTF to zero, one or the mean based on prior experience or as is appropriate to the policy site attributes. As described in the data collection procedure, the out-of-sample original WTP values are adjusted for inflation, thus all values are reported in 1996 U.S. dollars.

5.1. Out-of-Sample Test Using the U.S. Studies

We evaluated the predictive accuracy of national and regional BTFs using out-of-sample studies from the U.S. The predictive validity was analyzed using the percentage difference (λ), paired t-test, and correlation test. Because of the relatively small sample of predicted WTP values for each of the four regional BTFs, the WTP values for regional BTFs were aggregated for hypothesis testing.

It is evident from the results reported in Table 4 that the mean predicted WTP values of $47.10 and $40.91 using national and regional BTFs, respectively, are higher than the mean WTP estimates of the corresponding $34.40 and $32.48 from the original out-of-sample studies. This result indicates that on average our BTFs over-predict WTP values. The mean predicted value obtained from our national BTF is 37% larger, and that of aggregated regional BTF is 26% larger than the mean WTP estimates reported in original studies. When aggregated by out-of-sample studies, the absolute values of the mean λ between original and predicted WTP using national and regional BTFs are approximately 80%, ranging from 17.58% to 411.08% and from 23.20% to 357.44%, respectively.

The results of paired t-test reported in Table 4 show that the t-statistics are significant at or below probability value of 0.05 in both national and regional BTFs, implying the H_0 must be rejected. In other words, there is a significant difference

Table 4. Paired T-Test, Correlation Test, And Percentage Difference Using The U.S. Out-Of-Sample

Measurement	National	Regional
Original out-of-sample mean WTP $	34.04 (1.97–116.78)	32.48 (1.97–116.78)
BTF predicted mean WTP $	47.10 (6.40–80.09)	40.91 (12.12–84.55)
Percentage difference (λ) at mean %	37	26
Percentage difference (λ) by study %	80.88 (17.58–411.08)	79.95 (23.20–357.44)
Paired t-test (t-statistics)	3.832**	2.10*
Pearson correlation coefficient (r-statistics)	0.651**	0.191
Sample size	37	44

** significance at $p \leq 0.01$, * significant at $p \leq 0.05$.
Numbers in parentheses are ranges of percentage difference.

between the benefit function predicted and original WTP values, thus the transfer values do not converge based on the paired t-test.

The test statistics based on Pearson's correlation coefficient show a positive correlation between original and predicted WTP values. The correlation coefficient for the national BTF is 0.651, significant at $p \leq 0.01$, indicating that the original and predicted WTP estimates are positively correlated, that is, an evidence of consistency between estimates. The result suggests that the national BTF predicts higher benefit values for the site or resource where original WTP is larger. For regional BTF, the correlation is weakly positive ($r = 0.191$) and insignificant implying no strong relationship (Table 4). Thus, correlation between the national BTF predicted and original WTP estimates is the only evidence of a consistency of the meta-analytic benefit transfer using the U.S. out-of-sample studies.

5.2. Out-of-Sample Tests Using International Studies

A similar procedure is applied to test the validity of the BTF to estimate the value of international recreation sites. To estimate the BTF predicted values for the ROW out-of-sample studies, the national BTF was used. The validity test procedures were carried out in three separate treatments in order to account for the potential differences in recreation site values between high-income and low-income countries (World Bank, 1996). Thus, the international data is segregated into three groups–all international studies (ROW), low-income countries, and high-income countries. The mean WTP for the combined international out-of-sample studies are \$34.91, \$45.59, and \$38.90 for high-income, low-income, and ROW, respectively. The BTF predicted WTP values, however, were \$40.32, \$35.97, and \$38.70 for high-income, low-income, and ROW. Thus, absolute values of the prediction error represented by λ indicate 15.5%, 21.10%, and 0.51% for high-income counties, and low-income countries, and ROW, respectively. However, the percentage difference aggregated by studies show the λ value of 216.03%, 175.89%, and 197.39% for high-income, low-income, and ROW overall (Table 5).

The paired t-test reveals that all t-statistics are insignificant implying the BTF predicted values and the original WTP estimates across all countries together, and high-income and low-income countries separately, were not statistically different (Table 5), that is, the BTF predicted WTP values converge into the original out-of-sample WTP values drawn from the ROW sites.

The results of Pearson correlation coefficient analysis also revealed that in all cases there were significant positive correlations between the BTF predicted values and the original WTP values. The correlation coefficients across all groups were significant at $p < 0.05$, and the ROW and high-income countries significance at $p < 0.01$ (Table 5). This shows that our BTF predicted WTP values are highly correlated with the original ROW study values. The BTF tends to predict relatively higher values, when the original study values are higher and *vice versa*.

Altogether, using the U.S. out-of-sample studies the validity tests largely rejected the transferability of the BTF predicted values to estimate the WTP value of the resources at the new policy sites. The λ values show fairly large percentage differences between BTFs predicted and out-of-sample original WTP values, and paired t-test results reveal that BTF predicted WTP estimates are statistically different from out-of-sample original WTP estimates. Thus, the correlation test is the only evidence that show a minor support for benefit transfer using the U.S. out-of-sample studies. The same test procedures when applied to international out-of-sample studies reveal very different results. The paired t-test and correlation tests overwhelmingly demonstrated the consistency and convergence of the BTF predicted and original WTP estimates across the board. Similarly, the percentage errors at the mean remain fairly small (0.51 – 21.10%). However, the range of the percentage errors with respect to the individual study from both the U.S. and international out-of-sample studies show remarkable differences. The relatively better performance of the BTF for the international out of sample as compared to the national out of sample is unexpected result.

6. DISCUSSION

Benefit transfer with meta-analysis can be a more vigorous approach compared to the other benefit transfer methods. In particular, meta-analysis enables analysts to capture increased heterogeneity in the value estimates such that the transfer value is likely to be closer to the actual value of the resources at new policy sites. Our tests for convergent validity of meta-regression models or benefit transfer functions (BTF) predicted values and original out-of-sample WTP estimates, however, indicate mixed results. The correlation tests indicate consistency of BTF predicted values regardless of using the U.S. or international out-of-sample studies. The paired t-tests show the convergence of BTF predicted values only in case of international out-of-sample studies. While transfer error at the mean WTP estimates is fairly large (37%) in case of the U.S. out-of-sample studies, the mean error significantly

Table 5. Paired T-Test, Correlation Test, And Percentage Difference Using International Out-Of-Sample

Measurement	ROW	High income	Low income
Original out-of-sample mean WTP	38.90 (0.72–387.03)	34.91 (1.84–353.60)	45.59 (0.72–387.03)
BTF predicted mean WTP	38.70 (11.13–73.44)	40.32 (13.82–73.44)	35.97 (11.13–58.13)
% difference (λ) at mean	0.51	15.50	21.10
% difference (λ) by study	197.39 (12.76–1504.67)	216.03 (12.76–772.24)	175.89 (12.76–1504.7)
Paired t-test (t-statistics)	−0.031	0.730	−0.748
Pearson corr. coeff.	0.447**	0.547**	0.350*
Sample size	83	52	31

Note: Low income = countries with per capita income below US $8,955, High income = countries with per capita income above US $8,955 (World Bank, 1996).

** significance at $p \leq 0.01$, * significant at $p \leq 0.05$. Numbers in parentheses are ranges of percentage difference.

reduces (0.51%) in using international out-of-sample studies. A major caveat to these results is that the range of the transfer errors remains fairly large in all cases at the individual study level suggesting analysts to take cautions in real world benefit transfer.

Meta-analysis BTFs are limited by the quality of those studies they are derived from. We did not use any subjective judgments regarding the inclusion or exclusion of original study estimates from our BTFs. In addition, lack of reporting of site characteristics (resource quality, market characteristics, scope of affected area) in original studies further limits our ability to specify the underlying valuation function using meta-regression analysis. Limited simple sizes in the meta-analysis and out-of-sample tests may have also influenced our results. Given these limitations, our out-of-sample tests show that meta-analysis based BTFs can be a promising approach to benefit transfers.

In constructing our meta-analysis BTFs, we assumed the existence of a single, underlying valuation function for outdoor recreation activities, regardless of type of activity. Due to the heterogeneity of outdoor recreation activities, this assumption may be Herculean. Possibly a more realistic approach would be the assumption of meta-valuation functions for single or similar activities, such as winter sports, hunting, fishing, or sightseeing. This would potentially reduce the complexity and inherent heterogeneity of the data, leading to more accurate, robust, and practical meta-analytic BTFs. Rosenberger and Phipps (2002) show the distinct gains in accuracy of meta-analysis benefit transfers over traditional approaches when the specified valuation function is more tenable. Smith and Pattanayak (2002) and Smith et al. (2002) show how limited information on valuation studies can be used to calibrate models for specific uses.

Benefit transfer is largely more an art than a science. As McConnell (1992) suggests benefit transfer may never be mechanical because the value of natural resources and environmental services is estimated in a setting where the market does not clear and the 'true value' estimates are never observed. Thus, any application of benefit transfer requires researchers to choose the right number from range of possible value estimates, which involves a great deal of informed judgments. Meta-analysis provides a systematic basis for the final judgments on the accuracy of the transfer value estimates. Obviously, through meta-analysis the analysts are able to adjust for some of uncertainties related to the physical attributes, socioeconomic characteristics, and methodological differences in their transfer value estimates, which will narrow down the magnitude of value judgments they will have to superimpose.

Ram Shrestha is postdoctoral researcher, School of Forest Resources and Conservation, University of Florida, Gainesville, FL, USA. Randall Rosenberger is assistant professor, Department of Forest Resources, Oregon State University, Corvallis, OR, USA. John Loomis is professor, Department of Agricultural and Resource Economics, Colorado State University, Fort Collins, CO, USA.

7. REFERENCES

Bergl, O, Magnussen K, Navrud S (1995) Benefit Transfer: Testing for Accuracy and Reliability Discussion Paper #D-03/1995, Department of Economics and Social Sciences, Agricultural University of Norway

Bijmolt, THA, Pieters RGM (2001) Meta-Analysis in Marketing when Studies Contain Multiple Measurements, Marketing Letters 12(2):157–169

Boyle, K, Bergstrom J (1992) Benefit Transfer Studies: Myths, Pragmatism, and Idealism, Water Resource Research 28(3):657–663

Boyle, K, G Poe, Bergstrom J (1994) What Do We Know about Groundwater Values? Preliminary Implications from a Meta Analysis of Contingent-Valuation Studies, American Journal of Agricultural Economics 76:1055–1061

Brouwer R, Spaninks FA (1999) The Validity of Environmental Benefits Transfer: Further Empirical Testing, Environmental and Resource Economics 14:95–117

Brouwer R, Langford IH, Bateman IJ, Turner RK (1999) A Meta-Analysis of Wetland Contingent Valuation Studies, Regional Environmental Change 1(1):47–57

Carson RT, Flores NE, Meade NF (2001) Contingent Valuation: Controversies and Evidence, Environment and Resource Economics 19:173–210

Carson RT, Flores NE, Martin KM, Wright JL (1996) Contingent valuation and revealed preference methodologies: Comparing the estimates for quasi-public good, Land Economics 72(1):80–99

Desvousges WH, Naughton MC, Parsons GR (1992) Benefit Transfer: Conceptual Problems in Estimating Water Quality Benefits Using Existing Studies, Water Resource Research 28:675–683

Desvousges WH, Johnson FR, Banzhaf HS (1998) Environmental Policy Analysis with Limited Information: Principles and Applications of the Transfer Method Northampton, MA: Edward Elgar

Downing MT Jr, Ozuna (1996) Testing the Reliability of the Benefit Function Transfer Approach, Journal of Environmental Economics and Management 30:316–322

Glass GV (1976) Primary, Secondary, and Meta-Analysis of Research, Educational Researcher 5:3–8

Glass GV, McGraw B, Smith MS (1981) Meta-Analysis in Social Research, Sage Publications Inc, California & London

Greene WH (1997) Econometric Analysis (Third Edition), Prentice Hall, New Jersey

Kirchhoff, S, Colby BG, LaFrance JR (1997) Evaluating the Performance of Benefit Transfer: An Empirical Inquiry, Journal of Environmental Economics and Management 33:75–93

Loomis JB (1992) The Evolution of a More Rigorous Approach to Benefit Transfer: Benefit Function Transfer, Water Resources Research 28:701–705

Loomis J, Roach B, Ward F, Ready R (1995) Testing the Transferability of Recreation Demand Models Across Regions: A Study of Corps of Engineers Reservoirs, Water Resources Research 31:721–730

Loomis J D, White (1996) Economic Benefits of Rare and Endangered Species: Summary and Meta-analysis, Ecological Economics 18:197–206

Loomis JB, Walsh RG (1997) Recreation Economic Decisions: Comparing Benefits and Costs (Second Edition) Venture Publishing, Inc, Pennsylvania

Luken RA, Johnson FR, Kibler V (1992) Benefit and Costs of Pulp and Paper Effluent Control Under the Clean Water Act, Water Resources Research 28(3):665–674

MacNair D (1993) RPA Recreation Values Database USDA Forest Service, RPA Program, Washington DC

Markowski M, Unsworth R, Paterson R Boyle K (1997) A Database of Sport Fishing Values, Industrial Economics Inc prepared for the Economics Division, US Fish and Wildlife Service, Washington, DC

McConnell KE (1992) Model Building and Judgment: Implication for Benefit Transfer With Travel Cost Models, Water Resources Research 28(3):695–700

Morgan C N, Owens (2001) Benefits of Water Quality Policies: The Chesapeake Bay, Ecological Economics 39:271–284

Rosenberger RJ, Loomis (2000a) Panel Stratification of Meta-Analysis of Economic Studies: An Investigation of Its Effects in the Recreation Valuation Literature, Journal of Agricultural and Applied Economics 32(3):459–470

Rosenberger RJ, Loomis (2000b) Using Meta-Analysis for Benefit Transfer: In-Sample Convergent Validity Tests of an Outdoor Recreation Database, Water Resources Research 36(4):1097–1107
Rosenberger RS, Phipps TT (2002) Site Correspondence Effects in Benefit Transfers: A Meta-Analysis Transfer Function, Paper presented at 2nd World Congress of Environmental and Resource Economists, June 24–27, Monterey, California
Shrestha RK, Loomis JB (2001) Testing A Meta-Analysis Model for Benefit Transfer in International Outdoor Recreation, Ecological Economics 39:67–83
Sirkin RM (1995) Statistics for the Social Sciences SAGE Publications Thousand Oaks, California
Smith VK, Huang J (1995) Can Markets Value Air Quality? A Meta-Analysis of Hedonic Property Value Models, Journal of Political Economy 103:209–227
Smith VK, Kaoru Y (1990) Signals or Noise?: Explaining the Variation in Recreation Benefit Estimates, American Journal of Agricultural Economics 72:419–433
Smith VK, Osborne L (1996) Do Contingent Valuation Estimates Pass a Scope Test?: A Meta-analysis, Journal of Environmental Economics and Management 31:287–301
Smith VK, Pattanayak SK (2002) Is Meta-Analysis a Noah's Ark for Non-Market Valuation?, Environmental and Resource Economics 22:271–296
Smith VK, Houtven GV, Pattanayak SK (2002) Benefit Transfer via Preference Calibration: Prudential Algebra for Policy, Land Economics 78(1):132–152
Stanley TD (2001) Wheat from Chaff: Meta-Analysis as Quantitative Literature Review, Journal of Economic Perspectives 15(3):131–150
Sturtevant LA, Johnson FR, Desvousges WH (1998) A Meta-analysis of Recreational Fishing, Unpublished Manuscript, Durham, NC: Triangle Economic Research
US Water Resources Council (1973) Principles, Standards and Procedures for Water and Related Land Resource Planning, Federal Register 38(174), Part III
US Water Resources Council (1979) Procedures for Evaluation of National Economic Development (NED) Benefits and Costs in Water Resources Planning (Level C), Federal Register 44(243):72892–72976
US Water Resources Council (1983) Economic and Environmental Principles and Guidelines for Water and Related Land Resources Implementation Studies, US Government Printing Office, Washington DC
Valdes S (1995) Non-market Valuation of Recreational Resources: Testing Temporal Reliability in Contingent Valuation Studies and an Explanation of Benefit Transfer Procedures with Discrete Choice Models, An unpublished PhD dissertation, The University of Maryland
Walsh RG, Johnson DM, McKean JR (1989) Issues in Nonmarket Valuation and Policy Application: A Retrospective Glance, Western Journal of Agricultural Economics 14:78–188
Walsh RG, Johnson DM, McKean JR (1992) Benefit Transfer of Outdoor Recreation Demand Studies: 1968–1988, Water Resources Research 28:707–713
Woodward RT, Wui YS (2000) The Economic Value of Wetland Services: A Meta-analysis, Ecological Economics 37:257–270
World Bank (1996) World Development Report 1996: From Plan to Market, The World Bank, Oxford University Press, Washington DC
World Bank (1999) World Development Indicators: CD-ROM, The World Bank, Washington DC

R. SCARPA, W.G. HUTCHINSON, S.M. CHILTON
AND J. BUONGIORNO

BENEFIT VALUE TRANSFERS CONDITIONAL ON SITE ATTRIBUTES: SOME EVIDENCE OF RELIABILITY FROM FOREST RECREATION IN IRELAND

1. INTRODUCTION

The best way to investigate the magnitude of benefits from recreation at one natural resource site,[1] such as a forest park, is to conduct an original data collection on site. However, this is often costly[2] and time consuming. If information on the economic benefits produced by similar sites (the *study* sites) is already available, then one may consider the much cheaper option of 'transferring' these to the site of interest (the *policy* sites) after accounting for the idiosyncratic nature of the policy site. Then policy decisions can be based on these "second best" values. In the literature, this practice is referred to as 'benefit transfer' (BT), and it has been regarded as so important to natural resource management agencies that in 1992 a whole special issue of *Water Resources Research* was dedicated to the practical assessment of this technique by leading academics (issue n. 28, 1992).

More results from research on reliability of benefit transfer estimates, mostly developed in the context of North American water resources, were published by Loomis *et al.*(1995), Downing and Ozuna (1996), Kirchhoff *et al.* (1997), and Feather and Hellerstein (1997). More work from the nineties is reported in a book on water resource valuation by Bergstrom *et al.* (2000). Much of these reliability studies produced mixed recommendations on the use of BT, as reported in a review published by Bergstrom and De Civita (1999).

More recently the debate has rekindled with the publications of research papers reporting on applications evaluating new approaches, such as meta-analysis (Santos, 1998) choice modelling (Morrison *et al.* 2002), preference calibration (Smith *et al.* 2002) and alternative testing strategies for reliability (Kristofersson and Navrud this volume and 2002). However, most practitioners would probably agree that the state of knowledge has not substantially changed since Kirchhoff *et al.* (1997, p.75) wrote that:

> Although benefit value transfers are currently used in decision making by public agencies, the scientific debate over benefit transfer continues and many issues remain unresolved."

Benefit value transfers estimates are obviously of great potential interest to practitioners, provided they can be proven to be adequate surrogates of on-site estimates achievable by conducting more costly full-scale studies. In other words, to gain acceptance in the policy arena, BT estimates must show *convergent validity* (Bishop *et al.* 1995). That is, they must show theoretically meaningful and statistically

S. Navrud and R. Ready (eds.), Environmental Value Transfer: Issues and Methods, 179–205.

significant relationships with alternative measures of the same theoretical construct such as other site-specific estimates of the same welfare change. But the issue of how to evaluate this convergence remains contentious.

Validity assessment must recognise that both the BT value estimate for the policy site and the on-site value estimate to be surrogated by it are random variables. Hence, a measure of reliability must account for the probabilistic nature of these values. The major obstacle to this form of validity is argued to be represented by various source of measurement errors (Bergstrom and De Civita 1999). In benefit estimates from outdoor recreation one such source is represented by site-specific attributes, which are determinants of recreational value. When benefits are determined by site attributes their omission from the econometric specification of the benefit function causes mis-specification errors. In the context of maximum likelihood estimation this omission results in biased estimates.

On the other hand, the inclusion of attributes in the specification may induce collinearity because all observations from the same site are associated with the same set of value attributes. For this reason, BT function estimation with site-specific attributes can only be achieved with data from a sufficiently large number of sites and must bear the consequences of collinearity, such as wide standard error estimates. However, with large sample sizes these consequences are less deleterious than one may expect.

In this respect multi-attribute stated preference techniques based on orthogonal experimental design of choice contexts (choice modelling or choice experiments) can be considered superior from the view-point of statistical efficiency. Contingent valuation (CV) responses from samples collected at different sites and choice modelling responses from experimentally designed choice contexts both face the problem of the choice of value to employ to represent attribute levels in the econometric specification. However, in as much as choice contexts are separated from the experience of the effects of the site attributes of relevance – as is often the case in choice modelling – these techniques have another drawback. This is due to the fact that often attributes levels are *described to* respondents, rather than being *experienced by* respondents. Behavioural psychology and common sense suggest that the *experienced* and the *described* give rise to two different sets of value constructs by individuals. See for example the work of Adamowicz *et al.* (1997) on real versus perceived attributes.

In the present study willingness to pay (*WTP*) responses were obtained from respondents who had just completed a visit to the forest, and therefore had experienced the attributes of the site. Furthermore, in merging data from CV surveys at different sites, one must ensure the invariance of the survey instrument, and possibly of the preference structure over time. In our study this was achieved by using data from an identical survey instrument across all sites, conducted more or less simultaneously, over a time interval of a few weeks (summer 1992) (see details in Ni Dhubhain *et al.* 1994).

One novel contribution of the present study is therefore that of systematically addressing and controlling for what we think are the most common sources of measurement and specification errors, namely site-specific recreational attributes.

A second novel contribution is that, rather than water resources, we assess the reliability of BTs from benefit estimates for forest recreation. The social benefits drawn from the recreational value of forests is of growing interest in land use policy, particularly when assessing alternative user-friendly destinations for agricultural land, perhaps because of under-performance in current economic conditions, or for explicit objective of environmental policy (e.g. article 3 of Kyoto forests).

A third novelty is the focus of the statistical test on the differences between on-site value estimates and transferred value estimates, which allows for both the conventional test of no-difference and the more recent one based on equivalence testing (Kristofersson and Navrud, 2002), although in our case – for the sake of comparison with previous studies – we focus on no difference testing. To emphasize the diversity in conclusion between the reliability testing we also present results of a test equivalent to the often cited work of Downing and Ozuna (1996).

Finally, some regression results are reported in an attempt to explain the observed departure from transferred values and on-site values on the basis of site-attributes and estimation factors.

The unit of analysis is the compensating variation for a single visit to a forest and the valuation approach is that of a "*conventional*" discrete-choice CV survey with one follow-up. One can always use sophisticated econometrics, but because the focus of BT is mostly operational, and it concerns its applicability in policy advice within resource management agencies, we believe that tests of its reliability must be kept within reach of the skills available in these contexts, rather than those available in research universities. This approach also ensures wider generality of the results.

In our approach, benefit estimates from value functions estimated from CV data collected on *study* forests are considered *reliable* if their difference with analogue estimates obtained from a full-scale study in the *policy* forest is sufficiently small to be due to chance. If the difference is larger than a given threshold, then a full-blown on-site valuation study would be warranted to properly inform policy makers, rather than the use of a transferred value estimate from existing studies. The adequate size of the threshold depends on the amount of uncertainty tolerable in the decision process (see Navrud and Pruckner, 1997 and Kristofersson and Navrud, 2002 for a discussion of this point). This is clearly a matter that needs addressing in a case-specific manner, mainly on the basis of the cost and benefit of increased estimate accuracy in the particular policy decision at hand.

More specifically, we assess transfer reliability by obtaining estimates of commonly employed location parameters of willingness to pay (*WTP*) distributions, such as median and mean *WTP* for access to the policy forest from the benefit function estimated from data collected at the study forests. With reference to Bergstrom and De Civita (1999), we therefore use a value estimator BT method.

Generally, when the benefit value function is conditional on forest attributes, then the benefit value for the study forest can be predicted by "plugging-in" the values of the forest attributes of the policy site into the estimated conditional function. We formally test the hypothesis of no difference between the on-site and transferred estimates conditioning on forest-specific attributes. We do not check whether on site and transferred estimates have overlapping confidence intervals, but following recommendations by Poe *et al.* 1994 and similarly to Kristofersson and Navrud (2002), we focus on the distribution of differences between these two estimates and employ a test whose empirical significance is close to its nominal one. Then we report on the results of such a strategy to investigate the reliability of benefit value transfers when forest(site)-specific attributes are used as conditioning variables in the estimation of the probability of a positive response, and then used as predictors in the transfer of this benefit value estimate.

Our results would seem to indicate that *WTP* value transfers are frequently reliable when forest attributes are used as predictors. We speculate that previous negative results on transfer reliability obtained in similar contexts may have suffered from misspecification, preference instability, or small sample sizes. We believe our data and estimation procedure to be less prone to these problems mostly because of the simultaneity of data collection across forest sites, the fairly large number of sites involved and the relative homogeneity of the recreational experience object of valuation. However, we note that the transfer value practice in our empirical sample induces over-prediction compared to the on-site benefit estimates.

Our results also seem to suggest that when data sets are sufficiently large, and sufficient variation across site attributes is present, then discrete choice contingent valuation (DC-CV) can be effectively used for BTs without the need to incur the additional effort and potential "lack of experience" bias involved in alternative stated preference-methods, such as choice modelling.

The remainder of the paper is articulated as follows. The next section deals with the conceptual framework behind the study. Section 3 presents the formal reliability test employed. Section 4 reports about the data and present a discussion of the econometric results. Section 5 concludes and outlines a number of issues that require further research in this field.

2. CONDITIONAL AND UNCONDITIONAL BT FROM DISCRETE CHOICE CV RESPONSES

In the context of benefit estimation from DC-CV survey responses, the econometric task is the estimation of the probability of positive response conditional on the proposed bid amount t, and possibly on other covariates. In our opinion,[3] the probability function, and hence the conditional benefit function from which the transfer estimates are derived ought to accommodate a vector $\mathbf{q}$ of site-specific variables, as well as the conventional bid amount t and vector of socio-economic variables $\mathbf{s}$.

The policy site specific values of $\mathbf{q}$ can then be used as conditional predictors in the transfer phase. Similarly, the value estimates of location parameters for $\mathbf{s}$ can be obtained by pre-existing statistics on the potential population of visitors to the policy site.

For the *i*th forest visitor characterized by $\mathbf{s}_i$, who visits the *j*th forest with attributes $\mathbf{q}_j$, in a parametric probability estimation context, we are therefore postulating a *conditional* benefit function of the form:

$$(1) \qquad B(\mathbf{s}, \mathbf{q}; \hat{\boldsymbol{\theta}}) = f[\Pr(Yes \mid t, \mathbf{s}, \mathbf{q}; \hat{\boldsymbol{\theta}})]$$

instead of an *un*conditional one of the type:

$$(2) \qquad B(\hat{\boldsymbol{\theta}}) = f[\Pr(Yes \mid t; \hat{\boldsymbol{\theta}})]$$

as used in most previous studies.

Here $\hat{\boldsymbol{\theta}}$ may be the maximum likelihood estimates maximizing:

$$(3) \qquad \ln(L) = \ln[g(R_{ij} \mid t_{ij}, \mathbf{s}_i, \mathbf{q}_{j;} \hat{\boldsymbol{\theta}})], i = 1, \ldots, n; j = 1, \ldots, J$$

over the parameter space, where $\ln[\cdot]$ is the natural log transformation of the likelihood specification $g(\cdot)$, and R_{ij} is the recorded discrete-choice individual response in the random sample observation, which may be with or without a follow-up.

That forest attributes play a role in explaining the magnitude of the benefits enjoyed in forest recreation makes both intuitive and economic sense, no matter whether the theoretical paradigm underlying the analysis is a RUM (Hanemann, 1984, 1989) or a valuation function one (Cameron and James, 1987). It is also practically advantageous. In fact, the identification of a set of theoretically meaningful and statistically significant relationships with forest-specific attributes is of particular relevance when transferring estimates as it allows the analyst to make a conditional prediction by 'plugging-in' the benefit function the values of the attributes observed for the policy site. This produces a conditional value prediction, which is expected to be more precise – everything else equal – than an unconditional one.

Suppose one has DC-CV responses to the *same* survey for J sites and wishes to evaluate the performance of a benefit function conditional on site attributes. One may estimate the parameters of the conditional probability in the likelihood in equation 3 by using the DC-CV responses collected at $J - 1$ study sites, here indexed with $-j$. These parameter estimates can then be used to predict the benefit transfer value at the *j*th site as:

$$(4) \qquad \hat{B} = (\mathbf{s}_j, \mathbf{q}_j; \hat{\boldsymbol{\theta}}_{-j})$$

this value can then be compared in terms of its statistical properties with the unconditional estimate obtained only from the responses collected on policy site j.

The conventional *no difference* hypothesis can be formally tested by using the following formulation of the null:

$$(5) \qquad \mathrm{H_o}: \Delta\hat{B}_{j,-j} = \hat{B}(\mathbf{s}_j, \mathbf{q}_j; \hat{\boldsymbol{\theta}}_{-j}) - \hat{B}(\hat{\boldsymbol{\theta}}_{-j}) = 0,$$

$$\mathrm{H_a}: \Delta\hat{B}_{j,-j} = \hat{B}(\mathbf{s}_j, \mathbf{q}_j; \hat{\boldsymbol{\theta}}_{-j}) - \hat{B}(\hat{\boldsymbol{\theta}}_{j}) \neq 0,$$

We call the difference $\Delta\hat{B}_{j,-j}$ "value transfer error". Notice that the equivalence test proposed by Kristofersson and Navrud, (2002) can be easily tested in this context, by identifying a small positive value Δ, which will depend on the purpose of the policy. The null can then be formulated as:

$$(6) \qquad \mathrm{H_o}: \Delta\hat{B}_{j,-j} \leq \Delta, \mathrm{or} \Delta\hat{B}_{j,-j} \geq -\Delta$$

$$\mathrm{H_a}: -\Delta < \Delta\hat{B}_{j,-j} < \Delta$$

Although we share the motivation for the use of equivalence, rather than the no-difference testing approach in BT, in this paper we focus exclusively on the latter, which has been the most commonly employed so far. This allows a comparison with a larger number of studies and emphasizes the effect on transfer accuracy due to the reduction of measurement error that characterize our approach.

Under this approach a failure to reject the null in (5) will provide evidence of transferability of the benefit value estimate (though not of the benefit *function*, for which we did not test in this study).

This somewhat departs from the approached used by other authors, which is based on either overlapping confidence intervals (Downing and Ozuna) or on whether confidence intervals for study site (policy site) benefit estimates are contained in the estimated confidence interval of the policy site (study site) (Kirchhoff *et al.*), but is more in keeping with the findings in Poe *et al.* (1994), and ultimately more statistically sound.

The question we seek to answer in this study is the following: *how do benefit transfers perform when conducted conditional on site-specific attributes relevant for recreation*? In order to answer this question we systematically test the hypothesis of no difference between the transfer estimate and the on-site estimate. We conduct this systematic test across 26 forests and find encouraging results in that a number of mean and median *WTP* transfers fail to reject the null at conventional significance levels.

3. MEASUREMENT ERROR MINIMIZATION AND RELIABILITY TEST

The inclusion of covariates in the value estimator function plays an important role in systematically correcting various measurement errors. For example, amongst the various measurement error categories discussed by Bergstrom and De Civita (1999), the presence in the value estimator of forest-specific covariates relating to forest attributes and visitor characteristics should attenuate commodity measurement error, population characteristic measurement error, and physio-economic linkage measurement error.

Other important categories of measurement error are also controlled for by our approach. For example, welfare change measurement error is minimized by the use of an *identical* survey instrument at each site, while preference stability measurement error is addressed by the nearly *simultaneous* collection of the CVM data across all 26 forest sites (over a few weeks period).

The last category of measurement error that one must address is related to the estimation procedure. In general, benefit estimates are known to be crucially sensitive to many judgment calls the analyst must make in the process of estimation. Choice of distributional assumption, functional form of the deterministic component of the model, nature of the data and their fit to the estimating framework are all important. In particular, in DC-CV analysis, tests that allow a discrimination across alternative choices are known to have little power (Alberini, 1995). Hence, much rests with the "wisdom" incorporated in the choices made by the analyst.

The estimates obtained at the end of this nested decision process are conditional on these value judgments, which affect the measurement error introduced by the estimation procedure. In practice BTs are likely to be employed by resource management agency whose analysts are likely to employ off-the-shelf econometric approaches. For this reason in our benefit estimation procedure we kept things simple and adopted a well known and typically robust random utility specification: the log-in-the-bid logistic probability model. This model ensures non-negativity and asymmetry (left-skewness) of the distribution of *WTP* for access to forests for recreation[4]. Furthermore, it is a well-known approach easy to adopt and by now well established amongst analysts in resource management agencies.

3.1. Specification of the Benefit Function for Value Prediction

The generic form of the argument of the RHS of Equation (1) was therefore specialized in:

$$\text{(7)} \qquad \Pr(Yes \mid t, \mathbf{s}, \mathbf{q}; \theta) = \left[1 + e^{-[+\ln(t) + \gamma' \mathbf{s} + \delta' \mathbf{q}]}\right]^{-1}$$

While the first three elements of the vector $\boldsymbol{\theta}$ can always be estimated from individual data, those of $\boldsymbol{\delta}$ may be identified only when the sample pools individual responses from a sufficiently large number of sites. In the context of forest recreation as well as in other forms of outdoor recreation it is quite plausible that *WTP* be associated with site-specific attributes.

Under the response probability specification in equation 7 both median and mean *WTP* have close-form solutions, leading to two difference functions.

$$\text{(8)} \qquad \Delta \hat{M}(WTP) = e^{\left(\frac{\hat{\alpha}_{-j} + \hat{\boldsymbol{\gamma}}_{-j}' \mathbf{s}_j + \hat{\boldsymbol{\delta}}_{-j}' \mathbf{q}_j}{-\hat{\beta}_j}\right)} - e^{\left(\frac{\hat{\alpha}_j}{-\hat{\beta}_j}\right)}$$

The second is $\Delta\hat{E}(\boldsymbol{WTP})$, the difference between expected *WTP* estimates, which are obtained by multiplying the first and second terms of Equation 8 respectively by: $\frac{\pi}{\hat{\beta}_{-j}}[\sin(\pi / \hat{\beta}_{-j})]^{-1}$ and $\frac{\pi}{\hat{\beta}_j}[\sin(\pi / \hat{\beta}_j)]^{-1}$ to account for the asymmetry of the *WTP* distribution due to the log-bid transformation in the logistic specification.

Since benefit value estimates are highly non-linear functions of parameter estimates they have unknown sampling distributions, and so do their differences $(X - Y)$. Checking for the overlaps in confidence intervals is hence an inadequate procedure as the empirical size of the test was found to significantly differ from its theoretical one (see Poe *et al.* 1994 for discussion). We therefore proceed by parametrically bootstrapping directly the differences (Poe *et al.* 1995), and checking either for the presence of zeros within the confidence interval – which implies no difference – or for systematic under or over-prediction. This is done by checking the representative percentiles of the simulated distribution obtained using the Krinsky and Robb (1986) parametric bootstrap procedure, which is suitable to approximate confidence interval of highly non linear functions of maximum likelihood parameter estimates. If the lower percentile of the chosen confidence interval is negative and the higher one is positive then the H_o cannot be rejected and the transfer is reliable.

To sum-up, the test for validity of the BT estimate is conducted for each of the 26 forests, and it involves six steps:

1) Single out one forest as the policy site and use all the DC-CV data from the other forests as study site data;
2) Estimate the parameters α_{-j}, β_{-j} and $\boldsymbol{\gamma}_{-j}$, along with the relative variance-covariance matrix $\boldsymbol{\Omega}_{-j}$ of the BT function conditional on forest attributes in $\mathbf{q}_j$ from the $j-1$ study sites data (leave-one-out BT function). In our case these were data from 25 forest parks.
3) Estimate the parameters α_j and β_j along with their variance-covariance matrix $\boldsymbol{\Omega}_j$ from the data of the jth candidate policy site;
4) Parametrically bootstrap 10,000 times both $\Delta\hat{M}(\boldsymbol{WTP})$ and $\Delta\hat{E}(WTP)$ using the two estimated variance-covariance matrices and parameter vectors;
5) Check whether the bounds of the simulated confidence intervals contain zero, or whether they contain only over- or under-predictions;
6) When zero is contained, the BT value is defined to be *reliably transferable* because the difference between the forecast benefit value for the policy site obtained from the value estimator and the on-site estimate is small enough to be due to chance.

Model estimation and no-difference tests are repeated under single-bound, double-bound (interval-data) assumptions, with a restricted (national) and extended (entire island) set of forests. Altogether this procedure required the estimation of $(26 \times 2 + 26 \times 2 = 104)$ models with covariates[5] and $(26 \times 2 = 52)$ without covariates (on-site estimates), for a total of 156 models each with mean and median $W\hat{T}P$ estimates and three sets of simulated confidence intervals (one for each conventional level of probability for Type I Error) around the differences[6].

3.2. *Determinants of Value Transfer Errors*

Value transfer errors $\Delta\hat{B}_{j,-j}$ are the differences between the *WTP* value from transfer functions and on-site *WTP* estimates. Their magnitude may be systematically determined by estimation factors. For example, the particular value (Median versus Mean WTP), the use of double-bounded versus single-bounded estimates, sample size of the on-site estimates etc. They may also be linked to the site-specific attributes employed to represent site-specificity $\mathbf{q}$. In an attempt to explain the variation in the magnitude of the observed transfer errors we report an OLS regression whereby all the 200 absolute values of observed transfer errors $|\Delta\hat{B}_{j,-j}|$ are pooled and explained on the basis of such potential determinants.

4. DATA, RESULTS AND DISCUSSION

4.1. *Data*

In 1992, the Queen's University of Belfast conducted a CV survey as part of a larger forest recreation study (Ni Dhubhain *et al.*, 1994). The survey was administered by conducting on-site, face-to-face interviews in 14 forest parks in Northern Ireland[7] and 13 in the Republic of Ireland (Figures 1 and 2). The summary of the CV discrete-choice responses is presented in Table 1. Over 9400 visitors were interviewed by trained interviewers who completed the task in a period of a few weeks, short enough to ensure preference stability. All the CV surveys shared an identical design across forest sites. The question asked of all respondents at all sites was:

> If it were necessary to raise funds through an entry charge to ensure this forest or woodland remained open to the public and with no charge being made for parking, would you pay an entry charge of t for each person in your party (including young people under 18) rather than go without the experience?

One is therefore comparing two states, the first is the event of the outdoor visit to site j and the payment of the admission charge t which guarantees access to the forest-specific experience $f(\mathbf{q})$ and defines the state $u(m - cv, f(\mathbf{q}); \mathbf{s})$. The second is the forgoing of the outdoor visit to site j and intact income level m, which defines the state $u(m, f(\mathbf{q}); \mathbf{s})$. This money measure is a Hicksian compensating measure as it includes an income effect.

The initial (first bound) bid amounts t used were: {50, 100, 150, 250, 400} (in pence). They were uniformly distributed across visitors. Respondents who answered 'yes' were presented with a follow-up question that probed the *WTP* at a higher bid amount th: {100, 150, 250, 400, 700}, respectively. Instead, respondents who answered 'No' were asked the same question again, with a lower bid amount tl: {30, 60, 80, 150, 250}, respectively. Bid amounts were chosen on the basis of initial parameter estimates of the *WTP* distribution obtained from extensive pilot studies. Notice that they are so low that income can never be a binding constraint.

During the interview, other information was also obtained concerning the socio-economic profile of visitors, such as age, sex, household income, personal income, dominant reason for the visit, means of transport to the forest and other

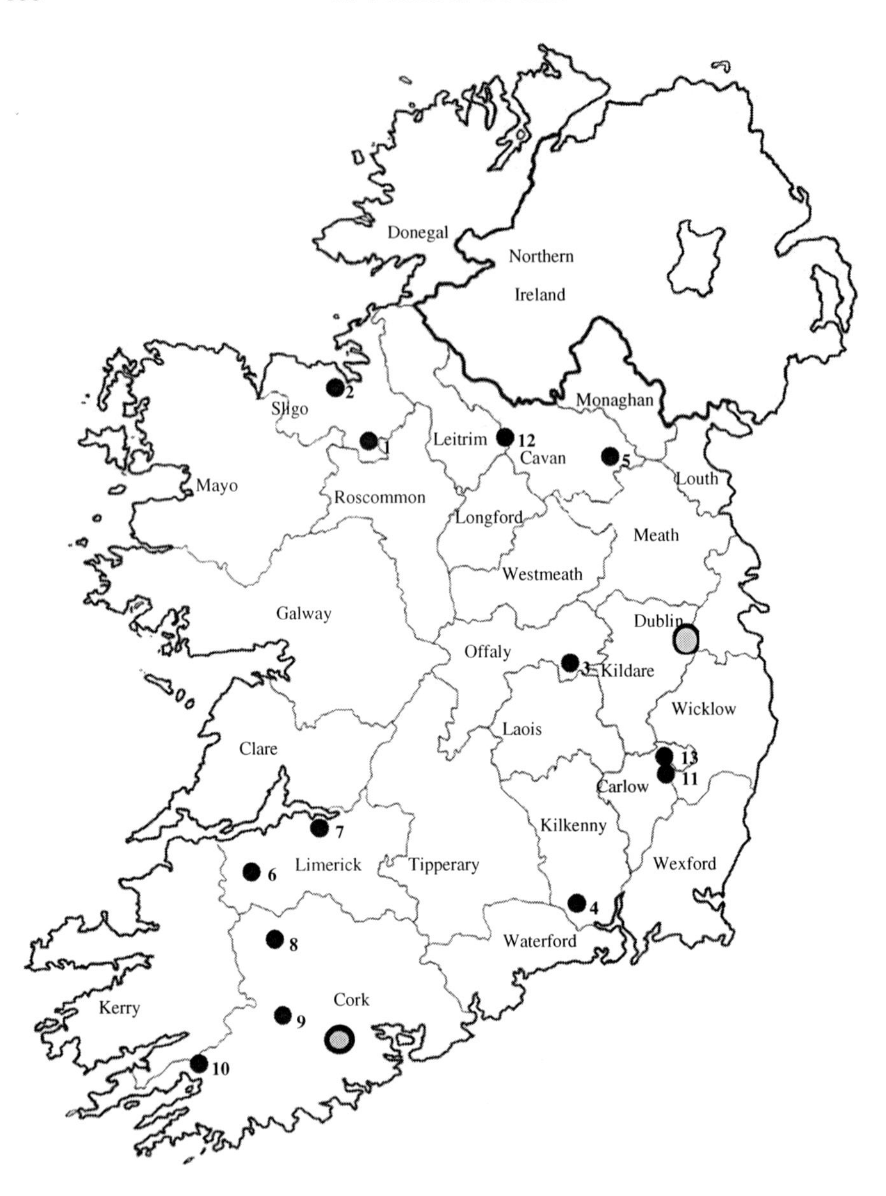

Figure 1. Forest parks and administrative districts in the Republic of Ireland

1. Lough Key.	8.Douneraile.
2.Hazelwood.	9.Farran.
3.Donadea.	10.Guaghan Barra.
4.John F. Kennedy.	11.Avondale.
5.Dun-a-Ri.	12.Killykeen.
6.Currachase.	13.Glendalough.
7.Cratloe.	

Figure 2. Forest parks and administrative districts in Northern Ireland

1.Tollymore	8.Glenariff
2.Castlewellan	9.Ballypatrick
3.Hillsborough	10.Somerset
4.Belvoir	11.Florencecourt
5.Gosford	12.Lough Navar
6.Drum Manor	13.CastleArchdale
7.Gortin Glen	14.Crawfordsburn

information characterizing the profile of the visitor. All of these were included in the s vector. However, only household income had a statistically significant negative effect and was stable for different functional forms[8]. This was combined with the available data on the site attributes deemed relevant for outdoor recreation, which made up the **q** vector. The forest attributes relevant for this paper are in Table 2. We note that the proxy for congestion is measured by the estimated number of visits divided by 100 over the number of cars that can be fitted in the car-park of the forest site. We do recognise that this is less than ideal and may be seen as contentious, but the crowding of the car-park is often taken as a signal of crowding of the resort, and may deter sensitive visitors from attempting an entry. The other attributes are measured in physical units and are quite standard.

Table 1a. Responses to CV elicitation questions by forest site in Northern Ireland

Tollymore, $N = 498$

£	0.5	1.0	1.5	2.5	4.0
yy	63	37	16	3	1
yn	29	42	37	29	11
ny	2	12	22	29	25
nn	5	7	25	40	63

Castlewellan, $N = 497$

£	0.5	1.0	1.5	2.5	4.0
yy	47	42	13	2	1
yn	45	32	41	18	4
ny	2	16	26	32	27
nn	6	9	20	47	67

Hillsborough, $N = 491$

£	0.5	1.0	1.5	2.5	4.0
yy	24	13	3	0	0
yn	39	28	16	7	3
ny	9	16	38	20	12
nn	26	42	42	71	82

Belvoir, $N = 476$

£	0.5	1.0	1.5	2.5	4.0
yy	20	16	5	2	1
yn	43	25	25	8	3
ny	5	17	28	12	13
nn	28	38	38	72	77

Gosford, $N = 489$

£	0.5	1.0	1.5	2.5	4.0
yy	57	22	9	0	1
yn	34	49	31	19	3
ny	6	23	45	40	19
nn	2	5	12	39	73

Drum Manor, $N = 370$

£	0.5	1.0	1.5	2.5	4.0
yy	40	23	2	1	1
yn	20	17	20	9	7
ny	5	13	25	23	8
nn	9	21	27	41	58

Gortin glen, $N = 341$

£	0.5	1.0	1.5	2.5	4.0
yy	34	21	4	0	0
yn	25	29	25	11	3
ny	7	14	24	31	15
nn	3	5	15	26	49

Glenariff, $N = 480$

£	0.5	1.0	1.5	2.5	4.0
yy	64	42	20	6	0
yn	28	37	51	20	15
ny	1	11	21	42	26
nn	2	5	5	29	55

Ballypatrick, $N = 90$

£	0.5	1.0	1.5	2.5	4.0
yy	11	4	1	0	0
yn	7	11	2	2	0
ny	0	3	9	1	1
nn	0	0	6	15	17

Somerset, $N = 243$

£	0.5	1.0	1.5	2.5	4.0
yy	9	3	1	0	0
yn	21	11	9	2	1
ny	5	5	12	3	4
nn	14	30	27	43	43

Florencecourt, $N = 167$

£	0.5	1.0	1.5	2.5	4.0
yy	14	9	4	0	0
yn	13	5	8	4	4
ny	3	9	7	4	6
nn	5	10	15	25	22

Lough Navar, $N = 265$

£	0.5	1.0	1.5	2.5	4.0
yy	23	27	6	0	0
yn	23	11	25	12	5
ny	1	10	10	17	15
nn	6	5	12	24	33

Castlearchdale, $N = 465$

£	0.5	1.0	1.5	2.5	4.0
yy	49	39	8	2	1
yn	34	30	34	22	4
ny	2	13	20	24	16
nn	6	10	30	47	74

Table 1b. Responses to CV elicitation questions by forest site in the Republic of Ireland

	Lough Key, $N = 482$					Hazelwood, $N = 493$					Dun a Dee, $N = 195$					J.F. Kennedy, $N = 498$				
£	0.5	1.0	1.5	2.5	4.0	0.5	1.0	1.5	2.5	4.0	0.5	1.0	1.5	2.5	4.0	0.5	1.0	1.5	2.5	4.0
yy	81	53	20	8	0	45	14	5	0	0	19	10	1	0	0	88	69	36	8	2
yn	12	34	49	17	1	33	34	26	16	4	15	10	15	7	1	11	23	45	41	16
ny	1	2	19	46	23	1	8	26	18	8	0	8	11	8	6	0	5	14	29	31
nn	3	8	9	24	72	18	44	42	63	88	5	11	12	24	32	1	2	5	22	50

	Dun a Ree, $N = 249$					Currachase, $N = 498$					Cratloe, $N = 160$					Douneraile, $N = 273$				
£	0.5	1.0	1.5	2.5	4.0	0.5	1.0	1.5	2.5	4.0	0.5	1.0	1.5	2.5	4.0	0.5	1.0	1.5	2.5	4.0
yy	41	24	3	0	0	63	36	7	2	2	1	3	0	0	0	29	17	4	1	0
yn	5	14	22	6	0	28	36	39	28	8	21	6	9	3	0	23	20	25	5	0
ny	0	6	14	15	12	5	19	41	29	20	3	7	8	5	2	2	10	15	20	10
nn	4	5	11	30	37	4	9	12	40	70	7	16	15	24	30	2	7	11	28	44

	Farran, $N = 491$					Guaghan Barra, $N = 135$					Avondale, $N = 318$					Killykeen, $N = 199$				
£	0.5	1.0	1.5	2.5	4.0	0.5	1.0	1.5	2.5	4.0	0.5	1.0	1.5	2.5	4.0	0.5	1.0	1.5	2.5	4.0
yy	49	30	10	5	0	20	13	6	2	0	40	21	4	0	0	21	15	5	1	0
yn	33	25	37	9	7	6	10	9	9	4	21	23	24	9	3	14	9	13	6	5
ny	2	15	34	32	15	0	4	6	7	4	3	11	24	24	8	0	7	12	12	8
nn	15	28	19	50	76	2	1	5	9	18	0	9	12	30	52	6	9	9	20	27

	Glendalough, $N = 496$				
£	0.5	1.0	1.5	2.5	4.0
yy	74	63	26	12	3
yn	15	24	42	33	18
ny	1	1	14	19	21
nn	9	11	17	35	58

Table 2a. Site attributes for Northern Ireland forests

Forest site	Total area (100 of hectares)	Congestion (100 visits per car park space)	Natural Reserve	Trees before 1940 (% of total)	Tree coverage (% of total forest area)			Median Household income bracket*
					Conifers	Broadleaves	Larch	Bracket
Tollymore	6.29	2.68	No	26	57	5	21	5
Castlewellan	6.41	1.38	No	12	44	7	17	5
Hillsborough	1.99	40.00	No	6	57	12	17	5
Belvoir	0.95	44.00	Yes	0	24	6	27	5
Gosford	2.51	1.39	No	2	40	21	0	4
Drum Manor	0.94	1.40	No	11	20	9	0	4
Gortin glen	14.60	1.17	No	3	70	2	3	4
Glenariff	11.82	1.75	Yes	2	67	1	7	5
Ballypatrick	14.61	0.85	No	0	81	0	3	4
Somerset	1.38	2.00	No	3	59	14	6	3
Florencecourt	13.93	0.50	Yes	1	32	5	0	5
Lough Navar	26.09	0.77	Yes	0	68	1	1	5
Castlearchdale	4.99	4.75	Yes	1	54	3	4	4

*Income bracket was: 1 = under £3,999; 2 = £4000 – £7,999; 3 = £8,000 – £11,999; 4 = £12,000 – 15,999; 5 = 16,000 – 19,999; 6 = 20,000 – 29,999; 7 = £30,000 – £39,999; 8 = higher than £40,000.

Table 2b. Site attributes for Republic of Ireland forests

Forest site	Total area (100 of hectares)	Congestion (100 visits per car park space)	Natural Reserve	Trees before 1940 (% of total)	Tree coverage (% of total forest area)			Median Household income bracket*
					Conifers	Broadleaves	Larch	
Lough Key	3.4	3.00	No	7.3	22	78	0	5
Hazelwood	0.7	20.00	No	0	7	93	0	6
Dun a Dee	2.4	5.00	No	2.6	51	48	1	6
J.F. Kennedy	2.52	1.70	No	0.4	35	60	5	5
Dun a Ree	2.29	3.00	No	2.2	64	36	0	6
Currachase	2	3.30	No	0.3	20	68	12	5
Cratloe	0.65	3.80	No	2.1	56	3	41	6
Douneraile	1.6	4.00	No	8.1	4	96	0	4
Farran	0.75	1.70	No	0.9	83	7	10	6
Guaghan Barra	1.4	5.00	No	4.2	46	12	42	6
Avondale	2.86	1.80	Yes	2.4	30	10	4	5
Killykeen	2.4	2.00	No	2.7	90	8	2	5
Glendalough	3.26	2.00	Yes	4.3	42	7	27	6

*Income bracket was: 1 = under £3,999; 2 = £4000 – £7,999; 3 = £8,000 – £11,999; 4 = £12,000 – 15,999; 5 = 16,000 – 19,999; 6 = 20,000 – 29,999; 7 = £30,000 – £39,999; 8 = higher than £40,000.

4.2. Results and discussion

Tests of the null hypothesis were conducted at the three conventional levels of significance. A higher value of the probability of Type I error is associated here with a more conservative assessment of reliability of the transfer, as the interval of the simulated sampling distribution which might include zero is smaller than when values for this probability are lower. For this reason, departing somewhat from convention, three stars were associated with those transfers that were reliable at 10%, two stars with those at 5% and 1 star for those at 1%. A plus sign indicates that the BT estimate was an over-prediction at all the three levels, while a minus sign an under-prediction. Table 3 shows the results of the convergence validity tests implemented using the no-difference test from single bound (SB) and double bound (DB) leave-one-out estimates from the entire pooled sample across all forest parks as well as from Northern Ireland and Republic of Ireland separately. This latter distinction allows us to evaluate the effect of restricting to the national context the set of sites from which the forest-attributes conditioning the response probabilities are drawn.

4.3. Results on the Reliability of Benefit Value Transfer

These results can be used to discuss five aspects of reliability of benefit value transfer:

1) median versus mean *WTP* transfer;
2) DB versus single bound SB effects on reliable BTs;
3) effects on reliability of BTs of estimates from an extended set of forests (all forests) versus those from a reduced set (only national forests);
4) the prevalence of over or under prediction of the transfer with respect to the on-site study estimates; and
5) the choice of values for probability of Type I error amongst the three conventional ones.

Out of 100 mean *WTP* and 100 median *WTP* comparisons,[9] there are 62 reliable mean transfers and 51 reliable median transfers at *any* significance level. Of these, 42 mean *WTP* estimates are transferable at a 10% significance level, 7 more are transferable at 5% and 13 more at 1%. For median *WTP* the transferable values are 41, 3 and 7 respectively. There are 12 under-predictions in mean transfers and 18 in median ones, while there are 26 and 31 over-predictions for mean and median respectively. Transfer values have therefore a tendency to over-predict, which indicates that—in this case—the BT practice would tend to induce the analyst to infer an excessively high volume of benefit compared to the on-site estimation.

Table 4a, reports the statistics for the on-site and transferred values estimates, while Table 4b reports the statistics for the observed widths of the simulated confidence intervals around $\Delta\hat{B}_{j,-j}$. The relative magnitudes of the latter are related as one would anticipate. That is, at a given level of significance confidence intervals are tighter for DB estimation than for SB, for the higher precision of the former. They are larger for mean WTP estimation than for median WTP, due to the ‘fat tail problem’ of the log-in-the-bid specification.

Table 3a. Tests for reliability of benefit transfer from CV in Northern Ireland Forest Parks

	SB estimates from NI		DB estimates from NI		SB estimates from all		DB estimates from all	
	Mean	Median	Mean	Median	Mean	Median	Mean	Median
Tollymore	***	***	*	**	+	+	+	+
Castlewellan	**	***	***	*	**	***	***	***
Hillsborough	***	***	+	+	**	+	+	+
Belvoir	***	***	+	+	***	*	+	+
Gosford	+	+	+	+	+	***	***	+
Drum Manor	***	***	***	***	***	***	**	***
Gortin glen	***	***	***	+	***	***	***	***
Glenariff	+	***	*	+	+	***	***	+
Ballypatrick	***	**	–	***	***	*	–	–
Somerset	–	*	+	+	–	**	+	+
Florencecourt	–	***	***	–	–	***	***	–
Lough Navar	*	***	*	–	–	***	*	–
Castlearchdale	***	***	***	***	***	***	***	***

Table 3b. Tests for reliability of benefit transfer from CV in the Republic of Ireland Forest Parks

	SB estimates from ROI		DB estimates from ROI		SB estimates from all		DB estimates from all	
	Mean	Median	Mean	Median	Mean	Median	Mean	Median
Lough Key	***	***	***	***	***	+	***	+
Hazelwood	+	+	+	+	*	–	+	+
Dun a Dee	***	–	**	–	***	*	**	–
J. F. Kennedy	+	+	+	+	+	+	+	+
Dun a Ree	–	*	*	***	*	***	***	***
Currachase	*	–	–	–	***	***	**	***
Cratloe	***	–	+	+	*	–	***	***
Douneraile	***	***	***	***	–	–	–	–
Farran	+	+	+	+	***	–	+	+
Guaghan Barra	*	–	+	+	***	***	***	***
Avondale	n.a.	n.a.	n.a.	n.a.	–	*	*	–
Killykeen	***	***	***	***	***	***	***	***
Glendalough	n.a.	n.a.	n.a.	n.a.	*	+	+	+

Table 4a. Statistics of estimated $\hat{B}_{j,-j}$ in pence

	Single bounded			
	$\hat{B}(\mathbf{s}_j, \mathbf{q}_j; \hat{\theta}_{-j})$		$\hat{B}(\hat{\theta}_j)$	
	$\hat{M}(WTP)$	$\hat{E}(WTP)$	$\hat{M}(WTP)$	$\hat{E}(WTP)$
Mean	139	198	134	189
St.deviation	44	63	44	52
Minimum	9	11	55	88
Lower quartile	125	182	104	159
Median	141	205	132	175
Upper quartile	161	232	155	214
Maximum	239	327	250	343
	Double bounded			
	$\hat{B}(\mathbf{s}_j, \mathbf{q}_j; \hat{\theta}_{-j})$		$\hat{B}(\hat{\theta}_j)$	
	$\hat{M}(WTP)$	$\hat{E}(WTP)$	$\hat{M}(WTP)$	$\hat{E}(WTP)$
Mean	133	179	154	209
St.deviation	39	53	36	62
Minimum	10	12	98	123
Lower quartile	122	166	132	162
Median	137	185	145	188
Upper quartile	154	206	179	245
Maximum	214	289	240	408

Table 4b. Statistics of the widths of the simulated confidence intervals around $|\Delta\hat{B}_{j,-j}|$ in pence

	Single bounded					
	$\lvert\Delta\hat{M}(WTP)\rvert$			$\lvert\Delta\hat{E}(WTP)\rvert$		
	$\alpha = 0.1$	$\alpha = 0.05$	$\alpha = 0.01$	$\alpha = 0.1$	$\alpha = 0.05$	$\alpha = 0.01$
Mean	44	53	70	108	149	490
St.deviation	17	21	28	69	116	819
Minimum	23	27	35	44	54	76
Lower quartile	32	39	50	65	82	119
Median	39	48	63	85	106	155
Upper quartile	48	57	78	128	176	354
Maximum	108	128	172	336	538	3,407
	Double bounded					
	$\lvert\Delta\hat{M}(WTP)\rvert$			$\lvert\Delta\hat{E}(WTP)\rvert$		
	$\alpha = 0.1$	$\alpha = 0.05$	$\alpha = 0.01$	$\alpha = 0.1$	$\alpha = 0.05$	$\alpha = 0.01$
Mean	37	45	61	92	119	229
St.deviation	17	21	32	94	133	381
Minimum	17	20	26	23	28	38
Lower quartile	26	31	42	42	50	67
Median	32	38	51	64	80	118
Upper quartile	40	48	65	100	126	173
Maximum	90	110	161	480	676	1,909

α indicates probability of type I error in the test for no difference.

Median *WTP* value transfers perform quite well in Northern Ireland forests from SB estimates obtained from both pooling all sites in the island, and from those in Northern Ireland alone. In the Republic of Ireland, instead, mean *WTP* value transfers seem to be more frequently reliable than median ones.

When DB estimates are used for the median BT the number of transferable estimates markedly decreases to only five forests in the case of estimates from Northern Ireland sites, and down to four for the median transfers from the pooled sample estimates. This is due to the increased efficiency of the DB over the SB estimates which translates into tighter confidence intervals around the point estimates at any given value of probability of Type I error, and hence into a lower transferability.

DB estimations produce gains in efficiency (Hanemann *et al.* 1991) by assuming *a priori* that first and second responses are drawn from the same distribution, as a result the confidence interval around the point welfare estimates is typically tighter than with the SB estimates. The implication for our reliability test is that fewer transfers should pass the test when moving from SB estimates to DB ones. This expectation is in agreement with the results shown in Table 3, especially in the case of the median transfers.

A similar pattern is observed for mean *WTP* value transfers, which are less transferable also due to their higher variability for the fat tail problem.

Estimates from the set of sites of the entire island do not seem to pass the transferability test more frequently than those from the national subsets, 57% versus 56. So, extending the sets of sites from which to draw an estimate of the benefit function does not seem to improve convergence validity of the BT in our case, although it clearly reduces estimation problems due to collinearity of the forest attributes. This is visible in that none of the 4 convergence problems found in the ROI sub-sample were present in the pooled sample.

Finally, the number of over-predictions dominates that of under-predictions in both mean and median *WTP* value transfers, but this might be due to the choice of specification which is log-linear in the bid amount. The choice of specification is partly dictated here by the desire to compare these results with those obtained in similar studies that employed this specification.

In Northern Ireland, the forests for which it is generally possible to transfer both median and mean CV estimates are Castlewellan, Drum Manor, Castlearchdale and – with one median transfer exception – Gortin Glen.

In the Republic of Ireland, only for Killykeen, although Dun a Ree is also always transferable, save in the case of mean transfer for the SB model estimated on national forests, and Lough Key is also transferable, save in the two median transfers from the pooled model.

On the other extreme, J. F. Kennedy forest is the only one for which there is never a reliable transfer, although benefit estimates for Hazelwood forest are transferable only in one case, while in Northern Ireland no forest produces estimates that are never transferable in at least some form.

Altogether these results are encouraging. Systematically addressing various categories of measurement error is obviously important. When the effect of important determinants of *WTP* for forest recreation are accounted for, such as site attributes and socio-economic characteristics, and an empirically correct convergence validity test is employed, CV estimates appear to be frequently transferable across sites, and therefore of reliable practical use. Transfers from SB estimates and from a larger set of forests tend to perform better than those from DB estimates and from a reduced set of forests. Conclusions with regards to mean/median *WTP* are more difficult to draw, although this particular set of results shows that mean *WTP* is more frequently transferable.

4.4. Comparison with the Downing and Ozuna Benefit Function Transferability Test

How does our data perform under previous benefit function reliability tests? We investigated this issue by carrying out a test for benefit *function* transfer asymptotically equivalent to the one used in Downing and Ozuna. In our Northern Ireland samples, a log-likelihood test conducted to check for the reliability of the unconditional benefit function – in a similar (they used *t*-ratios[10]) fashion to the one conducted by them (i.e. to test for the significance of site-specific slope and constant dummy variables in the pair-wise pooled samples) – *always* rejected the null of both dummies being zero. This exercise requires the estimation of $(K^2 - K)$ models, which for our 13 Northern Ireland sites gives 156 estimates. The result, as mentioned earlier, is fully consistent with the fact that forest-specific attributes play a role in determining *WTP* for the recreational experience. When not explicitly accounted for, such as in a mis-specified model, the differences across sites are captured by the slope and constant site-specific dummies, which therefore show significance. Notice that we argue that *this is not evidence of non-transferability of values per se*, but the consequence of having chosen an excessively parsimonious[11] benefit function specification, perhaps as a consequence of data inadequacy. The issue here is whether it makes sense to assess transferability of the benefit function on the basis of such a parsimonious and clearly mis-specified model, rather than focussing on the transferability of the *value* estimates, as we do here. While the link 'Transferable Benefit Function' → 'Transferable Welfare Estimates' may hold if the specification of the benefit function is correct, expecting to find transferability of the benefit function when its specification is evidently too parsimonious to be correct is an exercise bound to provide little insight.

4.5. Determinants of Transfer Errors

In Table 5 we report the OLS regression results explaining the effect of estimation factors on the observed magnitude of the absolute value transfer errors (mean = 50.84, st.dev. = 47.34) from both median and mean *WTP*.

Table 5. OLS regression of absolute transfer errors

Variable	Estimate	P-value	Sample average
Intercept***	85.96 (3 .45)	0.001	1.00
Total area	0.44 (0.64)	0.523	5.19
Congestion**	0.63 (1.98)	0.049	6.28
Natural Reserve	−4.07(−0.36)	0.722	0.24
Trees before 1940***	−1.87(−2.77)	0.006	4.05
Conifers	−0.05(−0.21)	0.833	47.48
Broadleaves*	0.40 (1.85)	0.067	24.14
Larch***	1.25 (3.41)	0.001	9.38
Income class	−7.09(−1.494)	0.137	4.94
Median *WTP****	−18.41(−3.05)	0.003	dummy
From all sites**	−13.65(−2.25)	0.026	dummy
Double bounded	9.47 (1.57)	0.118	dummy
Sample size at site *j*	−0.01(−0.58)	0.560	458.08

N = 200, Adjusted $- R^2 = 0.187$. Dimensions as in table 2.

The regression shows that congestion has a significant positive effect on the error. This is hardly surprising as this variable has a very large variability in the set of sites employed in this investigation, with a standard deviation to sample average ratio of 1.78.

The presence of old and plausibly large trees significantly diminish the effect of the value transfer error, while the presence of broadleaves and larch both significantly increase the magnitude of the value transfer errors. Again the last two show a significantly higher variance in the sample of sites, with a standard deviation to sample average ratios of 1.24 and 1.31, compared to a value of only 0.48 for conifer coverage.

The dummy variable for median *WTP* shows a significant negative effect confirming the higher stability of this measure of location with respect to mean *WTP*. Finally, the dummy variable for value transfer from estimates from all sites show lower error than those from the national sub-sets.

All other variables are not significant, included the dummy variable for value transfer from double-bounded estimates.

5. CONCLUSIONS

Forest attributes important for recreation are plausible determinants of forest recreation benefits from both intuitive and theoretical standpoints.

Starting from this observation we systematically investigate the effect of estimating a benefit function conditional on selected forest attributes, from a large scale discrete-choice CV study, and use these estimates for the purpose of benefit value transfer. The assessment of the value transfer reliability is made on the basis of a comparison with a forest-specific estimate obtained from on-site responses, which are the conceptually superior measure of the same value construct, yet represent a more costly alternative in applied policy analysis.

The null hypothesis of no difference between the on-site and transfer benefit estimates is formally tested by means of a test based on simulated distributions.

The study is conducted so as to allow the investigation of various effects, such as one extra bound in the discrete CV data, the number of forest sites from which the benefit function estimate is obtained, the use of the two most frequently used measures of welfare (mean and median *WTP*), and a comparison with a previous often cited study (namely Downing and Ozuna, 1996).

We find that forest attributes show significant and plausibly signed coefficients and value transfers based on a benefit function conditional on these attributes are often reliable. This finding produces evidence somewhat contrary to previous findings from similar research and to speculations of inadequacy of CV data to be used in this context. The data requirement, however, is quite high as a critical sample mass is required across numerous sites.

A number of issues remain to be explored in this area of research. For example, it remains unclear how powerful the statistical tests for ‘no difference’ are in these contexts. Furthermore, it may appear counter-intuitive that the more accurate an estimate is (e.g. a DB estimate *viz-a-viz* a SB one) the less transferable it is by the approach employed here. Taken to the extreme this may incorrectly be interpreted as less-accurate estimates being more desirable because more transferable. Of course, this line of reasoning is wrong because what determines the accuracy benchmark should be the accuracy level required for the on-site study the transfer estimates wishes to surrogate. In as much as the benefit value transfer for the policy site has similar accuracy, as it does here, it should be deemed an appropriate surrogate.

Finally, what appears to be important for the assessment of the benefit transfer practice is that provision be made for it to be included in the systematic data collection and data analysis that natural resource agencies conduct. Only by making provision for adequate data availability can one hope to have the necessary information to inform decision making in new policy contexts.

Riccardo Scarpa is Professor, Department of Economics, University of Waikato, Hamilton, New Zealand. W. George Hutchinson is Senior lecturer, Department of Agricultural and Food Economics, Queen's University – Belfast, U.K. Sue M. Chilton is Lecturer, Business School-Economics, University of Newcastle Upon Tyne, U.K. Joseph Buongiorno is the John N. McGovern Professor, Department of Forest Ecology and Management, University of Wisconsin-Madison, U.S.A. *Much of the results presented in this paper were previously circulated in the year 2000 as working paper n. 34 by Fondazione ENI Enrico Mattei. We wish to acknowledge useful comments on previous drafts from the editors of this volume, Richard Bishop and Erick Nordheim, and the graduate students taking the class in Advanced Forest Economics at the Department of Forest Ecology and Management at the University of Wisconsin-Madison in Fall 1999. The usual disclaimer applies.*

6. NOTES

[1] In the remaining part of the text the terms *site* and *forest* are used as synonymous.

[2] For example, conducting an on-site DC-CV exercise is currently costed by market research firms at £25–30 per completed survey. Adequate precision requires a sample size in excess of 350–400, hence survey administration alone would cost £8,750–12,000, plus the cost of the data analysis.

[3] The role of site attributes was also emphasized by Opaluch and Mazzotta (1992) and is ignored in other studies employing discrete choice CVM.

[4] It is worth pointing out that a different set of assumptions might affect the benefit transfer reliability tests conducted here. Investigating the extent to which these results are sensitive to alternative estimation assumptions is beyond the scope of this study, but certainly a worthwhile subject for further research.

[5] For 4 of these models convergence was problematic, presumably due to high collinearity, hence the total number of transfer estimates was 200, 100 for mean *WTP* and 100 for median.

[6] Details of each model estimates and a specification estimated on the whole set of 26 forests are available from the authors.

[7] One forest park from Northern Ireland was dropped from the BT study because of its anomalous pattern of recreation, due to the fact that it is a city forest park.

[8] The empirical observation that forest recreation is an inferior good may also be due to the fact that as income increases the choice set for recreation also increases, but leisure time tends to decrease. Therefore

shorter time is destined to any single activity, while the number of leisure activities increases. A more in-depth discussion of the derivation and implication of the model are reported in Scarpa *et al.* 2000.

[9] The procedure broke down computationally in 4 cases. They concerned the estimation of the double and single bounded benefit function from the ROI leave-one-out set of forests when data from Avondale and Glendalough forests in Carlow district were dropped, possibly because of the high collinearity present in they respective leave-one-out subsets.

[10] Personal communication.

[11] Constant slope specifications are in fact the *most* parsimonious in this context.

7. REFERENCES

Adamowicz, W, Swait, J, Boxall, P Louviere, J, M Williams (1997) Perception versus objective measures of environmental quality in combined revealed and stated preference models of environmental valuation *Journal of Environmental Economics and Management,* 32:65–84

Alberini, A (1995) Testing Willingness-to-Pay Models of Discrete Choice Contingent Valuation Survey Data, *Land Economics*, 71:83–95

Bergstrom, JC, De Civita P (1999) Satus of benefits transfer in the United States and Canada: a review, *Canadian Journal of Agricultural Economics*, 47:79–87

Bergstrom, JC, Boyle KJ, Poe GL (eds) (2000), *The Economic Valuation of Water Quality* Edward Elgar Publishing: Cheltenham, UK and Northhampton, MA

Bishop RC, Champ P, Mullarkey D (1995) Contingent Valuation, in *Handbook of Environmental Economics*, Bromley DW (ed.), Basil Blackwell Publisher, London

Cameron TA, James MD (1987) Efficient Estimation Methods for “closed-ended” Contingent Valuation Surveys, *Review of Economics and Statistics*, May, LXIX(2):269–276

Downing M, Ozuna T (1996) Testing the reliability of the benefit transfer approach, *Journal of Environmental Economics and Management*, 30:316–322

Feather P, Hellerstein D (1997) Calibrating Benefit function transfer to assess the conservation reserve program, *American Journal of Agricultural Economics*, 79:151–162

Hanemann MW, Loomis J, Kanninen B (1991) Statistical Efficiency of Double Bounded Dichotomous Choice Contingent Valuation, *American Journal of Agricultural Economics*, November, 73(4):1255–1263

Hanemann WM (1984), Welfare Evaluations in Contingent Valuations Experiments with Discrete Responses, *American Journal of Agricultural Economics*, 66:332–341

Hanemann WM, 1989, Welfare Evaluations in Contingent Valuations Experiments with Discrete Response data: a reply *American Journal of Agricultural Economics*, 71:1057–1061

Kirchhoff S, Colby BG, LaFrance JT (1997) Evaluating the performance of benefit transfer: An empirical inquiry, *Journal of Environmental Economics and Management,* 33(1):75–93

Krinsky I, Robb A (1986) Approximating the Statistical Properties of Elasticities, *Review of Economics and Statistics*, 68:715–719

Kristofersson D, Navrud S (2002) Validity Tests of Benefit Transfer: Are we performing the wrong tests? Paper presentated at the 2nd World Congress for Environmental and Resource Economists, June 24–27 2002 Monterey, California, USA

Loomis J, Roach B, Ward F, Ready R (1995) Testing transferability of recreation demand models across regions - A study of corps of engineer reservoirs, *Water Resources Research*, 31(3): 721–730

Morrison M, Bennett J, Blamey R, Louviere J (2002) Choice Modeling and Tests of Benefit Transfer, *American Journal of Agricultural Economics*, 84(1):161–170

Navrud S, Pruckner GJ (1997) Environmental valuation – To use or not to use? A comparative study of the United States and Europe, *Environmental and Resource Economics,* 10:1–26

Ni Dhubhain A, Gardiner J, Davis J, Hutchinson G, Chilton S, Thomson K, Psaltopoulos D, Anderson C (1994) Final Report EU CAMAR Contract No 8001-CT90-0008 *The Socio-Economic Impact of Afforestation on Rural Development*, by University College of Dublin, The Queens University of Belfast and University of Aberdeen

Opaluch JJ, Mazzotta MJ (1992) *Fundamental issues in benefit transfer and natural resource damage assessment, in Benefit transfer: procedures, problems and research needs* Workshop proceedings, AERE, Snowbird, UT, June

Poe GL, Severance-Lossin EK, Welsh MP (1994) Measuring the difference $(X - Y)$ of simulated distributions: a convolutions approach, *American Journal of Agricultural Economics*, 76: 904–915

Poe GL, Welsh MP, Champ PA (1995) Measuring the difference in mean WTP when dichotomous choice contingent valuation responses are not independent, *Land Economics*, 73:255–267

Santos JML (1998) *The economic valuation of landscape change,* Cheltenham, UK, Edward Elgar

Scarpa R, Hutchinson G, Chilton S, Buongiorno, J (2000) Reliability of Benefit Value Transfers from Contingent Valuation Data with Forest-Specific Attributes, FEEM Working Paper n 3400

Scarpa R, Hutchinson WG, Chilton SM, Buongiorno J (2000) Importance of Forest Attributes in the Willingness To Pay for Recreation: A Contingent Valuation Study of Irish Forests, *Forest Policy and Economics*, 1(3–4):315–329

Smith VK, Van Houtven G, Pattanayak SK (2002) Benefit Transfer via Preference Calibration: "Prudential Algebra" for Policy, *Land Economics*, 78(1):132–152

D. KRISTÓFERSSON AND S. NAVRUD

CAN USE AND NON-USE VALUES BE TRANSFERRED ACROSS COUNTRIES?

1. INTRODUCTION

An alternative to conducting a new environmental valuation study is to use existing values as an approximation. This method has been termed benefit transfer (or more general; value transfer) because estimates are transferred from a site where a valuation study has been conducted to a site of policy interest. Benefit transfer has become popular in practice due to the high costs and time associated with conducting original studies, regardless of the numerous difficulties associated with obtaining valid estimates. Numerous studies have tested the validity and magnitude of errors in benefit transfer for example Loomis (1992), Loomis et al. (1995), Bergland et al. (2002), Downing and Ozuna (1996), Kirchhoff et al. (1997), Brouwer and Spaninks (1999) and Ready et al. (2004) to name a few. It is easily argued that benefit transfer can only produce valid estimates in the few cases where the environmental good and the population are virtually identical. Otherwise there is no reason to believe that the willingness to pay (WTP) is the same for non-identical goods and populations. This raises a question about the way in which the validity is tested. Usually, a null hypothesis of no difference between the original and the transferred estimate is tested. Valid benefit transfer is reported in the cases where the null hypothesis has not been rejected at the chosen level of significance, most often $\alpha = 0.05$. However, non-rejection of a null hypothesis is not proof of its truth as a rejection is proof of its untruth (Lehman 1983, Hoenig and Heisey 2001). It is, therefore, not possible to prove equality when such a null hypothesis is not rejected. When we are interested in the validity of the null hypothesis itself, as is the case for studies of benefit transfer validity, it is appropriate to test for equivalence and not for difference. Such equivalence tests have been developed and used for quite some time in pharmaceutical research (Hauck and Anderson 1984, Schuirmann 1987, Welling et al. 1992 and Berger and Hsu 1996) and in psychological research, for example Stegner et al. (1996), but have no widespread use in economics.

We will here test the equivalence of benefit transfer values and original values for use and non-use values of freshwater fish populations between three Nordic countries; Norway, Sweden and Iceland. Identical contingent valuation surveys were conducted at the same time in these three countries. Observed differences in willingness-to-pay (WTP) should therefore be due to such factors as demographic differences between the populations, non-quantifiable differences in underlying preferences and the institutional organization of recreational fishing in different countries.

S. Navrud and R. Ready (eds.), Environmental Value Transfer: Issues and Methods, 207–225.

To our knowledge, this is both the first application of equivalency analysis to environmental value transfers, and the first study to compare the transferability of non-use values versus use values.

2. BENEFIT TRANSFER

Assume that all respondents in all countries have identical underlying preferences. Using country *m* as the baseline country, WTP for individual *i* living in country *m* for an improvement in environmental quality from Q_0 to Q_1 is defined using indirect utility functions by:

$$(1) \qquad V(p_m, I_i, Q_0) = V(p_m, I_i - WTP, Q_1)$$

where $\boldsymbol{p}_m$ is a vector of prices for goods and services in country *m* and I_i is the individuals income or wealth. Let us suppose that individual *j* has the same preferences as individual *i* but lives in country *n*. He faces prices $\boldsymbol{p}_n = \boldsymbol{p}_m$. Because the indirect utility function is homogeneous of degree 0 in prices and income, known as the absence of money illusion, it will not influence the result. The WTP of individual *k* will be $WTP_i = WTP_k$.

Benefit transfer methods can be divided into two major types: i) unit value transfer and ii) value function transfer.

Unit value transfer methods estimate total benefits at the policy site by aggregating exiting standard values per unit. These values are derived from study site data. For example, the total benefits of fishing at the policy site may be estimated as the product of some standard value for a fishing day at the study site and the number of fishing days at the policy site. The obvious problem with this method is that individuals at the policy site may not value the good in question in the same way as individuals at the study site. There are two principal reasons for this. First, the characteristics of the population may differ in terms of income, education, religion, demographic composition and so forth. Second, even if the individual preferences are the same, the supply of the good in question may differ (Kirchhoff et al. 1997).

A more sophisticated approach would be to adjust the value before transferring it to the policy location. There are two different types of adjustments that can be made. First, the analyst may regard the unit value available from the study site to be biased, or estimated inaccurately. This might be based on an evaluation of the methodology used in the original study. Second, the value may have to be adjusted to better reflect the conditions at the policy site. Four potential differences should be addressed in this kind of adjustments:

- The quality/quantity of the environmental good affected
- What caused the environmental change
- The socioeconomic characteristics of the households affected
- The availability of substitutes

In *value function transfer* methods, estimator models derived from study site data are used with explanatory variables collected at the policy site to estimate both value per unit and total units. For example, an estimated recreational demand function from the study site, may be used on data from the policy site to estimate both the value of a fishing day and total value of fishing (Brookshire and Neill 1992). Value function transfer is viewed as the best approach to benefit transfer as it relies on a better theoretical basis than unit transfers. However, the benefit estimates derived from contingent valuation studies are often a complex function of the site and user characteristics and the spatial and temporal setting. Not to take these into account is to make very strong assumptions about preferences, that is: preferences are universally stable over populations and time (Loomis 1992).

While rigorous guidelines exist on how to carry out original valuation studies (Arrow et al 1993), no such protocol exists for benefit transfer as of yet. However, as the number of studies on the validity of benefit transfer increases rules of practice emerge, see for example Desvousges et al. (1992), Bergland et al. (2002), Brouwer (2000) and Ready et al. (2004).

3. THE SURVEY

Toivonen et al. (2000) performed identical Contingent Valuation (CV) surveys in all Nordic countries during the period September–December 1999 to estimate the total economic value of the Nordic freshwater fish stocks. We have used the data from Norway and Sweden to test benefit transfer between two countries under close to ideal conditions. These countries are very similar with regards to geographical, ecological, cultural, and institutional context for this environmental good. We have then added Iceland as the "odd ball out" among the Nordic countries, with a larger degree of privatization of recreational fishing and having less threatened freshwater fish stocks.

Random samples of the national population (between age 18 and 69) of 2,500, 5,000 and 7,500 persons were selected for the mail surveys in Iceland, Norway and Sweden, respectively. The response rates were 34.2, 44.6 and 46.7 %, respectively.

The questionnaire had four main parts. In the first part, general attitudes towards wildlife and fishing were assessed. The second part, which was for anglers only, was aimed at identifying their angling activity, preferences for different types of recreational fishing, and annual expenses for recreational fishing. The third part contained four CV scenarios; three use-value scenarios aimed at anglers and one non-use scenario for all respondents. Only two of the use-value scenarios are relevant for all three countries analyzed here. Table 1 provides an overview of the three CV scenarios used in the comparative study reported here. The last part contained questions about socio-economic data, such as, sex, age, education and income.

A multiple bounded approach based on a payment card was used to elicit respondents' WTP. The payment card consisted of ten offered amounts and five levels of certainty for each offered amount. The amounts were the same in all countries, and converted to national currencies using Purchase Power Parity (PPP)-adjusted exchange rates. However, there is one exception. The highest amounts were higher

Table 1. Characteristics of the scenarios used in the contingent valuation study

	Scenario 1: Use Value Local River fishing	Scenario 2: Use Value Local Lake fishing	Scenario 3: Non-use Value Nordic fish stocks
Description	A stream is opened for fishing after being closed for many years. It is near your home. It has high water quality. It has a restricted number of anglers	A lake is opened for fishing after being closed for many years. It is near your home. It has high water quality. It has a restricted number of anglers	The natural fish stocks of the Nordic countries are threatened in several ways: Low water quality Regulation of water flow Eutrophication Acid rain Parasites Disease
Fish stock	Salmon and trout. Above average chance of catch	Grayling, brown trout and artic char. Above average chance of catch	All
Methods allowed	Rod and line	Rod and line	–
You buy	12 months access	12 months access	Preservation of natural fish stocks
Payment vehicle	Annual rent per person paid into a local fund, where the anglers get representation at the board of the fund.	Annual rent per person paid into a local fund, where the anglers get representation at the board of the fund	Increase in annual, personal income tax

in Iceland to account for the generally higher price level of fishing licenses. This was done to avoid a "thick tail" of the distribution of WTP, which can influence the estimated mean WTP. The respondents were asked to assign a certainty level to each amount. The certainty levels were "I would certainly pay", "I would almost certainly pay", "I am unsure", "I would almost certainly not pay" or "I would certainly not pay" the offered value. The problem with this approach is choosing the level of certainty that corresponds to the respondents "true" WTP. Welsh and Poe (1998) and Notaro and Signorello (1999) compared multiple bounded question format with open-ended and single bounded formats. Their levels of certainty correspond to those used in the study. Their results show that the first level of certainty produces lower mean WTP then an open-end format. In both cases the single bounded model produces mean WTP higher than the "not sure" level of certainty, in the multiple bounded format. The results of Ready et al. (2001) indicate that this difference is due to "yea-saying" resulting from uncertainty. They conclude that applying the level of "almost certain" to closed-ended questions reduces the estimated WTP to the same level as for the payment card response. Thus, in our study, we use the second level of certainty ("I would almost certainly pay") in the estimation of values, as the best approximation of "true" WTP.

The data is well suited for studies on the validity of benefit transfer as the surveys are identical, conducted at the same time with the same scenarios, using the same sampling procedure and in similar populations. The study was conducted in accordance with existing guidelines and the dataset is adequate in size. This should result in data that is free of bias resulting from temporal differences or differences in the offered good. The size of the data set ensures that parameter estimates are stable and statistical tests can be used.

However, the data also have some weaknesses. Although the scenarios are identical in all countries, they need not represent the same *relative* changes in environmental quality. That will ultimately depend on the initial level in each country. This may lead to differences in the relative change in environmental quality perceived by respondents in different countries. Sample selection is another potential problem. With the relatively low response rates in these mail surveys of the general public, there could be an over-representation of respondents with an interest in recreational fishing. However, if this takes place in *all* countries, it should not have any impact on the benefit transfer *validity tests* between countries.

4. MODEL AND TEST PROCEDURES

4.1. Statistical Model

Ready et al. (2004) stress the importance of correcting WTP for differences in purchase power when attempting benefit transfer between countries. Purchasing power parity (PPP) is a theory that states that exchange rates between currencies are in equilibrium when their purchasing power is the same in each of the two countries. This means that the exchange rate between two countries should equal the ratio of the two countries' price level of a fixed basket of goods and services. When a country's domestic price level is increasing (i.e., a country experiences inflation), that country's exchange rate must depreciate in order to return to PPP (Burda and Wyplosz 1997).

All monetary values had to be adjusted for different exchange rates as well as purchase power. It was decided to use Norwegian kroner (NOK) as the monetary unit of the study. The exchange rate used for Icelandic and Swedish kroner was the mean exchange rate of the Norwegian central bank for the duration of the survey (September – December 1999). Comparative price levels based on purchase power parity (PPP) were obtained from the OECD. The estimates that were used are shown in Table 2.

Multiple-bounded responses do not provide point estimates, but identify the intervals within which the "true" WTP lies. An upper and lower bound on maximum WTP is obtained from the payment card responses. Respondents with zero WTP can be identified using an open-end question following the payment card. These individuals have to be accounted for when estimating the expected WTP. Some respondents did not state their WTP, and were deleted from the data set.

Reiser and Shechter (1999) and Kriström (1997) showed the importance of not excluding true zero bids from the statistical analysis. The solution they suggest is to estimate a spike model. The first part of such a model identifies those with positive and those with zero WTP. The second part estimates a value function for the first

Table 2. Mean exchange rate and purchase power indices used for conversion into Norwegian kroner (NOK); 1 euro = 8.70 NOK (exchange rate January 2004)

	Mean exchange rate	PPP based comparative price levels	Overall adjustment
	To NOK	Norway=1,00	To NOK
Iceland	0,10962	1,01	0,11072
Sweden	0,94833	1,10	1,04317

Source: OECD, Norwegian Central Bank.

group and assigns zeros to the second. By the method of Reiser and Shechter (1999) the likelihood function breaks up into two parts, in correspondence to the two parts of the model. The log of the first part is:

$$\ln L = \sum_{i=1}^{n} [(1 - S_i)\, p + S_i\,(1 - p)] \tag{2}$$

where S_i is a dummy variable taking the value 1 if WTP<0, p is the percentage of zero bids and n is the number of observations.

For the second part we need to define the intervals within which each bid falls. The payment cards included ten offered values. Eleven intervals can be constructed, nine where upper and lower values are known, and two where only the upper or lower value is known. A population consisting of t observations can be indexed by the set T. T can then be portioned into eleven disjoint subsets depending on the interval into which the true WTP falls.

Observations belonging to subset T_1 are left censored, observations belonging to $T_2 - T_{10}$ are interval censored and observations belonging to T_{11} are right censored. The log-likelihood function for the second part can be written as:

$$\ln L = \sum_{i \in T_1} \ln \mathrm{F}(h_{1i}) + \sum_{k=2}^{10} \sum_{i \in T_k} \ln\left(\mathrm{F}(h_{ki}) - \mathrm{F}(l_{ki})\right) + \sum_{i \in T_{11}} \ln\left(1 - \mathrm{F}(l_{11i})\right) \tag{3}$$

where $F(.)$ is the cumulative density function of the distribution of WTP across the population and l, h are lower and upper levels of the intervals. This model can be estimated by some statistical packages, for example the PROC LIFEREG procedure in the SAS® statistical system. A Weibull probability density function was used (Greene 2000, Bergland et al. 2002, Allison 1995, Lindsey and Ryan 1998).

The maximum likelihood estimator for probability of zero WTP, p, is according to Reiser and Shechter (1999):

$$\hat{p} = \frac{n - \sum_{i=1}^{n} S_i}{n} \tag{4}$$

Combining the results from the estimation of zero bidders and those with positive WTP we can estimate the mean WTP. Let $f(x)$ be the probability density function associated with the distribution of the WTP. The mean WTP is given by:

$$[WTP] = (1 - \hat{p}) \int_{0}^{\infty} x f(x) dx \tag{5}$$

The integration was done using numerical integration and ∞ approximated with a large number.

This type of spike model uses more of the available information, but has the advantage of being completely combinable with existing estimation procedures (Reiser and Shechter 1999).

A comparison was made of three scenarios in all the countries, two use-values, and one nonuse-value scenario. The criteria used in constructing the models for each scenario was that the included parameters for the explanatory variables were significantly different from zero (10% level of significance was loosely applied), that they were not seriously correlated with each other (with $r = 0{,}3$ as upper limit) and that the model was applicable as a pooled model for all countries. Some variables were included for the sake of completeness, for example if one level of a classification variable was included, then all the other levels were automatically included as well.

The expected value of WTP is calculated by numerical integration. This value is then adjusted for missing values, as previously described.

Goodness of fit is calculated in a likelihood ratio index;

$$I_{LH} = -2(\ln L_0 - \ln L_1) \tag{6}$$

where $\ln L_0$ is the log likelihood for a model that only includes a constant and $\ln L_1$ is the log likelihood for a model with covariates. The index indicates how much the model improved with added information. The index can be tested for significance as it has a χ^2-distribution with degrees of freedom equal the number of covariates.

The calculated WTP function is highly nonlinear in the estimated parameters. This makes estimation of standard errors difficult. Therefore, bootstrap distributions of mean WTP are obtained from 1000 bootstrap iterations that resemble the original dataset with replacement (see e.g. Bergland et al. 1993). Each iteration includes a full model is estimation. These results are then used to estimate mean WTP for an original model and a value function transfer model. This is a preferable method when model residuals are not well defined. Cooper (1994) conducted a simulation study of the validity of four different approaches for calculating confidence intervals for closed-ended WTP models. He concluded that this method showed best overall performance when the true underlying distribution was Weibull.

Transfer error is used as a measure of the accuracy of the benefit transfer estimates the definition for transfer error used here is

$$\text{transfer error} = \frac{|WTP_E - WTP_T|}{WTP_T} \tag{7}$$

where the subscript T stands for true value as estimated by an original study and the subscript E for estimated value as given by benefit transfer.

4.2. Equivalence Testing and Acceptable Transfer Errors

In equivalence testing one reverses the roles of the null hypothesis (H_0) and the alternative hypothesis (H_A). Equivalence is demonstrated by testing a set of these reversed hypotheses with a predetermined significance level. It is not sufficient to fail to show a difference one must be fairly certain that a large difference does not exist. Suppose that we are willing to conclude that a difference is negligible if its absolute value is no greater than a small positive value Δ. In contrast to the traditional casting of the null hypothesis, the null hypothesis becomes[1]

$$H_0 : D \leq \Delta \text{ or } D \geq -\Delta$$

$$H_A : \Delta < D < -\Delta$$

where D is the absolute transfer error. The structure of the statistical hypothesis is determined by the objective of the analysis. The null hypothesis states that D is not within the interval that has been determined as equivalent. The alternative hypothesis states that D is within the interval, which implies that the two parameters are equivalent. If we can reject the null hypothesis on the basis of a study result, then we conclude that H_A is true, i.e. the two are equivalent. If the null hypothesis is not rejected we do not conclude that H_0 is true. Rather, we say that H_A has not been shown to be true. This procedure is exactly the same as the classical methodology of testing, only with reversed null and alternative hypothesis.

Hauck and Anderson (1984) and Schuirmann (1987) showed that if a $1–2\alpha$ confidence interval lies entirely between $-\Delta$ and Δ, we could reject the null hypothesis of non-equivalence in favour of equivalence at the α level. The equivalence test is at the α level because it involves two one tailed tests, which together describe the $1–2\alpha$ level confidence interval. A simple version of this kind of test is the two one-sided test or TOST. In its simplest form it involves conducting two one-sided t-tests at the α level of significance.

$$t_1 = \frac{D - \Delta}{s_p\sqrt{1/n_1 + 1/n_2}} \geq t_{1-}$$

and

$$t_2 = \frac{\Delta - D}{s_p\sqrt{1/n_1 + 1/n_2}} \geq t_{1-} \tag{8}$$

where t_{1-} is the t-value associated with the chosen significance level and degrees of freedom, s_p is an estimate of pooled standard deviation and n_1 and n_2 are the number of observations in the two samples used to achieve the estimates that are being compared. Equivalence tests have some appealing properties over traditional non-rejections. It becomes increasingly difficult to reject the null hypothesis with increasing variance. A well-preformed CV study is therefore more likely to show equivalence given that it is the true state of nature while the reverse is true for non-rejections of classical null hypothesis.

The TOST is only one of the ways in which equivalence can be tested. Other, more powerful parametric test exist as well as non-parametric tests, see for example Berger and Hsu (1996). The simplicity and widespread application of the TOST in, for example pharmaceutical research, and its basis in the well-known t-test makes it a good choice for the purpose of our study.

In order to assess the equivalence of two test groups we must first define what would be considered equivalent. In the pharmaceutical industry the agreed upon standard is that the population mean tested must be within 20% of the mean of the reference group (Δ=0, 2μ_{ref}). Such a standard must be set for each application. This standard must be based on what is considered theoretically relevant. This should not impose a problem since the theoretical interpretation of statistical results is contingent upon that such a definition exists. Kristofersson and Navrud (2005) suggest that in the case of benefit transfer, it should be left to the users of the estimates to determine the acceptable level. Thus, the results of the equivalence test could be presented in a table where one shows the highest transfer error level that can be applied in order to show equality. The rationale is that the acceptable level varies dependent on the policy use of the benefit transfer estimate. Generally speaking; a lower level of accuracy is needed for benefit estimates in cost-benefit analyses (CBAs) of projects and policies than for environmental costing and green accounting. However, even for CBAs the level of accuracy needed would be high if benefits and costs are very close. The highest level of accuracy is needed in Natural Resource Damage Assessments (NRDA), which are used to derive the compensation the polluter should pay to compensate damages to the general public from the pollution incident. (see also Navrud and Pruckner 1997). The acceptable transfer error should depend on the costs of making the wrong decision if the decision is based on benefit transfer instead of a new valuation study. We suggest testing at two levels, 20% transfer error and 40% transfer error. The 20% level identifies the cases where benefit transfer could produce estimates that could be used in a similar way as original estimates. The second level of 40% identifies those cases where the benefit transfer estimate gives approximation that could be used in applications where the need for cost effectiveness outweighs the demand for accuracy, as is the case for some CBAs.

4.3. T-tests and Hypotheses

Hypothesis of benefit transfer will be tested using both the classical procedure of non-rejection of a null hypothesis and the two one-sided equivalence test. The pooled estimator for standard deviation used is:

$$s_p^2 = \frac{(n_a - 1)\,s_a^2 + (n_b - 1)\,s_b^2}{n_a + n_b - 2} \tag{9}$$

The t-statistic used for comparison was the pooled version:

$$(10) \qquad t = \frac{|WTP_a - WTP_b|}{s_p\sqrt{1/n_a + 1/n_b}}$$

where *WTP* are willingness to pay estimates from the two sites, s_p is pooled standard deviation and n is sample size.

The following hypotheses were tested:

Hypothesis 1 (Transferability of unit values)

Under the null hypothesis estimated mean values can be transferred between sites and countries, and the WTP estimates from the study and policy sites are not significantly different and/or directly equivalent at $\Delta=\theta$ where $\theta \in \{0.2\text{WTP}_\text{P}, 0.4\text{WTP}_\text{P}\}$ and WTP_P is the estimate from the policy site.

A more sophisticated approach is to use the value function from the study site with sample information from the policy site to predict the mean willingness to pay at the policy site:

Hypothesis 2 (Transferability of value functions).

Under the null hypothesis of transferability, the predicted WTP at the policy site using the parameters from the study site is not significantly different and/or directly equivalent to the WTP at the policy site for $\Delta=\theta$.

5. RESULTS

Table 3 presents the descriptive statistics for each country, and for anglers, non-anglers and the combined sample, respectively. Table 3 shows that the descriptive statistics of the samples are quite similar across countries. The only large differences found are fishing expenses and lower and upper bounds of WTP between countries, which reflect the much higher value of fishing licenses in Iceland compared to Norway and Sweden.

It is worth noting that the mean cost falls within the higher and lower bounds of stated WTP (over and above costs) for fishing in Iceland and Norway, and for river fishing in Sweden. This shows that WTP is large compared to current fishing expenses.

Results from the model estimation, presented in tables 4, show that income has the expected positive sign in all models. Parameters for both personal and other household income have positive signs. Considerable information is contained in the covariates, as can be seen from the likelihood ratio indices.

Men seem to have significantly higher WTP for both fishing and conservation of fish stocks. This is clearest for the Swedish and Norwegian respondents. Men also dominate the group that says they are anglers (Table 3).

Age has a negative parameter where it is significant, suggesting that WTP decreases with age. Years of education do not seem to have any clear effect except in the non-use value scenario. Here, the respondents with higher education have

Table 3. Descriptive statistics for anglers, non-anglers and the combined sample from the surveys in Iceland, Norway and Sweden. Mean values of all variables, upper and lower bounds of willingness-to-pay (WTP) for the three scenarios (Table 1), and number of observations are reported. (The mean values for a dummy variables are percentages)

	Iceland			Norway			Sweden		
Variables	Non anglers	Anglers	All	Non anglers	Anglers	All	Non anglers	Anglers	All
Number of observations	571	268	839	1010	1148	2158	2166	1280	3446
Sex (male = 1, female = 0)	0,32	0,69	0,44	0,37	0,66	0,52	0,37	0,69	0,49
Age	41,7	40,9	41,4	43,4	42,1	42,8	44,5	42,9	43,9
Personal income (in NOK)	157 982	193 680	169 385	207 943	245 253	227 791	171 535	194 158	179 938
Other household income (in NOK)	196 511	193 904	195 678	156 991	161 467	159 372	168 922	157 062	164 516
Number of persons in household.	3,27	3,45	3,32	2,48	2,82	2,66	2,65	2,81	2,71
11–13 years of education [a]	0,31	0,31	0,31	0,30	0,32	0,31	0,38	0,41	0,39
> 14 years of education	0,50	0,48	0,49	0,39	0,41	0,40	0,34	0,31	0,33
Subsistence fisherman [b]	N.A. [d]	0,04	0,04	N.A.	0,06	0,06	N.A.	0,05	0,05
Generalist	N.A.	0,10	0,10	N.A.	0,15	0,15	N.A.	0,15	0,15
Occasional fisherman	N.A.	0,48	0,48	N.A.	0,53	0,53	N.A.	N.A.	N.A.
Semi-urban residential area [c]	0,27	0,27	0,27	0,30	0,29	0,30	0,27	0,27	0,27
Rural residential area	0,13	0,12	0,13	0,21	0,29	0,25	0,21	0,29	0,24
Fisherman	N.A.	N.A.	0,32	N.A.	N.A.	0,53	N.A.	N.A.	0,37
Total number of fishing days in the last 12 months.	N.A.	7,67	13,78	N.A.	13,78	13,78	N.A.	13,17	13,17
Fishing expenses for the last 12 months	N.A.	3 770	3 770	N.A.	1 307	1 307	N.A.	1 433	1 433

(Continued)

Table 3. *(Continued)*

Variables	Iceland			Norway			Sweden		
	Non anglers	Anglers	All	Non anglers	Anglers	All	Non anglers	Anglers	All
No stated fishing costs	0,94	0,03	0,65	0,89	0,05	0,44	0,97	0,08	0,64
Likes river-fishing best	N.A.	0,56	0,56	N.A.	0,22	0,22	N.A.	0,26	0,26
Likes lake-fishing best	N.A.	0,32	0,32	N.A.	0,26	0,26	N.A.	0,36	0,36
Lower bound of WTP for scenario 1	N.A.	3 325	3 325	N.A.	1 014	1 014	N.A.	816	816
Upper bound of WTP for scenario 1	N.A.	5 697	5 697	N.A.	1 762	1 762	N.A.	1 483	1 483
Lower bound of WTP for scenario 2	N.A.	2 402	2 402	N.A.	957	957	N.A.	796	796
Upper bound of WTP for scenario 2	N.A.	4 444	4 444	N.A.	1 794	1 794	N.A.	1 419	1 419
Upper bound of WTP for scenario 3	1 656	1 808	1 709	684	946	826	631	859	722
Lower bound of WTP for scenario 3	2 765	3 444	2 999	1 139	1 604	1 391	1 084	1 475	1 233

[a] The "hidden" education level is the respondents with less then 10 years of education.
[b] The "hidden" category of fishermen was sports fishermen.
[c] The "hidden " variable is respondents living in urban residential areas.
[d] N.A. indicates that the statistic is not available.

Table 4. Sign and significance of estimated parameters. Three signs indicate significance at the 0,1% level, two signs indicate 5% and one sign indicates significance at 10% level. No sign indicates a non-significant parameter while N.A. indicates that the variable was not included in the estimated model

	River fishing scenario			Lake fishing scenario			Non-use value scenario		
	Isl	Nor	Swe	Isl	Nor	Swe	Isl	Nor	Swe
Intercept	+++	+++	+++	+++	+++	+++	+++	+++	+++
Sex, man = 1		+++	++		++	+++		+++	+
Age	N.A.[c]	N.A.	N.A.	–		–		–	–
Fishing expenses last 12 months (in NOK)	+++	+++	+++	+++	+++	+++		+++	+++
No stated fishing costs			++		–			++	+++
Personal income (in NOK)	+	++		+			+++	+	++
Other income (in NOK)	+	++			++	+	+++	+++	+
11–13 years of education	N.A.	N.A.	N.A.	++	++				++
> 14 years of education	N.A.	N.A.	N.A.	+	+		++	+++	+++
Likes river-fishing best	++	+++		N.A.	N.A.	N.A.		++	++
Likes lake-fishing best	N.A.	N.A.	N.A.		+		N.A.	N.A.	N.A.
Fisherman	N.A.	N.A.	N.A.	N.A.	N.A.	N.A.		++	+++
Subsistence fisherman	–	++		N.A.	N.A.	N.A.	N.A.	N.A.	N.A.
Generalist				N.A.	N.A.	N.A.	N.A.	N.A.	N.A.
Occasional angler		–	N.A.	N.A.	N.A.	N.A.	N.A.	N.A.	N.A.
Semi-urban residential area				N.A.	N.A.	N.A.	N.A.	N.A.	N.A.
Rural residential area	–			N.A.	N.A.	N.A.	N.A.	N.A.	N.A.
Log-likelihood of estimated model	−414	−1911	−2251	−416	−1879	−2171	−1145	−3554	−5554
$-2^*(\text{log-likelihood}_1 - \text{log-likelihood}_0)$ [a]	87	215	114	34	98	94	78	165	225
Zero WTP [b]	0,034	0,117	0,077	0,026	0,105	0,080	0,162	0,133	0,125
The number of observation with positive WTP	247	955	1117	237	952	1077	641	1735	2760

[a] −2 times the difference in log likelihood between a model with no covariates and this model. The resulting number is χ^2 distributed with degrees of freedom equal the number of covariates.

[b] An estimate for the perectage of true zero bidders obtained from the effected population. For the treatment of zero bidders see equations 4 and 5 and relevant discussion in text.

[c] N.A. indicates that the variable was not included in the relevant model.

significantly higher WTP in all countries. This could indicate a larger level of environmental awareness among the respondents with higher education.

Residential area produces some significant parameters. The signs indicate that respondents in rural Iceland and Norway have lower WTP than semi-urban and urban respondents.

Fishing expenses clearly influences WTP for both use and non-use value. This shows that anglers have a clear idea of the actual cost of fishing and their bids are strongly influenced by this.

A stated preference for one type of fishing increases WTP for that type of fishing, as expected. This effect is largest for those who prefer river fishing compared to those that prefer lake fishing. This might be caused by the fact that river fishing is more exclusive and often more expensive than lake fishing. The hidden variable is a stated preference for sea fishing.

Mean WTP per person per year with bootstrap standard errors are reported in Table 5. Table 5 shows that there is a strikingly large overall difference in WTP between Iceland and the other two countries. This reflects the large differences in costs associated with recreational fishing as reported in table 3. This seems also to affect the WTP for preserving the Nordic fish stocks (i.e. non-use value scenario). However, the relative difference in WTP between Iceland and the two other countries is smaller for the non-use values than the two use value scenarios. The difference in WTP is consistent between countries. It is always largest for Iceland, then Norway and smallest for Sweden.

The two separate hypothesis tested were equality and non-equivalence. Equality was tested by a classical t-test while non-equivalence was tested by the two one-sided procedure (TOST) at 20% and 40% transfer error. The tests were performed both for unit value transfer (hypothesis 1), Table 6, and value function transfer (hypothesis 2), Table 7.

Table 6 shows the results for all possible transfer combinations. The first column defines the country used for estimation while the rows define the country the value is transferred to. For example a transfer of values from Norway to Iceland for the river fishing scenario (row two, column one to three) would result in a

Table 5. Results of mean WTP per person per year with bootstrap standard errors and test for the normality of the bootstrap distribution

Scenario	Country	Mean WTP	Bootstrap standard error	Shapiro-Wilk statistic (W)[a]	Prob < W[b]
	Iceland	4 251	352	0,980	0,00
River	Norway	1 120	64	0,931	0,00
	Sweden	1 014	61	0,987	0,62
Lake	Iceland	3 119	245	0,984	0,15
	Norway	1 090	49	0,989	0,87
	Sweden	927	48	0,984	0,10
	Iceland	1 772	149	0,983	0,05
Non-use value	Norway	870	42	0,931	0,00
	Sweden	761	31	0,985	0,21

Notes:
[a] A test for normality applicable in small samples. The test is described in section 4.
[b] The probability of making a type I error by rejecting normality.

Table 6. Transfer error (Error) as given by equation 7, t-values from a classical test of equality and smallest level of significant equivalence (TOST) for unit value transfer

River fishing scenario

		To								
		Iceland			Norway			Sweden		
		Error [a]	t-value [b]	TOST [c]	Error	t-value	TOST	Error	t-value	TOST
From	Iceland	–	–	–	280%	16.21***	N.Eq.	319%	17.24***	N.Eq.
	Norway	−74%	16.21***	N.Eq.[d]	–	–	–	10%	1.20	40%***
	Sweden	−76%	17.24***	N.Eq.	−9%	1.20	40%***	–	–	–

Lake fishing scenario

		To								
		Iceland			Norway			Sweden		
		Error	t-value	TOST	Error	t-value	TOST	Error	t-value	TOST
From	Iceland	–	–	–	186%	14.25***	N.Eq.	236%	15.51***	N.Eq.
	Norway	−65%	14.25***	N.Eq.	–	–	–	18%	2.37*	40%**
	Sweden	−70%	15.51***	N.Eq.	−15%	2.37*	40%***	–	–	–

Non-use value scenario

		To								
		Iceland			Norway			Sweden		
		Error	t-value	TOST	Error	t-value	TOST	Error	t-value	TOST
From	Iceland	–	–	–	104%	8.50***	N.Eq.	133%	11.35***	N.Eq.
	Norway	−51%	8.50***	N.Eq.	–	–	–	14%	2.12*	40%***
	Sweden	−57%	11.35***	N.Eq.	−13%	2.12*	40%***	–	–	–

* = 5%, ** = 1% and *** =< 0.1% level of significance.

[a] Transfer error associated with the benefit transfer.

[b] The t-value for the hypothesis that the benefit transfer estimate and the original estimate are equal.

[c] The lowest level of significant equivalence. The levels tested are 10% 20% and 40% of the original study estimate of mean WTP.

[d] Not equivalent at any tested level.

transfer error of −51%. This transfer error is statistically different from zero and the hypothesis of non-equivalence cannot be rejected. The conclusion is therefore that such a transfer is not acceptable. The differences seen in Table 5 are also seen in the result in Table 6. Benefit transfers to and from Iceland result in very large transfer error that is highly significant. Both the t-test and the equivalence test clearly suggest that such a transfer is not acceptable. On the other hand, none of the t-tests of equality are rejected when transferring between Norway and Sweden. This, in addition to the small transfer errors, would have been interpreted as evidence of valid benefit transfer. The equivalence tests reveal a weakness of this argumentation. Non-equivalence is only significantly rejected at a 40% level of transfer error suggesting that the variance of the WTP makes it impossible to accurately predict a small transfer error. The estimates should only be used in cases where approximate figures are required, for example in a CBA where the estimated benefits and costs are far apart.

Table 7. Transfer error (Error), t-values from a classical test of equality, and smallest level of significant equivalence (TOST) for value function transfer

River fishing scenario

		To								
		Iceland			Norway			Sweden		
		Error [a]	t-value [b]	TOST [c]	Error	t-value	TOST	Error	t-value	TOST
From	Iceland	–	–	–	147%	9.09***	N.Eq.	210%	7.22***	N.Eq.
	Norway	−35%	3.28**	N.Eq.[d]	–	–	–	33%	3.53***	N.Eq.
	Sweden	-64%	9.4***	N.Eq.	-22%	3.27**	40%**	–	–	–

Lake fishing scenario

		To								
		Iceland			Norway			Sweden		
		Error	t-value	TOST	Error	t-value	TOST	Error	t-value	TOST
From	Iceland	–	–	–	123%	9.61***	N.Eq.	165%	12.98***	N.Eq.
	Norway	−33%	2.34*	N.Eq.	–	–	–	34%	4.10***	N.Eq.
	Sweden	−61%	15.22***	N.Eq.	−16%	2.37*	40%***	–	–	–

Non-use value scenario- full sample

		To								
		Iceland			Norway			Sweden		
		Error	t-value	TOST	Error	t-value	TOST	Error	t-value	TOST
From	Iceland	–	–	–	149%	9.40***	N.Eq.	137%	8.77***	N.Eq.
	Norway	−43%	3.91***	N.Eq.	–	–	–	11%	1.60	40%***
	Sweden	−52%	8.06***	N.Eq.	−8%	1.29	20%*	–	–	–

Non-use value scenario- non-anglers only

		To								
		Iceland			Norway			Sweden		
		Error	t-value	TOST	Error	t-value	TOST	Error	t-value	TOST
From	Iceland	–	–	–	162%	6.48***	N.Eq.	148%	8.77***	N.Eq.
	Norway	−55%	5.45***	N.Eq.	–	–	–	7%	0.82	40%***
	Sweden	−60%	9.28***	–	−8%	0.91	40%***	–	–	–

* = 5%, ** = 1% and *** =< 0.1% level of significance.

[a] Transfer error associated with the benefit transfer.

[b] The t-value for the hypothesis that the benefit transfer estimate and the original estimate are equal.

[c] The lowest level of significant equivalence. The levels tested are 10% 20% and 40% of the original study estimate of mean WTP.

[d] Not equivalent at any tested level.

Table 7 shows the results of the tests of value function transfer (hypothesis 2). The results for value function transfer in Table 7 are slightly different than the ones for unit value transfer in Table 6. Transfer error between Iceland and Norway is considerably reduced for the use value scenarios. The estimated models explain a considerable amount of the differences in mean WTP. Transfer error is however not reduced to a level that results in valid benefit transfer. For Norway and Sweden, on the other hand, using the value function transfer approach increases transfer errors for the use value scenarios. This indicates that Icelanders and Norwegians have similar preferences when it comes to fishing, while Swedes and Norwegians do not. All the t-tests for the use value scenarios reject equality of the benefit transfer and original estimates of mean WTP at a 5% significance level. The results of the equivalence tests are not symmetric. Non-equivalence is rejected when transferring from Sweden to Norway at a 40% level of the original mean WTP. If the transfer is from Norway to Sweden the non-equivalence is not rejected at any tested level. Swedish values could potentially be used in Norway but nor vice versa.

The results for the non-use value scenario are quite different. Transfer error is reduced in all cases, except when transferring from Iceland to Norway. The largest reduction is achieved when only a sub sample of non-anglers is used. Then the transfer error between Norway and Sweden is reduced to 7–8%. This produces the smallest t-values, taken as an indication of validity of the null hypothesis. Similarly non-equivalence is generally rejected at 40% transfer error, indicating significant equivalence. Further, non-equivalence is rejected at 20% transfer error for a transfer from Sweden to Norway using the full sample estimate. This clearly indicates a fully acceptable benefit transfer. The results for the non-use value suggest that preferences for that environmental good are similar in Norway and Sweden.

6. CONCLUSIONS

Several tests of both unit value transfer and value function transfer have been conducted for cases where the study site and policy site are in a different country. The general conclusion is that the accuracy of benefit transfer relies heavily on the similarity of populations and described scenarios in respect to environmental conditions in each country. This corresponds to the results of previous studies. It is evident that the more similar the populations and environmental goods are the smaller the transfer errors. This similarity must also include the perceived price levels of the resource at hand.

The WTP estimates are consistent between countries for all tested scenarios. The Icelandic values are largest, followed by the Norwegian and then Swedish as the smallest. Added information generally reduces transfer error. This is most obvious when the errors are large. This underlines the necessity of well defined and correctly specified value functions.

The suggested method of using equivalence tests instead of non-rejection of a classical test of equality seems to result in more consistent outcomes. The rejection of non-equivalence only happens in cases where both transfer error is small and WTP estimates are stable while non-rejections of classical hypothesis may happen when transfer error is large and WTP estimates are unstable. This is clearly seen in tables 6 and 7. The equivalence tests clearly indicate that Swedish WTP estimates can be used in Norway in cases where the acceptable transfer error is large, especially for non-use values. Norwegian results can to a lesser extent be used in Sweden. No clear-cut outcome results from the classical test procedure because the results vary form scenario to scenario.

The transfer errors are consistently smaller for the non-use value scenario. The results for the non-use value scenario by non-anglers in Norway and Sweden produce the only statistically equivalent benefit transfer at the strictest level of 20% transfer error. Anglers being excluded, the estimated WTP should be a strict non-use value estimate. Recreational fishing is a fairly well defined good in comparison to the non-use value of preserving natural fish stocks. The purely hypothetical nature of the latter might make the preferences more general, and less influenced by specific factors regarding recreational fishing in each country. Brouwer (2000) suggests that non-use values reflect some kind of overall moral commitment to environmental causes. He hypothesizes that such values can be expected to stay more or less constant across social groups and environmental domain. Our results seem to support his hypothesis.

Daði Kristófersson is researcher, Agricultural University of Iceland, Borgarnes, Iceland. Ståle Navrud is associate professor, Department of Economics and Resource Management, Norwegian University of Life Sciences, Ås, Norway.

7. NOTE

[1] It is assumed that Δ is symmetrical. That implies that only the size of the error is important not its sign. It is equally simple to work with two Δ, one for the possessive error and another for the negative error, denoted here by $-\Delta$.

8. REFERENCES

Allison PD, 1995 Survival analysis using the SAS® system: a practical guide SAS Institute Inc, Cary, NC 292 pp ISBN: 1-55544-279-X

Arrow KJ, Solow R, Portney PR, Leamer EE, Radner R, ShumanH, 1993 Report of the NOAA panel on contingent valuation Federal register, 58:4601–4614

Berger RL, Hsu JC (1996) Bioequivalence trials, intersection-union tests and equivalence, Statistical Science, 11(4):283–302

Bergland O, Magnussen K, Navrud S (2002) Benefit transfer: testing for accuracy and reliability In: Florax, RJGM, P Nijkamp and K Willis (eds) 2002: Comparative Environmental Economic Assessment, Edward Elgar, Cheltenham, UK and Northampton, MA, USA 117–132

Bergland O, Romstad E, Kim S, McLeod D (1993) The use of bootstrapping in contingent valuation studies Unpublished working paper Department of Economics and Social Sciences, Agricultural University of Norway, Ås

Brookshire DS, Neill HR (1992) Benefit transfers—conceptual and empirical issues Water resources research, 28(3):651–655

Brouwer R, Spaninks A (1999) The validity of environmental benefit transfer: further empirical testing Environmental and resource economics, 14:95–117

Brouwer R (2000) Environmental value transfer: state of the art and future prospects Ecological Economics, 32(1):137–152

Burda M, Wyplosz C (1997) Macroeconomics: a European text – 2nd edn Oxford University Press, Oxford, p. 613, ISBN: 0-19-877468-0

Cooper JC (1994) A comparison of approaches to calculating confidence-intervals for benefit measures from dichotomous choice contingent valuation surveys Land economics,70(1):111–122

Desvousges WH, Naughton MC, Parsons GR (1992) Benefit Transfer: Conceptual Problems in Estimating Water Quality Benefits Using Existing Studies. Water Resources Research, 28(3): 675–683

Downing M, Ozuna T (1996) Testing the reliability of the benefit transfer approach Journal of Environmental Economics and management, 30:316–322

Greene WH (2000) Econometric Analysis Prentice Hall International New Jersey ISBN: 0-13-015679-5

Hauck WW, Anderson S (1984) A new statistical procedure for testing equivalence in two-group comparative bioavailability trials, Journal of Pharmacokinetics and Biopharmaceutics 12(1): 83–91

Hoenig JM, Heisey DM (2001) The abuse of power: The pervasive fallacy of power calculations for data analysis American Statistician 55(1): 19–24

Kirchhoff S, Colby BG, LaFrance JT (1997) Evaluating the performance of benefit transfer: An empirical inquiry Journal of environmental economics and management, 33(1):75–93

Kristofersson D, Navrud S (2005) Validity Tests of Benefit Transfer: Are we performing the wrong tests? Environmental and Resource Economics 30(3):279–286

Kriström B (1997) Spike models in contingent valuation American Journal of Agricultural Economics, 79:1013–1023

Lehman EL (1983) Theory of point estimation, Wiley: New York

Lindsey JC, Ryan LM (1998) Tutorial in biostatistics–Methods for interval-censored data Statistics in medicine, 17(2):219–238

Loomis JB (1992) The evolution of a more rigorous approach to benefit transfer—Benefit function transfer Water resources research, 28(3):701–705

Loomis J, Roach B, Ward F, Ready R (1995) Testing transferability of recreation demand models across regions—A study of corps of engineer reservoirs Water Resources Research, 31(3):721–730

Navrud S, Pruckner GJ (1997) Environmental valuation - To use or not to use? A comparative study of the United States and Europe Environmental and Resource Economics 10:1–26

Notaro S, Signorello G (1999) Elicitation effects in contingent valuation: a comparison among multiple bounded, double bounded, single bounded and open-ended formats EAERA 9th annual conference, Oslo 25–27, June 1999

Ready RC, Navrud S, Dubourg R (2001) How Do Respondents with Uncertain Willingness To Pay Answer Contingent Valuation Questions? Land Economics, 77 (3):315–326

Ready R, Navrud S, Day B, Dubourg R, Machado F, Mourato S, Spanninks F,Rodriquez MXV (2004) Benefit Transfer in Europe How Reliable Are Transfers Between Countries? Environmental and Resource Economics 29:67–82

Reiser B, Shechter M (1999) Incorporating zero values in the economic valuation of environmental program benefits, Environmetrics, 10:87–101

Schuirmann DJ (1987) A comparison of the two one-sided procedure and the power approach for assessing equivalence of average bioavailability, Journal of Pharmacokinetics and Biopharmaceutics 15(6):657–680

Stegner, BL, A G Bostrom and T K Greenfield 1996 Equivalence testing for use in psychosocial and services research: An introduction with examples, Evaluation and Program Planning 19(3):193–98

Toivonen, Anna-Liisa; Appelblad, Håkan; Bengtsson, Bo; Geertz-Hansen, Peter; Gudbergsson, Gudni; Kristofersson, Dadi; Kyrkjebø, Hilde; Navrud, Ståle; Roth, Eva; Tuunainen, Pekka & Weissglas, Gösta (2000) The Economic Value of Recreational Fisheries in the Nordic Countries TEMA Nord Rapport 2000:604, 68 pp ISSN: 0908-6692; Nordisk Ministerråd; København

Welling, PG, FLS Tse and SV Dighe (1992) Pharmaceutical Bioequivalence, Marcel Dekker: New York

C. LEÓN, R. LEÓN AND F. VÁZQUEZ-POLO

THE APPLICATION OF BAYESIAN METHODS IN BENEFIT TRANSFER

1. INTRODUCTION

Benefit transfer is a technique for measuring the benefits of environmental goods based on past information. In this chapter we focus on the application of Bayesian methods to the transfer method. Bayes' theorem involves the combination of prior information with sample information in order to derive a posterior distribution from which any inference can be made. Whereas classical approaches discard past statistical information on a random variable, the Bayesian approach updates prior beliefs and experiences in the light of new empirical information. Thus, Bayesian methods are particularly suitable to the transfer method. They can be utilized as a general framework in which prior information can be handled to obtain predictions on the value of new environmental goods or policy sites.

The application of a Bayesian approach to benefit transfer requires the definition of a prior distribution. This can be obtained from expert opinion or from past studies. Expert opinion is very valuable when there is little information on the potential benefits of a particular policy site. Prior distributions can be elicited from experts using elicitation methods. These methods are intended to derive the parameter of some specified model by relying on expert judgment and experience. Benefit transfer could be based on the elicited prior distribution. However, the application of Bayes theorem allows the researcher to update the prior distribution by utilizing some sample data from the new policy site. Thus, in a Bayesian framework, benefit transfer can be improved by considering the role of sample information in complementing the lack of past information. This means that sample information can be seen as substitute for past information on study sites. Therefore, on-site empirical studies could be reduced by the use of a Bayesian approach which incorporates past information or expert opinion.

In the next section we first outline the principal concepts involved in applying Bayesian methods to benefit transfer. Section 3 discusses some elicitation methods for obtaining the prior distribution from expert opinion. Sections 4 and 5 develops the likelihood function and the simulation methods utilized to obtain the posterior distribution. Section 6 outlines an experiment intended to elicit a prior distribution for a National Park in Spain. Section 7 presents the results of the elicitation experiment and compares them with those obtained from sample information. Finally, Section 8 discusses the main conclusions and the prospects for future research.

S. Navrud and R. Ready (eds.), Environmental Value Transfer: Issues and Methods, 227–239.

2. BAYESIAN METHODS

Bayesian methods provide an alternative statistical analysis of empirical data. The principal difference with respect to classical methods is the consideration of prior beliefs in the estimation of statistical and econometric models. That is, the application of Bayes' theorem (Bayes, 1763) allows the researcher to combine prior information with the likelihood obtained from empirical data. Whereas classical analysis discards prior beliefs and information, the Bayesian approach emphasizes the usefulness of past experiences in the estimation process. In this setting, prior beliefs are revised when new empirical observations are obtained.

Prior beliefs from expert opinion or past experience are summarized in the prior distribution. Let us consider that the researcher is interested in estimating parameter θ, which can be considered as the consumer surplus to be obtained from a new policy site. This can also be seen as a function of unknown parameters defining the distribution of willingness to pay. If there is some knowledge on the possible values to be obtained from the empirical study, this information can be represented with the specification of a prior density distribution $\pi(\theta)$. This contains the probability of observing the parameter θ before empirical data is collected, based on all available evidence from past experience. The prior distribution could also incorporate beliefs from expert opinion.

If data is collected from the new policy site, this will be useful to define a likelihood function $f(x \mid \theta)$, which represents the likelihood of obtaining the sample x given that the population behaves according to parameter θ. This sample information allows the researcher to update her prior beliefs by applying Bayes Theorem. That is:

$$(1) \qquad \pi(\theta \mid x) = \frac{\pi(\theta) f(x \mid \theta)}{\int_{\Theta} \pi(\theta) f(x \mid \theta) d\theta} \propto \pi(\theta) f(x \mid \theta).$$

This is the expression for the posterior distribution, which is derived by combining the prior distribution and the likelihood function, and where $\propto$ denotes proportionality. If sample information on the new policy site does not induce relevant changes in the prior distribution, then there is no need for on-site data collection. In this case the prior distribution could be compatible with the posterior. However, new data would be required if the prior distribution does not accurately predict the value of the new policy site. Thus, there is a trade-off between prior and sample information. The collection of on-site data can be particularly useful for increasing the accuracy and efficiency of benefit transfer, while reducing the costs of environmental valuation.

In benefit transfer, it is most useful to consider the predictive distribution. This gives the probability of observing new sample data, given past experience and sample observations. That is, the predictive distribution can be expressed as

$$(2) \qquad \mathrm{m}(\mathrm{y} \mid \mathrm{x}) = \int_{\Theta} \mathrm{f}(\mathrm{y} \mid \theta) \pi(\theta \mid \mathrm{x}) d\theta,$$

where $f(y \mid \theta)$ is the likelihood for the sample observations which would be generated from a specific study for the new policy site, given parameter θ. This likelihood does not need to be the same as the one generating past observations.

Bayesian inference is carried out based on the properties and characteristics of either the posterior distribution when sample data is available, or the predictive distribution when there is no need to conduct further research. These characteristics include the mean, the mode or the median, and dispersion measures such as the variance, standard deviation, and interquartile range. A graphical representation of the posterior or predictive distributions could be useful in evaluating the relevant statistics.

The parameters defining either the posterior or the predictive distributions can be estimated by using a decision theory approach. This involves minimization of the expected value of a loss function, defined as the incurred loss when the real value of the parameter is θ but the research thinks is θ'. Alternative specifications of the loss function can be utilized depending on the particular statistics to be estimated (Berger, 1985). If we assume a quadratic loss function then the point estimate minimizing the expected loss is the mean. For an absolute value loss function it would lead to the median. The mode can be seen as the most likely value for the parameter θ in the distribution. That is, the mode $\tilde{\theta}$ is obtained from $\pi(\tilde{\theta} \mid x) = \max_{\theta \in \Theta} \pi(\theta \mid x)$. This is the reason why the posterior mode is considered the Bayesian maximum likelihood estimator.

Interval estimation is also convenient, in order to derive the bounds containing the parameter with a given probability. The interval (L, U) is a $(1-\alpha)100\%$ Bayesian interval for parameter θ if

$$\text{(3)} \qquad \text{Prob}\{L < \theta < U \mid x\} = \int_L^U \pi(\theta \mid x) d\theta = 1 - \alpha$$

Intervals could analogously be defined for the prior and predictive distributions. These are known as credible intervals, which should not be confused with the concept of confidence intervals in classical statistics. In the latter, the confidence level cannot be regarded as a probability. A credible interval is defined as the probability of the parameter lying within its bounds given the prior and the data, whereas a confidence interval gives the probability of the interval containing the parameter after innumerable repetitions of the experiment (Box and Tiao, 1965 and Press, 1989). Since Bayesian credible intervals are not unique, it is commonly chosen the one with the smallest length. It can be shown that for a unimodal distribution, this interval satisfies both (3) and $\pi(L \mid x) = \pi(U \mid x)$.

3. ELICITATION

Application of Bayes' Theorem requires a prior distribution to be specified by the analyst. The researcher could either specify her own prior distribution or rely on expert judgement. In any case, the prior distribution has to be elicited from expert opinion, which can be based on the study of the information available. It is convenient that experts know basic concepts of statistics, since the elicitation process involves asking questions about the parameters of a fully specified statistical model. In addition, experts should be familiar with the field for which further empirical data could be obtained. Thus, there must be previous experience in similar or identical situations. The researcher should also consider the potential bias from the elicitation process.

In benefit transfer, the specification of the prior could be based on the results from past studies on identical or similar sites. It is supposed that researchers can form an opinion about the potential results for the parameters of the empirical distribution which would be obtained from a new policy site. The elicitation of the prior distribution could be directed to generate a predictive distribution. That is, the information obtained from expert opinion is suitable to produce the probability of observing a new sample distribution, given the observation of past empirical data. In this context, the Bayesian approach would not require new sample information on the policy site to form predictions for the new study site. However, when previous information is lacking in quantity and/or quality, sample information could be combined with the elicited prior distribution, in order to update expert's beliefs on the benefits of a new policy site. Thus, benefit transfer could be complemented with sample data when the amount and characteristics of past experiments are not considered satisfactory.

There is a large diversity in the methods for eliciting prior distributions, most of them specific for the parametric statistical model to be considered (Kadane and Wolfson, 1998). Let us consider that we want to elicit the parameters of a shifted Beta distribution. This is a very flexible distribution, which allows for a variety of shapes. Thus, the prior density is

$$\pi(\theta; \alpha, \beta, a, b) = \frac{\Gamma(\alpha+\beta)}{\Gamma(\alpha)\Gamma(\beta)(b-a)^{\alpha-\beta-1}}(\theta-a)^{\alpha-1}(b-\theta)^{\beta-1} \tag{4}$$

where θ stands for consumer surplus or willingness to pay as obtained from the responses to a CV survey, and is assumed to be bounded in the interval (a, b). Therefore, the mean and the mode of θ are $a+(b-a)\frac{\alpha}{\alpha+\beta}$, and $a+(b-a)\frac{\alpha-1}{\alpha+\beta-2}$. There are two free parameters: $\ell = (\alpha, \beta)$, making possible a substantial variety of shapes, and introducing flexibility in the prior model of expert opinion.

The elicitation process starts by requiring from experts to assess the median (q_2) or second quartile, the third quartile (q_3), and the mode of the distribution for θ. Let F be a Beta cumulative distribution. Thus, we have the following two equations:

$$F(q_2; \alpha, \beta) = 0.5,$$
$$F(q_3; \alpha, \beta) = 0.75.$$

Parameters α and β are obtained by solving these two equations for the quantities q_2 and q_3. The procedure is iterative. In a first stage, interval $[q_2, q_3]$ is checked for a 50% high density region assuming a Beta distribution with parameters $\alpha = \beta = 1$.

If this condition is not satisfied, then the final parameters are found by considering the elicited mode for the prior distribution. That is, in the second stage parameter α is successively increased by 0.01, and the corresponding β is obtained by the relationship mode $= a + (b-a)\frac{\alpha-1}{\alpha+\beta-2}$, until interval $[q_2, q_3]$ does define a 50% high density region. When convergence is achieved, the final parameters $(\tilde{\alpha}, \tilde{\beta})$ define the prior distribution.

In order to check consistency, we also consider the first quartile $q_1(25\%)$ and the prior mean. Thus, parameters $\tilde{\alpha}$ and $\tilde{\beta}$ should satisfy the following two equations.

$$F(q_1, \tilde{\alpha}, \tilde{\beta}) \approx 0.25,$$
$$\frac{\tilde{\alpha}}{\tilde{\alpha}+\tilde{\beta}} \approx \text{mean}.$$

If the elicited mean and first quartile are inconsistent with parameters $\tilde{\alpha}$ and $\tilde{\beta}$, then the expert is asked to reassess the elicited quantities, until consistency is achieved. This can be defined as a 30% maximum deviation between the elicited quantities and those derived in the elicitation process (Wolfson, 1995).

4. THE LIKELIHOOD

A Bayesian approach should consider the role of empirical data in predicting the value of new policy sites. Some amount of data could help to improve the predictions obtained from the elicited prior distribution, and could be used as a consistency test between the prior and the posterior distributions. The likelihood function adopted to model the data generation process has to be appropriate for the empirical data. Let us consider a log logistic likelihood function, which is commonly utilized in valuing environmental goods through the contingent valuation method. We assume that willingness to pay in the population follows the distribution function

$$G(x) = \frac{(x \cdot \exp(-\delta))^{1/\sigma}}{1 + (x \cdot \exp(-\delta))^{1/\sigma}}. \tag{5}$$

This implies that the logarithm of willingness to pay follows a logistic distribution with location and scale parameters δ and σ, respectively. In the case of the double bounded elicitation format, the probability that an individual accepts to pay the first price (A_i) but refuses to pay a higher bid (A_i^u) offered subsequently is $\pi^{yn} = G(A_i^u) - G(A_i^d)$. Denoting by A_i^d the price that is offered to individual i after she refused to pay the price offered in the first place, the probability of the other possible answers is

$$\pi^{ny} = G(A_i) - G(A_i^d).$$

$$\pi^{yy} = 1 - G(A_i^u).$$

$$\pi^{nn} = G(A_i^d).$$

Hence, the likelihood of the model, for data $\tilde{x}$, is

$$f(\tilde{x} \mid \delta, \sigma) = \prod_{i=1}^{n} (\pi_i^{yy})^{v_1 v_2} (\pi_i^{yn})^{v_1(1-v_2)} (\pi_i^{ny})^{(1-v_1)v_2} (\pi_i^{nn})^{(1-v_1)(1-v_2)}. \tag{6}$$

where v_1 takes the value one if the individual answered "yes" to the first question and zero otherwise, and v_2 takes the same values but for the second question. Since willingness to pay follows a loglogistic distribution, the mean can be expressed as

$$\theta = \exp(\delta)\Gamma(1+\sigma)\Gamma(1-\sigma). \tag{7}$$

5. SIMULATION

The combination of a shifted Beta distribution with a loglogistic distribution through Bayes theorem is not straightforward. The posterior distribution can be simulated using the algorithm proposed by Metropolis et al. (1953). This algorithm belongs to an area of statistics known as Markov Chain Monte Carlo, usually referred to as MCMC. MCMC techniques enable simulation from a distribution by constructing a succession of random values with an appropriate algorithm. As the number of values in the succession grows, the probability of obtaining a value belonging to an interval (a, b) is approximately the probability determined by the distribution that we want to simulate. In other words, once convergence is reached, the values in the succession can be considered as approximate draws from the posterior distribution. These values can be used to calculate the mean and interquartile range of the posterior distribution. The succession that is constructed is a Markov Chain.

Let $(\delta, \sigma)^i$ be the ith element in the succession of values constructed with the Metropolis algorithm. The ergodic average of a real-valued function $t(\delta, \sigma)$ is the average $\bar{t}_n = \frac{1}{n}\sum_{i=1}^{n} t((\delta, \sigma)^i)$. If the expected value of $t(\delta, \sigma)(E_\pi[t(\delta, \sigma)])$ exists, then $\bar{t}_n \to E_\pi[t(\delta, \sigma)]$ as $n \to \infty$, with probability 1. This result is known as the ergodic theorem. One implication is that the average of the values in the succession is a strongly consistent estimate of the posterior mean. If we let $t(\delta, \sigma) = \exp(\delta)\Gamma(1+\sigma)\Gamma(1-\sigma)$ we obtain a more important implication for our model. The average of this transformation $t(\delta, \sigma)$ of the parameter values in the succession is a strongly consistent estimate of the posterior mean of mean willingness to pay. Thus, we can obtain the posterior mean of mean *WTP* without calculating the posterior distribution of mean *WTP*. Credibility intervals for the parameters or for mean *WTP* are similarly obtained by estimating the interval limits by the respective sample quartiles. By the ergodic theorem, the estimators are also consistent.

The values of the constructed succession are not independent draws from the posterior distribution. We could obtained a sample of independent draws by picking up one value every k iterations. For instance, for $k = 4$, and assuming that convergence is reached at the $1000th$ iteration, we could form the sample with the values $1000th$, $1004th$, $1008th$,... in the succession. However, discarding elements increases the variance of the estimators (O'Hagan, 1994; MacEachern and Berliner, 1994). This means that independence sampling comes at the expense of reduced efficiency.

One important aspect of MCMC is to determine when convergence is reached. Values from the posterior distribution are only obtained when the number of iterations of the chain approaches infinity. In practice this is not attainable and values obtained at a sufficiently large iteration are taken instead of being drawn from the posterior distribution. Convergence can be analyzed by observing the simulated chain. We could discard the first D values of the simulated chain and divide the rest of the chain in several parts. If the properties of these sub-samples are similar we can conclude that convergence is reached at least after the Dth iteration. This analysis can be reinforced when we also analyze sub-samples from chains started at different initial values.

The algorithm proposed by Metropolis et al. (1953) belongs to a family of algorithms known as Metropolis-Hastings algorithms. This name stems from papers by Metropolis et al. (1953) and Hastings (1970). There are large classes of schemes to simulate a distribution using MCMC techniques. There are as well different versions of the Metropolis algorithm. A review of these methods can be found in Gamerman (1997). The following is just one possible version.

Let $\pi(\delta, \sigma \mid x)$ be the posterior distribution that we want to simulate and

$$\lambda(h, i, j, k) = \min\left\{1, \frac{\pi(h, i \mid x)}{\pi(j, k \mid x)}\right\}.$$

Consider f_1 and f_2 are two random variables with distributions $N(0, \tau_1)$ and $N(0, \tau_2)$, respectively. These distributions are usually referred to as proposal densities. The initial values in the chain are fixed arbitrarily. The ith value of the chain (δ_i, σ_i) is obtained from its predecessor in the following way. A value x is drawn from the normal variable f_1. Consider the value $\delta_i = \delta_{i-1} + x \cdot \delta_i$ will be equal to a_i with probability $\lambda(a_i, \sigma_{i-1}, \delta_{i-1}, \sigma_{i-1})$ and equal to δ_{i-1} with probability $1 - \lambda(\delta_i, b_i, \delta_i, \sigma_{i-1})$. Similarly for the other coefficient. A value x is drawn from the normal variable f_2. Consider the value $b_i = \sigma_{i-1} + xb_i$ is accepted with probability $\lambda(\delta_i, b_i, \delta_i, \sigma_{i-1})$. If b_i is rejected, σ_i has the same value as σ_{i-1}.

The distribution of f does not have to be a normal distribution. It could be any other distribution symmetric around 0. The role of this distribution is to propose different values along the parameter space. The variance of the distribution must be large enough so that any value of the parameter space can be proposed in some iteration. In this version of the algorithm, the proposed values follow from the old values like in a random walk. There are multiple ways of proposing new values, but they affect the definition of λ. There are other versions of the Metropolis algorithm in which all the parameters are updated at the same time and not one by one. Note that we only need the ratio of the value of the posterior distribution in two different points. Hence, we do not need to calculate the constant that goes in the denominator of the posterior distribution.

6. EXPERIMENTAL DESIGN

The application was directed at eliciting the predicted benefits of Teide National Park, in Spain. This park contains 15,000 hectares placed in the island of Tenerife (Canary Islands). Annual visits are about 3 million, most of them by foreign tourists staying in the island for holidays. Visits are attracted by the magnificent sceneries in the park, and the volcano at the centre of it, which is the highest peak in Spain at 3714 meters. Tourists tend to walk around the volcano area, and travel in the carriage to the top of the peak, where impressive vistas can be enjoyed. The park also preserves a representation of the high mountain flora of endemic species from the Canary Islands. The visits last for a couple of hours, or one day at most, although there is a small hotel inside the park.

In the experiment we selected a group of students to carry out an elicitation practice on the recreational value of the park. The students were final year undergraduate students in Economics, who were trained as experts in environmental valuation and statistics, and were given information on the value of other parks in Spain. Each individual answered anonymously and independently to a structured questionnaire in three stages. Previous to the elicitation process, there was a stage in which subjects were asked technical statistical questions, as well as questions on the information package received, in order to screen for expertise and knowledge. Only those subjects who succeeded in passing this screening stage were considered for the elicitation process. There were 11 individuals who could be accepted as experts, out of 22 subjects initially considered. Following the procedure outlined in Section 3, subjects were asked to state the quartiles, the mean and the mode of the prior distribution for the potential results to be obtained through a contingent valuation survey on the value of the Teide. In a final stage, the experts were asked to revise their opinions in the light of the average values and the shapes of the distributions obtained form the pooled results.

In order to check the predictive ability of the elicitation process, results were compared with unpublished data on the recreational value of Teide National Park. These data were obtained from a contingent valuation survey carried out in summer 1997. A sample of 1133 tourists were interviewed on-site after experiencing recreational activities in the park. The interviews were conducted in six different languages by professional interviewers trained in the specifics of contingent valuation questionnaires. The design of the final version of the questionnaire followed a process of careful pretesting and work with focus groups. The payment vehicle was a hypothetical entrance-fee to the park. The elicitation format was dichotomous choice, with two binary questions making the double bounded format. The bid vector design was chosen by merging the results from the open ended pre-test surveys with past results on the value of other natural areas in Spain. The joint empirical distribution was then evaluated for four quartiles making the four bid vector design.

7. RESULTS

The responses from the elicitation procedure allow us to derive the prior distribution for the mean willingness to pay. Table 1 shows the characteristics of the elicited prior distribution, averaged across the sample of experts. It can be observed that experts differed considerably in their opinions in the early stage of the elicitation process. These differences can be attributed to variations in the interpretation of the information at hand, since none of the experts had previous experience in conducting empirical studies. Even though, a fair degree of consensus was reached on the average values for the specific quantities, starting from a wide range of values. The agreement was facilitated by the use of graphical techniques (histograms and probability functions), which allowed experts to compare their assessments with those derived from the average values.

The parameters for the prior Beta distribution were derived utilizing the iterative procedure outlined above. Thus, the prior distribution for θ reflects the beliefs of the experts, and takes the form

$$\pi(\theta) \propto \theta^{1.91}(7500-\theta)^{12.95}, \quad 0 \leq \theta \leq 7500. \tag{8}$$

Let us consider that the prior distribution of σ is non-informative and independent of θ, that is, $\pi(\delta, \sigma) = \pi(\delta)\pi(\sigma)$. Taking into account the relationship between θ and (δ, σ) described by equation (7), the prior distribution of (δ, σ) can be derived from prior (8).

$$\begin{aligned} \pi(\delta, \sigma) \propto & \left(e^{\delta}\Gamma(1+\sigma)\Gamma(1-\sigma)\right)^{1.91} \bullet \\ & \left(7500 - e^{\delta}\Gamma(1+\sigma)\Gamma(1-\sigma)\right)^{12.95} e^{\delta}\Gamma(1+\sigma)\Gamma(1-\sigma)\pi(\sigma) \end{aligned} \tag{9}$$

Prior (9) combined with likelihood (6) will determine the posterior distribution. The posterior distribution, with a constant prior for σ, was simulated in Gauss

Table 1. Characteristic values of the elicited prior information

	Teide national park			
Characteristics	Smallest	Average	Largest	Std. deviation
Mean	875	1379.55	2500	443.82
Mode	750	963.64	1300	145.07
First quartile	300	639.77	1200	283.44
Median	600	1137.73	2200	484.96
Third quartile	800	1567.05	3300	694.22
Maximum *WTP*	1000	3031.82	7500	1849.90

using the Metropolis algorithm. Three chains, each of length 20001, were run starting at different initial values for (δ, σ). In order to avoid very large numbers in the computations, the posterior distribution was specified for willingness to pay divided by 10,000. The proposed density for σ had a standard deviation of 0.08. The standard deviation of the proposed density for δ was different in each of the three chains (0.03,0.5,1). The first 5000 values of each chain were discarded. The remaining values were grouped into sub-samples of 5000 draws. Table 2 shows the mean values of δ and σ in each of these sub-samples. The similarity of the means in each group suggests that convergence was reached at least after the 5000th iteration. Hence, the values generated after the 5000th iteration form a sample of 45006 draws from the posterior distribution.

The final posterior results are presented in Table 3. Two alternative specifications for the prior of the precision ($\tau = 1/\sigma$) are considered. The "informative" case is given by DuMouchel (1995). This is:

$$\pi(\tau) = \frac{S_0}{(S_0 + \tau)^2}, \tag{10}$$

Table 2. Mean value of δ and σ in each sub-samples

Mean(δ)	Mean(σ)	Initial values
−2.751	0.668	$\delta_0 = -2, \sigma_0 = 0.8$
−2.752	0.668	$\delta_0 = -2, \sigma_0 = 0.8$
−2.745	0.666	$\delta_0 = -2, \sigma_0 = 0.8$
−2.736	0.664	$\delta_0 = -1.3, \sigma_0 = 0.5$
−2.757	0.670	$\delta_0 = -1.3, \sigma_0 = 0.5$
−2.757	0.670	$\delta_0 = -1.3, \sigma_0 = 0.5$
−2.736	0.0664	$\delta_0 = -2.4, \sigma_0 = 0.2$
−2.736	0.665	$\delta_0 = -2.4, \sigma_0 = 0.2$
−2.738	0.665	$\delta_0 = -2.4, \sigma_0 = 0.2$

Table 3. Posterior mean, median and 80% credible region for δ, σ, and mean WTP

	Posterior mean	Posterior median	80% Credible interval
		Flat prior for σ	
δ	−2.7460	−2.7441	(−2.8070, −2.6866)
σ	0.6671	0.6672	(0.6495, 0.6841)
Mean *WTP*	1556	1557	(1552, 1588)
		Informative prior for σ	
δ	−1.8424	−1.8419	(−0.18633, −1.8217)
σ	0.02050	0.02048	(0.01901, 0.02196)
Mean *WTP*	1585	1586	(1552, 1618)

Table 4. Loglogistic maximum likelihood

Parameter	Estimate (Std. error)
δ	7.3357 (0.02408)
θ	0.3608 (0.0152)
Median	1534
Mean	1919
log L	−1016.21
n	848

where S_0 is the standard deviation in the group of experts. It is shown that although the prior for the precision affects the posterior distribution of α and σ, it does not significantly affect the posterior distribution of mean willingness to pay.

The incorporation of sample data has raised the estimated mean consumer surplus from the expert assessments provided by the elicitation experience. This means that the elicited prior does not accurately predict the outcome from the contingent valuation study. Table 4 presents the results from modelling the sample data according to a log logistic distribution with maximum likelihood estimation. The results for the elicited prior based on the consensus mean distribution deviate somewhat from the observed data. However, by looking specifically at the individual experts, it can be found some subjects producing rather accurate predictions. The fact that we observe a large dispersion in expert opinion indicates a fair degree of uncertainty about the final outcome of an on-site contingent valuation study on Teide National Park. This uncertainty can be explained because of lack of previous experience in dealing with field work, and the absence of a large amount of past information on the value of specific parks.

8. CONCLUSIONS

Environmental valuation is a costly activity and faces the challenge of efficiently predicting the value of new environmental goods. The benefit transfer method could be seen as a way of reducing the costs of valuing new policy sites, by making use of the pool of available experience and information. However, there is a need to correct for all the potential biases involved in appraising new benefits on the base of past benefits. The development of new techniques is required in order to address predictions, both from a theoretical and from an empirical point of view.

Bayesian methods provide a set of statistical techniques which are appropriate to model the transfer process. In fact, the Bayesian approach can be seen as a theoretical framework to be utilized for tracing predictions on new policy sites. The decision-maker is assumed to follow Bayes' theorem when combining the prior distribution with the sample information to be obtained from the policy site. This is the most efficient way of combining the two sources of information. When further information is not available, the researcher could based his predictions on the predictive distribution, which gives the probability of observing new data, given the set of past information. However, on-site sample data might increase the accuracy of the predictions, and could serve the researcher to update the prior distribution in the light of new evidence. This is most necessary when the set of past studies is limited and lack in quality.

A crucial step in the application of Bayesian methods is the elicitation of the prior distribution. This prior contains all available information, and is elicited from experts on environmental valuation. It might be expected that experts will be increasingly capable of forming an opinion about the characteristics of the prior for a particular good as the pool of specific studies increases. There are diverse methods for eliciting prior distributions, most of them specific to the particular statistical model. In benefit transfer, the incorporation of expert opinion in the estimation process could serve to improve the predictions on the new policy site. In some cases, it might be the only feasible way to appraise these benefits.

In the example shown in this chapter, experts were asked to elicit their prior distribution for the benefits of a National Park in Spain. From the results to the elicitation questions, we developed the parameters of a shifted Beta distribution. When the elicited results are compared with the sample observations, predictions are significantly different from the actual results, although they are in the right direction. This could raise some doubts for the straightforward use of elicitation mechanisms in benefit transfer. However, further research is needed in order to find out what are the appropriate sources of prior information to be utilized in forming predictions, and how sensitive are the results to the consensus procedure.

Carmelo León and Francisco José Vázquez-Polo are professors, Department of Quantitative Methods in Economics, University of Las Palmas. Roberto León is lecturer, Department of Economics, University of Leicester. *Roberto León acknowledges funding by Zona Especial Canaria and Fundación Universitaria de Las Palmas.*

9. REFERENCES

Bayes T (1763) An Essay Towards Solving a Problem in the Doctrine of Chances, Philos Trans R Soc London, 53:370–418 Reprinted in Biometrika, 1958, 45:293–315.

Berger JO (1985) Statistical Decision Theory and Bayesian Analysis (2nd edn) New York: John Wiley and Sons.

Box GEP, Tiao GC (1965) Multiparameter Problems From a Bayesian Point of View, Annals of Mathematical Statistics, 36:5, 1468–1482.

DuMouchel W (1995) Predictive Cross-Validation of Bayesian Meta-Analyses In Bayesian Statistics JM Bernardo, JO Berger, AP Dawid, AFM Smith (Eds), Oxford University Press, Oxford, 5:107–127.

Gamerman D (1997) Markov Chain Monte Carlo: Stochastic simulation for Bayesian inference Chapman & Hall.

Hastings WK (1970) Monte Carlo Sampling Methods Using Markov Chains and Their Applications, Biometrika, 57:97–109.

Kadane JB, Wolfson LJ (1998) Experiences in Elicitation, Journal of the Royal Statistical Society, Series D (The Statistician), 47:1, 3–19.

MacEachern, SN and Berliner, LM (1994) Subsampling the Gibbs Sampler, The American Statistician, 48:188–190.

Metropolis N, Rosenbluth AW, Rosenbluth MN, Teller AH, Teller E (1953) Equation of State Calculations by Fast Computing Machine, Journal of Chemical Physics, 21:1087–1091.

O'Hagan A (1994) Bayesian Inference, Volume 2B of Kendall's Advanced Theory of Statistics Edward Arnold, London.

Press SJ (1989) Bayesian Statistics: Principles, Models, and Applications New York: John Wiley and Sons.

Wolfson L (1995) Elicitation of Priors and Utilities for Bayesian Analysis PhD Dissertation Dpt Of Statistics, Cornegie Mellon University.

S. PATTANAYAK, V. K. SMITH AND G. VAN HOUTVEN

IMPROVING THE PRACTICE OF BENEFITS TRANSFER: A PREFERENCE CALIBRATION APPROACH

1. INTRODUCTION

Most applied welfare analyses for environmental policy evaluations, whether benefit-cost or natural resource damage assessments, rely on adaptations of existing benefit estimates in what is described as benefits transfer rather than new research. Over 10 years ago, David Brookshire organized a set of papers in *Water Resources Research* to focus attention on the practice of benefit transfer (see Brookshire and Neill [1992]. Since then, interest in research on the potential for improvement in these techniques has exploded and this volume reports on a number of innovations relying on refinements in the statistical methods used in meta analyses that often provide empirical benefit functions for transfer. Nonetheless, where evaluations of benefits transfer have taken place, current practice is generally regarded as *very unreliable*![1]

This paper considers a different perspective on the practice of benefit transfers. It is one that interprets the benefit transfer problem as an identification problem. That is, the analyst must calibrate individual preferences for the environmental resources of interest based on the available empirical benefit estimates. Our proposed methodology is general. Here we apply it to one example – the development of consistent measures of the benefits of water quality improvements. To develop this logic we begin with a historical perspective, interpreting Harberger's [1971] approximation using indifference curves and then suggesting this logic seems to have been a conceptual antecedent to the logic used with unit benefit transfers. However, in this case the same desirable properties cannot be assured. Section 2 presents a detailed description of conventional benefit transfer practices and these antecedents. Section 3 illustrates the limitations of such practices through a simple example. In Section 4, we provide a detailed description of our proposed methodology in six steps. Using a case study from the Chesapeake Bay our proposed approach is illustrated in Section 5 with travel cost and contingent valuation (CV) data. We then demonstrate how the calibrated functions can be used to construct benefit estimates for a separate situation. We discuss how the resulting benefit estimates differ from those of a more traditional benefit transfer practice, hereafter labeled "simple approximation." Finally, in Section 6 we present a few methodological conclusions.

S. Navrud and R. Ready (eds.), Environmental Value Transfer: Issues and Methods, 241–260.

Table 1. Recent Evaluations of Conventional Benefit Transfers

Commodity/Service	Authors	Year	Source
Recreation water quality	Loomis, Roach, Ward, and Ready	1995	*Water Resources Research*
Fishing water quality	Downing and Ozuna	1996	*JEEM*
Recreation water quality	Kirchhoff, Colby, and LaFrance	1997	*JEEM*
Health water quality	Barton	1999a	Working Paper
Waste water treatment benefits	Barton	1999b	Working Paper
Rural farm landscape	Santos	1999	Working Paper
Peat meadow amenities	Brouwer and Spaninks	1999	*Environmental and Resource Economics*
Overview	Brouwer	2000	*Ecological Economics*
City air quality	Rozan	2000	Working paper
Forest amenities	Scarpa, Hutchinson, Chilton, and Buongiorno	2000	Working paper

[a] The numbers correspond to the RFF water quality ladder and index (boatable = 2.5, fishable = 5.1, and swimmable = 7).

2. BENEFIT TRANSFER: CONVENTIONAL PRACTICE AND CONCEPTUAL ANTECEDENTS

Benefit transfer adapts available estimates of the economic value for a change in environmental quality (or quantity) to evaluate a proposed policy-induced change in the same or a "similar" resource. In these situations, the analyst is typically taking the results from one or more existing studies (defined in terms of their time frame, the location, the environmental resource, or quality change, and the affected population), and using them to evaluate a different context that is relevant for a specific policy. The new policy context can require changes in both the features of the resource and the characteristics of the people who care about it.

Most benefit transfer methods use either the *benefit value* or the *benefit function* approaches. In the case of a benefit value approach, a single point estimate (usually a mean willingness to pay (WTP) estimate) or value range is typically used to summarize the results of one or more studies that have been developed for another purpose. For example, an average consumer surplus per fishing trip might be taken from a recreational travel cost study, or a mean WTP estimate for an incremental change in water quality may be estimated from a CV study. These values are then used to evaluate the benefits from proposed policies that change water quality at different locations. In these applications, the transfers are intended to assess the economic value of fishing trips or changes in water quality in new areas. In the case of a benefit function transfer, a model has been estimated to describe how benefit measures (from one or more existing studies) change with the characteristics of the study population or the resource being evaluated.[2] Often this function is derived from a meta analysis. With this second approach, the function is "transferred" to the policy context, and the benefit estimate is then "tailored", based on the arguments of the function, to meet the new population's characteristics and the new resource's features. For example, a travel cost demand model from one site might be used with the income, average travel costs, and quality conditions for a policy site to estimate the consumer surplus under different conditions.

Benefit transfers usually proceed in four steps:

1. Translate the policy change into one or more quantity changes related to the affected environmental resource, such as the resulting change in level of use of the resource for a typical user.[3]
2. Estimate the number of individuals linked to the affected resource (e.g., the number of typical users) before and after the policy change.
3. Transfer a per "unit" value (e.g., consumer surplus) measure, with the unit measured in terms of the quantity index selected in Step 1.
4. Combine estimates in Steps 1 through 3 for each year considered in the analysis and compute the discounted aggregate benefit measures.

A typical benefit transfer of this type can also be summarized in a simple relationship, such as the following equation.

$$(1) \qquad CS_p = \frac{CS_T}{\Delta d_T}(d_1 \bullet N_1 - d_0 \bullet N_0)$$

where

CS_P = the estimate of consumer surplus gain for policy being evaluated;

CS_T = the consumer surplus gain (for a representative individual) measured in other literature for a change judged to be comparable to that of the policy;

d_i = the quantity index, such as the level of use of the affected resource by a typical user (e.g., visits per year), in the presence of the policy change (i = 1) and in the absence of the policy change (i = 0);

Δd_T = the change in the quantity index that corresponds to the CS_T measures in the literature; and

N_i = the number of individuals linked to the affected resource with the policy change (i = 1) and without it (i = 0).

This approach focuses the analysis on individual-specific quantity measures, which are characterized by the d term. In the recreation context, d is usually measured as a trip to or visitor-day at a recreation site. However, in the health context it could be measured as an episode of illness avoided or the risk of some acute condition (e.g., asthma). N would then be the number of affected or exposed individuals (through the affected resource, such as contaminated surface water), and Δd_T would be expressed as a *reduction* in the health effect or risk.[4]

CS_P can also be expressed for a quality change. That is, d can be expressed as a quantitative measure of the quality of the resource (e.g., water quality) that is affected by the policy. In this case, CS_T would have to be measured and interpreted as the consumer surplus gain from a specified quality change (Δd_T) for a similar resource that is experienced by a typical individual. In this case, $d_1 - d_0$ would be interpreted as the quality change resulting from the policy number, and N would be the number of individuals experiencing the change ($N_1 = N_0$).

In any adaptation of equation (1), the term $CS_T/\Delta d_T$, serves as a unit value or "price." In recreation-based applications, it might be expressed as a consumer surplus per visitor-day. With the benefit value approach to benefit transfer, the unit value is treated as a constant, regardless of the characteristics of the individual or the size of the change experienced by each individual. The use of a benefit function approach can provide more flexibility in how $CS_T/\Delta d_T$ is defined for different individuals and, in some cases, for different sizes or types of change; however, in practice, nearly all existing transfers follow the basic format of equation (1).

Each adaptation to the basic format described in equation (1) changes the mix of assumptions required to evaluate their consistency with respect to the basic benefit concepts. Although such adaptations are likely to affect the performance of the benefit estimation approach, none of these adjustments is directly linked to the concept of Hicksian willingness to pay (WTP).

When the benefit transfer process is described by the relationship in equation (1), it resembles the methods used to approximate WTP measures for price changes. Figures 1a and 1b illustrate the basic logic. In figure 1a we assume a simple linear demand function and seek to estimate the consumer surplus from P_0 to P_1 (with $P_0 < P_1$). This is given in equation (2) and simplifies to the expression given in equation (3).

$$CS = (P_1 - P_0)q_1 + 1/2(P_1 - P_0)(q_0 - q_1) \tag{2}$$

$$CS = 1/2(P_1 - P_0)(q_1 + q_0) \tag{3}$$

Figure 1b illustrates how it relates to expenditure changes.

$$CS = 1/2(CD + AB) \tag{4}$$

This result is readily established when we recognize that:

$$\begin{aligned} CD &= (m_0 - P_0 q_1) - (m_0 - P_1 q_1) = (P_1 - P_0)q_1 \\ AB &= (m_0 - P_0 q_0) - (m_0 - P_1 q_0) = (P_1 - P_0)q_0 \end{aligned} \tag{5}$$

Thus, the consumer surplus is an average of the change in expenditures on all other goods when q_1 is evaluated at P_0 and P_1 versus evaluating q_0 at these prices. Therefore, the logic for transfer provides a direct parallel to the expenditure changes that Harberger considered as approximating consumer surplus. The average quantity is valued by the price change to compute the consumer surplus.

In practice, benefit transfers have used the same general approach to estimate WTP for a quantity or quality change, rather than a price change. The transfer values are used as if they were virtual prices (i.e., marginal WTP measures) for the quantity or quality change associated with the policy. An important limitation of this approach is that it treats these virtual prices as if they were constant, when in

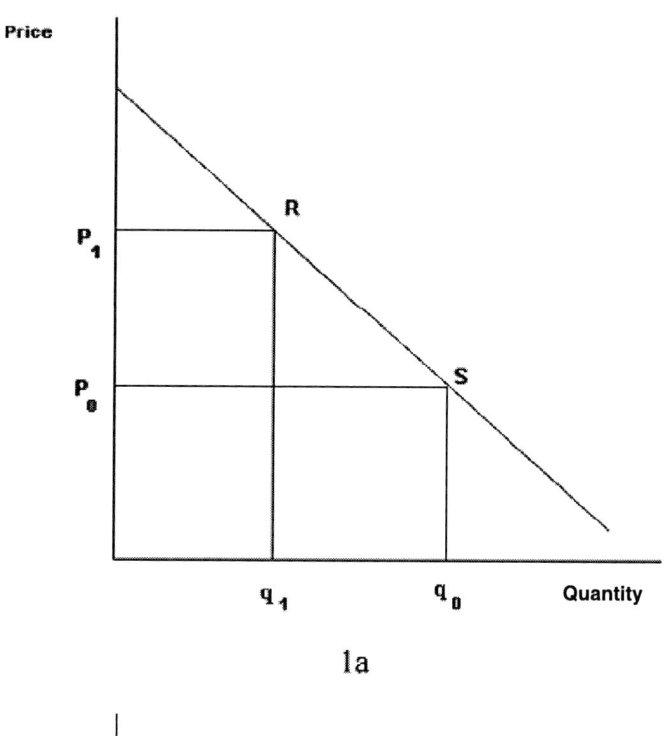

1a

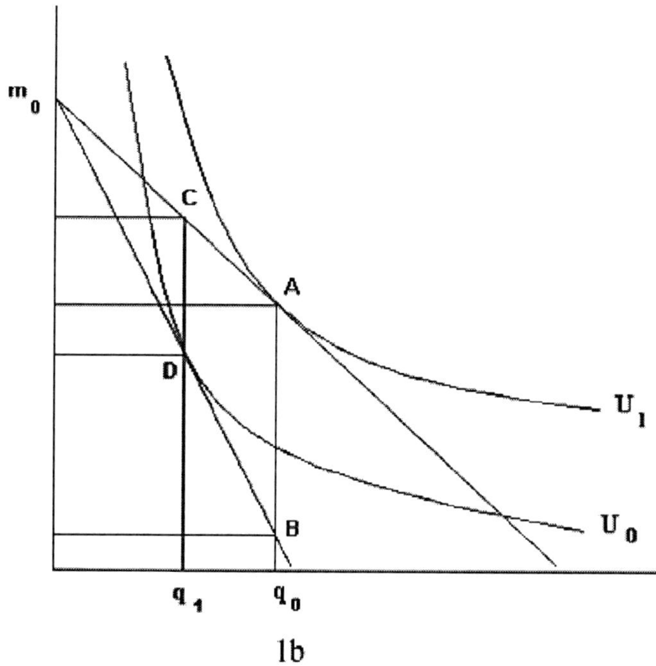

1b

Figure 1. Illustration of Harberger Approximation

fact they are not. This outcome contrasts with approximations intended to measure the value of *price* changes, where we can assume prices are constant. This can be illustrated with a simple version of a Hicksian expenditure function, e(.), with priced goods, (and $\bar{P}$ a vector of prices); one nonpriced good, Z; and a quality feature for Z, measured by s. The Hicksian consumer surplus (WTP) for a change in quality from s_0 to s_1 is given in equation (5a) with U_0 the initial utility level and Z_0 the level of the nonmarketed good:

$$\text{(5a)} \qquad WTP = e(\bar{P}, Z_0, s_0, U_0) - e(\bar{P}, Z_0, s_1, U_0) > 0$$

We expect this difference to be positive. Z is a desirable environmental service and its quality, s, is measured such that $s_1 > s_0$. Improvements in quality allow an individual to spend less on marketed goods but be able to attain the same level of well being.

Using the virtual price of Z, ρ_z, to approximate a parametric price, the first order approximation for consumer surplus, $\tilde{CS}_c$, for a change from s_0 to s_1 might be written as equation (5b).[5]

$$\text{(5b)} \qquad \tilde{CS}_c = \rho_z \bullet [Z(s_0) - Z(s_1)]$$

with $\rho_z = \frac{\partial e}{\partial z}$, which is evaluated at a value for Z (i.e., $Z[s_0]$ or $Z[s_1]$).

Unfortunately, any set of values we would propose to use for p_z and Z can be expected to be interrelated through e(.) and will be functions of s. The virtual price for the quality change can therefore, strictly speaking, not be treated as exogenous or constant.

3. DIFFICULTIES WITH CURRENT BENEFIT TRANSFER PRACTICES

Some of the limitations associated with conventional approaches to benefit transfer can be illustrated through an example involving water quality changes and recreation-based benefits. This example serves to explain the difficulties in using price-based approximations to describe the benefits due to the quality and quantity changes. In this case, the objective of the transfer is to assess the benefits of policies that have reduced pollutant discharges to a major river system in the U.S. The capacity of this river system to support different types of fishing activities is assumed to increase because of improvements in the water quality.

Figure 2 illustrates the implicit logic underlying a simple benefit transfer approach. D_0 describes the pre-policy demand for fishing and D_1 the post-policy demand, assuming that the change in water quality leads to a parallel shift in the demand function. The benefits from a quality improvement that shifts the demand from D_0 to D_1 would be DFCG, treating OE as the average travel cost to use the site for fishing.

In this algebraic example, it is assumed that $CS_T/\Delta d_T$ is a "perfect match" with conditions at the improved site (i.e., *after* the policy change) so that the unit

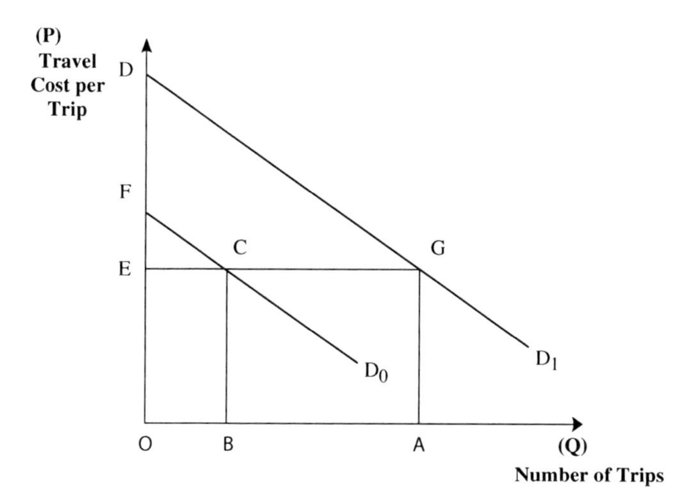

Figure 2. Quality as a Quantity Change

values would be estimates of DEG/OA, consumer surplus per trip for the desired fishing experience.[6] The benefit measure implied when equation (1) is adapted to approximate the value of quality change that has been linked to use $\tilde{C}S_P$ is then given by equation (6).

$$(6) \qquad \tilde{C}S_P = \left(\frac{DEG}{OA}\right) \bullet BA$$

BA is the postulated increase in fishing trips taken because of the water quality improvement. In contrast to this measure, the appropriate estimate of the value of the water quality improvement would be DFCG. We can use the geometry from Figure 2 to illustrate the extent of the mistake. The logic used in this transfer assumes OB is a constant multiple, α, of the activities currently observed:

$$(7) \qquad OB = \alpha \bullet OA$$

The "correct" benefit measure, given our assumptions, is $DFCG = DEG - FEC$, with quality leading to a parallel shift in FD_0 to DD_1. We can simplify matters using the following relationships for areas of the two triangles:

$$(8) \qquad DEG == \frac{1}{2} DE \bullet OA$$

$$(9) \qquad FEC == \frac{1}{2} DE \bullet OB = \frac{1}{2}(\alpha\ DE)(\alpha\ OA)$$

Simplifying the expression for DFCG, we have equation (10) describing the desired benefit measure:

$$(10) \qquad DFCG == \frac{1}{2} DE \bullet OA(1 - \alpha^2)$$

The expression given in equation (6) for the usual benefit transfer method can be expressed in terms of DE, OA, and α as

$$(11)\qquad CS_P = \left(\frac{DEG}{OA}\right)\bullet BA = \left(\frac{\frac{1}{2}DE\cdot OA}{OA}\right)\bullet(1-\alpha)OA = \frac{1}{2}DE\bullet OA(1-\alpha)$$

This geometry implies we have a relationship between the "correct" benefit measure and the simple approximation. The ratio of equation (10) to (11) suggests that the correct measure is $(1+\alpha)$ times the approximation. The approximation in equation (6) would therefore only be correct if the quantity measure is assumed to be zero with the pre-policy water quality conditions.

Of course, this example is just one possible way water quality could change the demand for recreation. Nonetheless it illustrates why the details that are often not made explicit in deriving the components of equation (1) from existing estimates are important for assessing the reliability of results from simple benefit transfers. Adjusting unit values to reflect different demographic characteristics (as might be done with a typical benefit functions transfer approach) would not address this issue because assumptions regarding how environmental quality relates to the observable quantity measure (in our example days or trips using the river for different types of recreation) would still not be made explicit.[7]

Thus, this graphical example illustrates two important problems with the current practices of benefit transfer. The first arises because the procedures are not consistently linked to the concept we wish to measure. The Harberger formulations for approximating the consumer surplus attributed to price changes are developed within a consistent economic framework. In contrast, conventional benefit transfer practices are not derived from a framework that links the quality change to be valued to the quantity of use associated with the quality change and to the economic values that people would place on the change. Morey's [1994] critique of using consumer surplus per trip in recreation applications is an example of this larger problem.

The second problem is much less apparent from most of the applications of benefit transfer techniques. Nothing in the methods assures the measures of WTP will be consistently related to household income. In other words, the transfers do not necessarily incorporate the restrictions implied by "ability to pay." For the most part, virtual prices (marginal WTP) are treated as constants, regardless of the scale of the changes being evaluated. With small, localized changes, the income effects may not be large. In other cases, such as Costanza et al.'s [1997] effort to measure the annual value of the earth's ecosystems and EPA's [1997]] retrospective analysis of the net benefits of the Clean Air Act, the large scope of the changes evaluated raises serious questions about the economic consistency of the results.[8]

Harberger's approach anticipated the potential problems associated with evaluating such large-scale changes. His approximation for price changes (equation (3)) can also be interpreted (when more than one good's price is changing) as building in an assumption that total expenditures on the commodities affected by the price changes do not themselves change. As Hines [1999] has suggested, Harberger's alternative to ordinary and compensated demands was an effort to form a simple general equilibrium demand function that recognized the importance of the income effects for large policy-induced changes in prices. He sought to evaluate policies after accounting for the income effects of the transfers that can arise from policy. In his applications the issue was with the disposition of the tax revenues. For our applications, where improvements in previously unpriced goods are to be paid for, the central issue underlying this type of closure condition is where will the money come from?[9]

4. A PREFERENCE CALIBRATION PROPOSAL

Preference calibration refers to a specific logic that specifies a preference function (usually an indirect utility function) and then defines each of the available benefit measures in terms of this function. The challenge of calibration is to determine whether the available information is sufficient to identify the parameters of the preference function and derive numerically calibrated values for them. The most direct practical insight from the approach is a requirement that each source of benefit estimates and each desired decomposition of these estimates should, in principle, link to a common specification for individual preferences. This type of overall framework describes how the environmental resources and their quality contribute to individual well-being. Moreover, it also summarizes how other changes in an individual's (or a household's) circumstances might change their economic valuation of the resource change.

In practice, this means that the analyst must first be willing to make explicit assumptions about the functional form of an individual's utility function (or indirect utility function). If V represents the maximum level of utility achievable, given the income, relative prices for marked goods (P), and Q level of environmental quality faced by the individual, κ represent parameters describing the "shape" of the function. WTP for a change in environmental quality can therefore be expressed as the reduction in income that would exactly offset the improvement in Q (i.e., $Q_1 > Q_0$) and leave utility unchanged.

$$V(m, P, Q_0; \kappa) = V(m - WTP, Q_1; \kappa) \quad (12)$$

Assumptions about the functional form of utility allow the analyst to express WTP as a function of the change in environmental quality, income, prices, and κ as in equation (13).

$$WTP = f(Q_1, Q_0, m, P; \kappa) \quad (13)$$

This function can be treated as a benefit transfer function. However, a key feature distinguishing it from statistical benefit functions is that it is derived from, and thus consistent with, the specification of preferences.

The second element of this approach is that it uses these existing benefit measures to estimate the parameters in κ. That is, the measured values for WTP and/or

Marshallian consumer surplus are used to "calibrate" the specified preference structure. The WTP function can then be derived from the preference function.

The process described above summarizes the basic logic of preference calibration as a strategy for benefit transfer. One way to offer more detail on the approach is through an illustration. Thus, we demonstrate the process for a specific example and contrast it with more traditional benefit transfer practices.

5. AN APPLICATION TO WATER QUALITY IN THE CHESAPEAKE BAY

To illustrate the logic we selected a contingent valuation study and a recreation demand analysis for water quality changes in the Chesapeake Bay area. The Chesapeake Bay application offers a situation with a well-defined water resource where large-scale changes in water quality are of policy interest and different nonmarket valuation methods have been applied previously to evaluate water quality-related benefits. The policy relevance of the Bay arises from the fact that its water quality has improved significantly over the last three decades because of Clean Water Act (CWA) initiatives. These changes have prompted EPA and federal and state agencies to consider evaluating the achievements of water pollution control policies in the region by comparing their benefits and costs (Owens and Morgan [2000]). Unfortunately, as described below, the regulatory processes giving rise to the cleanup did *not* include a systematic effort to collect the economic data over this time that would allow the evaluation to take place using observed behavioral responses to the improved conditions. Thus, because of the data limitations described below, this application of the benefits calibration approach to the Chesapeake Bay should be viewed primarily as an illustration of our methodology rather than as a rigorous assessment of CWA policies in the Bay.

To calibrate a preference function using the benefit estimates available in the literature, we propose six steps:

1. Specify a "representative" individual's preference function.
2. Specify information required to ensure that the parameters of this preference function can be identified from the estimates.
3. Define explicitly the relationships between the available benefits measures and the specified preference function.
4. Adapt the available information to assure cross-study compatibility.
5. Calibrate the preference function using these benefit measures.
6. Estimate benefits using calibrated WTP function.

5.1. Specify a "Representative" Preference Function

The first step in preference calibration requires selecting a function to describe the representative individual's preferences. This decision is itself a tradeoff. Complex functions may well capture a wider range of behavioral responses, but they will also be difficult to calibrate with limited data. Simple expressions may be so restrictive that they are inconsistent with findings in the existing literature. As in the case of the functions used in computable general equilibrium models, the forms that will ultimately gain acceptance (if our proposed strategy for transfer is adopted) will no doubt balance these types of considerations. For this illustration we selected a simple form with fairly transparent implications.

To introduce quality in a simple way that is linked to the use of a resource, we follow Willig [1978] and Hanemann [1984] and adopt a preference specification that is consistent with what Willig labels "cross-product repackaging." This implies that the indirect utility function is structured so that the water quality measure reduces the effective "price" of using the recreation site, as in equation (14).

$$(14) \qquad V = [(P - h(d))^{-\alpha} m]^{b}$$

P is the round-trip travel costs, d is the water quality measure, m is household income, and α and b are parameters. h(d) is a function that describes how increases in water quality reduce the effective price of a trip.

5.2. Specify Information to Identify Preference Parameter

Consider the task of estimating the recreational fishing benefits from water quality improvements with two sources of benefit information for the Chesapeake. One source uses a contingent valuation (CV) estimate that includes all possible uses in the Bay, and the second focuses on recreational fishing. The specific CV study we selected to illustrate this example is Lindsey et al. 1989 survey (reported in Lindsey, Paterson, and Luger [1995]). This survey sought to estimate people's WTP to undertake storm water control programs to help achieve Chesapeake Bay nutrient-reduction objectives.

The second study by Bockstael et al. [1989] relies on a travel cost model, which has two components. One model links water quality, measured using nitrogen and phosphorous loadings, to striped bass catch levels in the Chesapeake Bay. These catch models were then linked to Maryland fisher's demand for fishing trips during 1980. The authors report the average consumer surplus for a 20% improvement in nitrogen and phosphorous loadings in the Chesapeake Bay.

We selected the Lindsey et al. estimate for a fourpercent improvement from fishable conditions (i.e., conditions suitable to support game fish). The estimates from the Bockstael study can be adapted to consider an approximately equal change in water quality. Therefore, the two studies can be interpreted as providing comparable water quality improvements in ways that are relevant to users. Bockstael et al. [1989] measure the Marshallian consumer surplus based on fishing trips, and Lindsey et al. estimates the total Hicksian WTP for the water quality improvement.

5.3. Define the Relationships between the Available Benefit Measures and the Specified Preference Function

To calibrate the preferences with the empirical record in the literature, each benefit measure must be related to this common preference structure. Using Roy's identity, the demand for trips, X_1, can be expressed as equation (15).[10]

$$(15) \qquad X_1 = -\frac{V_p}{V_m} = \frac{\alpha m}{(P - h(d))}$$

The Marshallian consumer surplus, MCS, associated with access to the recreation sites providing these fishing opportunities at travel costs corresponding to P_0 can be found from the area under this demand between P_0 and the choke price (labeled here as P_C). This is given in equation (16).

$$\text{(16)} \qquad MCS = \alpha m \int_{P_0}^{P_c} \frac{1}{(P - h(d))} dP = [\alpha m \ln(P - h(d))]_{P_0}^{P_c}$$

Evaluating the integral yields equation (17).

$$\text{(17)} \qquad MCS = \alpha m \, [\ln(P_C - h(d)) - \ln(P_0 - h(d))]$$

The Bockstael et al. [1989] analysis implicitly evaluates how MCS changes with d. This relationship is described, for this preference specification, with equation (18).

$$\text{(18)} \qquad \frac{\partial MCS}{\partial d} = \alpha m \left[-\frac{h'(d)}{(P_C - h(d))} + \frac{h'(d)}{(P_0 - h(d))} \right]$$

where $h'(d) = dh/dd$.

Simplifying terms, the first term is the product of demand for fishing trips at the choke price (P_C) multiplied by $(-h'(d))$ and the second is the demand at P_0 multiplied by $h'(d)$. The definition of the choke price (even if it cannot be expressed in closed form) implies the first of the terms on the right side of equation (15) should be zero. The second term offers one approach to linking Bockstael et al.'s [1989] measures to our preference specification, but first we must specify h(d). For our example, assume h(d) follows a logarithmic function with a declining marginal effect of d on the price: $h(d) = \beta \ln(d)$, where β is assumed to be constant.[11] More specifically, the increase in Marshallian consumer surplus per fishing trip is exactly $\beta/d(= h'(d))$, as in equation (19).

$$\text{(19)} \qquad \frac{\dfrac{\partial MCS}{\partial d}}{\dfrac{\alpha m}{(P_0 - h(d))}} = \frac{\dfrac{\partial MCS}{\partial d}}{X_1} = h'(d) = \frac{\beta}{d}$$

Bockstael et al.'s [1989] consumer surplus estimates of quality improvements, scaled by their estimated number of trips and the ratio of water quality changes in the two studies, offers an estimate of the left side of equation (16). This is the effect of a quality adjustment on incremental consumer surplus per trip, $h'(d)$. We interpret it as the Marshallian surplus estimate for the water quality change as described by Bockstael et al. (i.e., improving water quality by four percent in the Chesapeake Bay). Thus, their estimate allows us to develop a calibrated estimate for β.

The Lindsey et al. study uses his CV question to provide an estimate of the WTP for a four percent improvement from fishable water quality (d_F). Specifying this question using the indirect utility function we have WTP defined by the indifference condition in equation (20a).

$$(20a) \quad [(P - h(1.04d_F))^{\alpha}(m - WTP)]^{b} = [(P - h(d_F))^{\alpha} m]^{b}$$

Simplifying the equations and solving for WTP we have equation (20b). This relation is the WTP for a four percent change in water quality from a baseline of fishable conditions that is implied by our preference function.

$$(20b) \quad WTP = m - \left(\frac{P - h(1.04^{*}d_F)}{P - h(d_F)}\right)^{\alpha} m$$

Values from the Lindsey et al. study for WTP can be combined with values for m, P, and $h(d_F)$ to calibrate the remaining parameter, α, as described below.

5.4. Adapt the Available Information to Assure Cross-Study Compatibility

With our benefit measures defined within a consistent economic framework, the next task is to convert the data relevant to each study's estimates in compatible units. We begin by adjusting the Bockstael et al.'s [1989] per trip consumer surplus estimates of the gain due to water quality improvements. The estimates of travel cost reflect 1980 dollars and a time cost corresponding to income levels that are substantially lower than what was observed with the Lindsey et al. study. Therefore, the travel cost should be adjusted to reflect a higher opportunity cost of time.[12] Second, the adjusted measures of travel cost and per-trip-consumer surplus ($2) need to be placed in comparable dollars with Lindsey et al., reflecting the effect of the overall price level. This is consistent with an implicit restriction in the preference function—homogeneity of degree zero in *all* prices and income. The consumer price index (CPI) is used to adjust monetary measures of prices, incomes, and consumer surplus from 1980 to 1998 dollars. We make a similar adjustment to the WTP amount of $42 in Lindsey et al., converting 1989 to 1998 dollars.

Finally, water quality, d, is characterized so that it is consistent between the two studies. The RFF water quality ladder and index (Vaughn [1981]) is used to establish this correspondence in the water quality measures.[13] The Bockstael et al. study measures water quality in terms of nitrogen and phosphorous. Recognizing that the Bay waters must be of fishable condition for the Maryland recreationists to be fishing for bass, we use the water quality ladder to assign this condition a water quality index value of 5.1. Similarly, we represent the baseline water quality in the Lindsey et al. study—again of fishable conditions—using a water quality index value of 5.1. Thus, the four4 percent improvement in fishable water quality described in Lindsey et al. study corresponds to improving the index value from 5.1 to 5.3, and the 20% improvement described in the Bockstael et al. study corresponds to improving the index from 5.1 to 6.1.

5.5. Calibrate the Preference Function Using these Benefit Measures

Equations(19) and (20b) allow us to calibrate the two unknown parameters—β, and α. Here we describe how these results can be reproduced. Specifically, from equation (19), we calibrate β to be

$$\hat{\beta} = \left(\frac{\frac{\partial MCS}{\partial d}}{X_1} \right) \bullet d. \tag{21}$$

We can take this calibrated value of β and use it in equation (20b), to calibrate α as

$$\hat{\alpha} = \frac{\ln\left(\frac{m - W\hat{T}P}{m}\right)}{\ln\left(\frac{P - \hat{\beta}(1.04 \bullet d_F)}{P - \hat{\beta}(d_F)}\right)}. \tag{22}$$

Using the estimates for travel cost, income, consumer surplus change, and baseline water quality from the Lindsey et al. and Bockstael et al. studies, we calculated the calibrated parameters as $\alpha = 0.29$ and $\beta = 10.33$.

5.6. Estimate Policy Benefits in the Chesapeake Bay Using the Calibrated WTP Function

To illustrate how the calibrated preference function can be used, consider the task of measuring the per household benefits of water quality improvements in the Chesapeake Bay as the result of CWA policies. An assessment of water quality models for the Chesapeake Bay indicates that the CWA improved bay-wide water quality by 60% (Owens and Morgan [2000]). Using two different reference points to characterize conditions without the CWA, the calibrated values from equations (21) and (22) are used in equation (20b) to estimate the per household benefits for

- a 60% improvement in water quality that *results* in fishable water quality (i.e., a change in the water quality index from 3.2 to 5.1), and
- a 60% improvement in water quality that *starts* from fishable water quality (i.e., a change in the water quality index from 5 to 8.2).

As reported in column 5 of Table 2, the calibrated per-household WTP estimates for the two improvements are $627 and $696 respectively.

For the sake of comparison, Table 2 also presents per household values that would be used in a simple, conventional benefit transfer: a travel cost-based value and a CV-based value. These values were estimated in two stages. First, per-unit (of the water quality index) values were estimated for each case (each expressed in 1998 dollars). Travel cost estimates were calculated by taking the adjusted per-trip consumer surplus estimate from Bockstael et al., multiplying it by the average

Table 2. Individual Benefit Estimates for a 60 percent Improvement in Chesapeake Bay Water Quality Using Simple Approximations and the Proposed Preference Calibration Approach[a] (1998 dollars)

Water Quality[b]		Approximation Using Travel Cost Study[c]	Approximation Using Contingent Valuation Study[d]	Calibrated WTP
Baseline	Final			
3.2	5.1	\$213	\$527	\$627
5.1	8.2	\$339	\$840	\$696

[a] Calibrated parameters are β=10.33, and α=.29. The travel cost information relies on Bockstael et al. [1989]. The CV estimate applies an adjustment of the Lindsey et al. estimate (reported in Lindsey, Paterson, and Luger [1995]). The consumer price index was used to convert into 1998 dollars.

[b] The numbers correspond to the RFF water quality ladder and index (boatable=2.5, fishable=5.1, and swimmable=7).

[c] The consumer surplus approximation is calculated by multiplying the per-trip per unit of dissolved oxygen consumer surplus estimate (from Bockstael et al. [1989]) by the size of the proposed policy change in water quality and by the average number of trips.

[d] The CV approximation is calculated by dividing the estimated WTP by the change in water quality (from Lindsey, Paterson, and Luger [1995]) for the original estimate and then multiplying by the size of the proposed policy change.

number of trips, and dividing it by the water quality increment evaluated in the study. The result is \$111 per unit of water quality improvement. The CV values were calculated by dividing the Lindsey et al. WTP estimate by the water quality increment evaluated in his study. The result in this case is \$275 per unit. In the second stage, each of these per unit values was multiplied by the 60% water quality improvements (1.9 and 3.1 index units, respectively). The per person value estimates from the travel cost and CV-based approaches are reported in Table 2 in columns 3 and 4 respectively.

The travel cost-based values can be interpreted as incomplete measures of the per household benefits of the water quality improvements (nonuse value, for example, are not captured in these values). The CV-based estimates would be expected to be more comprehensive. Thus, it should not be surprising that the CV-based approximations are closer than the travel cost-based approximations to the calibrated WTP estimates.[14] For the water quality change starting below the fishable level, the calibrated WTP value exceeds the CV-based value. The opposite is true above the fishable level. A further point that is not shown explicitly by the results in Table 2 should be noted. The differences between the calibrated and CV-based values increase in absolute value as one evaluates water quality changes further away from our point of calibration (in the 5.1 to 5.3 range). This distinction arises from the fact that the WTP function can be approximated with a linear function around the calibration point. The marginal welfare gains using the calibrated approach are larger at lower water quality levels (below fishable) and smaller at higher water quality level (above fishable). This pattern also is consistent with the logic of declining marginal WTP for improved water quality, but it would not be reflected in the linear estimates derived using the simpler benefit transfer approaches.

6. DISCUSSION

Conventional benefit transfer practices at best are approximations that have been developed to measure the consumer surplus associated with price changes. They have been used in ways that do not ensure they will be consistent with the economic concepts underlying the definition of WTP for quantity or quality changes. The larger the change being evaluated, the greater the likelihood of serious biases. Moreover, it is possible to find inconsistencies in smaller-scale, simple transfers. We illustrate this possibility with our graphical example in Section 3. Harberger approximations to control for income effects do not apply when the goods involved are not priced. To meet these shortcomings, we have proposed treating benefit transfer as a generalized estimation task—where the available information is linked to a preference function. Benefit estimates can be derived when the maintained assumptions and available empirical information together are sufficient to identify the preference parameter. This approach implies that consistent transfers require sufficient information to recover the parameters of a WTP function. This strategy was illustrated with an example application for water quality improvements in the Chesapeake Bay.

6.1. What Are the Main Advantages of this Proposed Alternative to Conventional Benefit Transfer?

The proposed approach offers a more systematic way to construct benefit measures under the time and resource constraints typically facing policy analysts. The primary argument for preference calibration is that it imposes economic consistency conditions on the ways the existing information is used. Experience and experiments (with computable general equilibrium models), comparable to what has taken place for nearly 50 years with Harberger approximations for the deadweight losses, provide the only basis for judging whether this strategy will be uniformly better. This conclusion does not imply we need to wait for 50 years of experience to consider revising current practices. While we wait for such numerical experiments, we can use preference calibration to judge whether imposing these types of consistency conditions would change conclusions.

The Chesapeake example highlights three underappreciated aspects of benefit transfers. First, the task associated with developing benefit estimates to evaluate a new policy should be interpreted as a type of *identification problem*. That is, when transferring benefits we must judge whether there is sufficient information to develop a *theoretically consistent* measure of the benefits for the changes being considered. This process makes explicit the roles of analyst judgment in developing the links between what has been measured and what is needed for each policy task. The analyst must also specify the structure of preferences in such a way that the critical preference parameters can be inferred from existing data and studies. There is also an important strategic element to selecting the functional form for preferences so that it is tractable. Second, benefit estimates assembled from studies that used different methods will often require that the same aspect of environmental quality be represented with different technical measures. Consistent use of these benefit estimates requires indexes of environmental quality that can be made compatible.

Differences in how this is accomplished may well be as important to discrepancies in transferred estimates as any distinctions in economic assumptions underlying those estimates.[15] Finally, the observed economic tradeoffs that people make to obtain increases in nonmarket resources are constrained by their available incomes. None of the existing approaches to benefit transfer meet this simple ability to pay test. That is, when transferring benefits, we must ensure that measured WTP values are affordable (i.e., well within people's disposable income).

6.2. *How Can We Evaluate Benefit Transfers?*

Most efforts to evaluate transfer methods have compared "direct estimates" of the benefits provided by some improvement in environmental quality in one location to a "transferred value." The latter is simply a different estimate. Random error alone would imply discrepancies. While sampling studies offer the prospect to control the standard used in evaluation, the assumptions required for describing preferences, true parameter values, characteristics of available data, etc., seem to offer so many combinations of alternatives that this also seems unlikely to offer many practical insights for evaluating transfer practices.

Because benefit measures are never observed, their estimates are unlikely to be evaluated in a context that will be fully satisfactory. That is, there is no "true benefit estimate" that could be found to serve as a measuring stick for the transferred estimates.

Preference calibration offers some advantages for evaluating benefit transfer. With a numerical characterization of the preference function, it is possible to consider estimating other observable "quantities" at the same time as the benefits are measured. For example, we could predict the number of recreation trips per household, the expenditure shares, and price and income elasticities. Such "indexes" may be easier to use in gauging the plausibility of a benefit function than a consumer surplus measure for an unobserved quality change. These types of estimates are not available with other transfer methods because they are not consistently linked to preferences. Large discrepancies between the predictions for the linked private good or elasticities that are judged to be implausible signal a need to evaluate the assumptions being used.[16]

6.3. *What Next?*

Clearly, what has been proposed here was done in the context of simple specifications to illustrate the logic of a different strategy for conducting benefit transfers. More complex functional forms are possible and numerical calibration analogous to what is used with numerical computable general equilibrium models is also possible. However, the desirability of pursuing such larger-scale efforts depends on the success of experimentation with smaller applications of the method and comparisons with current practice. It would be relatively easy to consider an exercise where recent benefit transfers were "redone" using the calibrated preference logic and compared with the approach used in the policy analysis. This would seem to offer a next step in evaluating the usefulness of the logic and could easily precede more extensive efforts at numerical calibration.

The possibilities for using the calibration logic do not stop with alternative spreadsheet computations under a wide array of judgments about which combination of point estimates to use to identify the preference parameters. Preference calibration also offers a strategy for using the existing literature as data to "estimate" the preference function used for benefit transfer. Instead of using reduced-form response functions in meta summaries, the logic of preference calibration implies that sufficient information can exist from the available literature to *identify* the parameters of preferences. When there are multiple sets of information, it is possible to treat the theoretical conditions to identify the parameters (e.g., marginal WTP from hedonic studies) as a system of equations that define moment conditions. That is, preference parameters could be estimated using the multiple sets of benefit measures as "data" in the form of moment conditions.

To summarize, mandates in the U.S., European Union, and in Development Agencies are increasingly calling for benefit-cost analyses to evaluate the performance of regulatory programs. In the absence of the time and resources needed for full studies of each new policy or a "non-market" general equilibrium model, analysts are likely to rely extensively on current transfer practices and off-the-shelf estimates. We have suggested that these practices pose real concerns. If responses to the mandates for economic information about policy tradeoffs are to avoid discrediting the practice of benefit cost analyses, they must recognize the need for imposing internal consistency measures of the gains (and losses) attributed to interrelated (from the consumer's perspective) but independently administered regulatory policies. Calibration offers a first step for avoiding inconsistent benefit estimates and for more completely accounting for the effects of large-scale policies.

Subhrendu Pattanayak and George Van Houtven are Senior Economists, RTI International, Research Triangle Park, NC, USA.. V. Kerry Smith is University Distinguished Professor, Department of Agricultural and Resource Economics, North Carolina State University, Raleigh, NC, USA and University Fellow, Resources for the Future, Washginton, DC, USA..

7. NOTES

[1] The basic approach for evaluation is to conduct benefit analyses at two or more sites and compare the transferred benefits (based on a function derived for Site 1) with actual benefits at the other site. See Table 1 for a list of recent studies in this category.

[2] These benefit functions usually come from one of two sources. The first is from CV studies. In these cases, the analyst uses a statistical summary of the design variations included in the primary CV study. The second type involves meta analysis summaries. In this case, the models are based on data that generally use summary statistics from the original sources and include characteristics of the resources, individuals (whose WTP is being estimated), and methods used. The three recent evaluations of benefit transfer have relied on the first type of benefit function and have considered benefit measures based on CV studies only.

[3] As we illustrate in the next section, this simplification can be misleading. The unit values derived from some approaches to nonmarket valuation are actually transformations of the consumer surplus attributed to price changes. In other situations, the model implicitly forms a specific "price equivalent" of a quality change so that the consumer surplus attributed to the quality change would have a price

change equivalent. For example, in the case of reductions in mortality risks, it is the *ex ante* marginal rate of substitution.

[4] The benefit measure in the health context can be different from a consumer surplus gain.

[5] The assumption implies Z and s are in a separable subfunction from the other arguments in the expenditure function.

[6] By assuming the demand is known, our example ignores this source of error and focuses instead on the error introduced by what the analyst does in constructing a transferred benefit. This error arises from using the approximation for a price change and treating consumer surplus per unit as the equivalent of a price.

[7] We could include environmental quality in the model and explicitly adjust for the responsiveness of demand to site quality.

[8] For Costanza et al. the estimated global annual WTP for these ecosystem services exceeded the global gross domestic product. For the EPA report, the per capita benefits of air quality improvements relative to average household income strain credibility.

[9] One could use a logic comparable to Harberger, requiring that the share of a fixed income attributed to the amenity was fixed before and after a changed (i.e., $\rho_z^0 Z_0 = \rho_z^1 Z_1$). $\rho_z^0 Z_0$ does not correspond to the expenditures on Z_0. It is an arbitrary concept that uses virtual prices to mimic what a consumer's expenditures would be under two different conditions. Using them, together with the conventional measure of incremental WTP for a change from Z_0 to Z_1, the Harberger restriction would imply that this measure should be approximated as $1/2(Z_1 + Z_2) \bullet (\rho_z^0 - \rho_z^1)$ with $\rho_z^0 \rangle \rho_z^1$.

[10] Note that in the simplified case without the quality term $V = [P^{-\alpha}{}_m]^b$, $X_1 = \alpha\left(\frac{m}{p}\right)$, and $\alpha = \frac{PX_1}{m}$, the share spent on X_1. With cross-product repackaging, $\alpha = (P - h(d)) \bullet X_1/m$. For small values of h(d), the two measures will be close to each other.

[11] The functional form for h(d) by itself is not critical, even though it represents another judgment by the analyst. We did consider an alternative form such as a power function with a declining marginal effect of d on the price and found different quantitative and qualitative solutions using $h(d) = d^{\square}$.

[12] The adjustment to travel cost attempts to take account of differences in the opportunity cost of time as a result of differences in income between the Bockstael et al. [1989] and the Lindsey et al. studies. This is accomplished by scaling travel cost by an adjustment for CPI differences and the relative income from the two studies.

[13] We do not claim that this water quality conversion is entirely accurate or ideal. It is used in our example as a simple and convenient approximation. It also highlights the general need for defensible conversion procedures and for consistent measures of water quality, which are essential to linking results from different benefit studies.

[14] By design of the calibration approach, the calibrated WTP estimate and the CV-based WTP estimate are the same for the water quality increment evaluated in the CV study. In other words, both approaches generate a WTP estimat of $56 for the water quality index increment examined in the Lindsey et al. study (from 5.0 to 5.2).

[15] This conclusion is supported by the recent Desvousges et al. [1998] meta analysis for environmental costing.

[16] The logic resembles the use of calibration in marketing research where the results of stated preference or conjoint surveys are calibrated based on a variety of other types of information before they are then considered relevant for a market analysis task.

8. REFERENCES

Barton D (1999a) "Quick and Dirty: Transferring the Benefits of Avoided Health Effects from Water Pollution between Developed and Developing Countries." Discussion Paper #D9/1999. Department of Economics and Social Sciences, Agricultural University of Norway

Barton D (1999b) "The Transferability of Benefit Transfer: An Experiment in Varying the Context of Willingness-to-Pay for Water Quality Improvements." Discussion Paper #D-10/1999. Department of Economics and Social Sciences, Agricultural University of Norway

Bockstael N, McConnell K, Strand I (1989) "Measuring the Benefits from Improvements in Water Quality: The Chesapeake Bay." Marine Resource Economics 6:1–18

Brouwer R, Spaninks F (1999) "The Validity of Environmental Benefits Transfer: Further Empirical Testing." Environmental and Resource Economics 14:95–117

Brouwer R (2000) "Environmental Value Transfer: State of the Art and Future Prospects." Ecological Economics 32:137–152

Brookshire, David S, Helen R. Neill (1992) "Benefit Transfers: Conceptual and Engineered Issues." Water Resources Research 28 (March):651–655

Costanza, Robert, Ralph d'Arge, Rudolf de Groot, Stephen Farber, Monica Grasso, Bruce Hannon, Karin Limburg, Shahid Naeem, Robert V. O'Neill, Jose Paruelo, Robert G. Raskin, Paul Sutton, Marjan van den Belt (1997) "The Value of the World's Ecosystem Services and Natural Capital." Nature 387 (May):253–260

Desvousges, William H, Johnson FR, Spencer Banzhaf H (1998) Environmental Policy Analysis with Limited Information: Principles and Applications of the Transfer Method. Cheltenham, UK: Edward Elgar

Downing, Mark, Teofilo Ozuna, Jr. (1996) "Testing the Reliability of the Benefit Function Transfer Approach." Journal of Environmental Economics and Management 30 (May):316–322

Hanemann, Michael W (1984) "Discrete/Continuous Models of Consumer Demand." Econometrica 52 (3):541–563

Harberger, Arnold C (1971) "Three Basic Postulates for Applied Welfare Economics." Journal of Economic Literature 9 (3:785–797

Hicks JR (1941) "The Rehabilitation of Consumer Surplus." Review of Economic Studies 8 (2):108–116

Hicks JR (1943) "The Four Consumer's Surpluses." Review of Economic Studies 11 (1):31–41

Hines Jr., James R (1999) "Three Sides of Harberger Triangles." Journal of Economic Perspectives 13(Spring):167–188

Kirchhoff S, Bonnie G. Colby, Jeffrey LaFrance (1997) "Evaluating the Performance of Benefit Transfer: An Empirical Inquiry." Journal of Environmental Economics and Management 33(May):75–93

Lindsey G, Paterson R, Luger M (1995) "Using Contingent Valuation in Environmental Planning." Journal of The American Planning Association 61 (2):252–262

Loomis, John, Brian Roach, Frank Ward, Richard Ready (1995) "Testing Transferability of Recreation Demand Models Across Regions: A Study of Corps of Engineers Reservoirs." Water Resources Research 31 (March):721–730

Morey, Edward (1994) "What is Consumer's Surplus per Day of Use, When is it a Constant Independent of the Number of Days of Use, and What Does it Tell Us About Consumer's Surplus?" Journal of Environmental Economics and Management 26 (May):257–270

Office of Management and Budget (1998) Draft Report to Congress on the Costs and Benefits of Federal Regulations. Federal Register 63 (158:44034–44099

Owens N, Morgan C (2000) Benefits and Cost of Water Quality Policies: The Chesapeake Bay. Washington, DC: U.S. Environmental Protection Agency, National Center for Environmental Economics

Rozan A (2000) "Benefit Transfer : A Comparison of WTP for Air Quality between France and Germany." University Louis Pasteur, Strasbourg

Santos J (1999) "Transferring Landscape Values: How and How Accurately." Technical University of Lisbon, Portugal

Scarpa R, Hutchinson G, Chilton S, Buongiorno J (2000) "Reliability of Benefit Value Transfers from Contingent Valuation Data with Forest-Specific Attributes." University of Newcastle upon Tyne, U.K.

U.S. Environmental Protection Agency (1997) The Benefits and Costs of the Clean Air Act 1970 to 1990. Office of Air and Radiation (October)

Vaughan, William J (1981) "The Water Quality Ladder." Appendix to R.C. Mitchell and R.T. Carson, A Contingent Valuation Estimate of National Freshwater Benefits. Prepared for the U.S. Environmental Protection Agency, Resources for the Future

Willig, Robert D (1978) "Incremental Consumer's Surplus and Hedonic Price Adjustment." Journal of Economic Theory 17 (2):227–253

D. BARTON

HOW MUCH IS ENOUGH? THE VALUE OF INFORMATION FROM BENEFIT TRANSFERS IN A POLICY CONTEXT

1. INTRODUCTION

With the growing body of non-market valuation studies, benefits transfer using single studies or meta-analyses have become increasingly popular. While a common justification for these approaches is the hope of being cheaper than conducting original on-site studies, while also providing *sufficiently* reliable welfare estimates, few have evaluated this cost-effectiveness argument. Extending the question of a recent paper, the answer to "how dirty?" needs to be divided by the answer to "how quick?" (van de Walle and Gunewardena 1998). Practical applications of welfare loss functions for information collection which can be used in the course of a benefit-cost analysis are needed for benefit transfer to become a familiar aspect of environmental policy evaluation. In this paper, a stepwise Bayesian updating approach to valuation is suggested where a practitioner evaluates whether to go ahead with a large survey based on such a welfare loss function, weighing study cost against the value of expected reduction in policy uncertainty.

This study uses an "intuitive" first-cut approach to Bayesian updating of the benefit estimate using a computationally simple approach which assumes normal distributions. An extension of this paper is the updating of the whole variance-covariance matrix, allowing for flexible distributional assumptions. The approach is evaluated based on the criteria of convergent validity and the expected net benefits of uncertainty reduction. A series of datasets are employed as they would realistically arise over time, including meta-analysis of a number of contingent valuation (CV) studies, benefit function transfer using census and pilot data, and a main CV survey at the policy site. The time-sequenced decision-making process on further information gathering is simulated, using actual study cost information, set in the policy context of a benefit-cost analysis of waste water treatment in Costa Rica. While none of these ideas are new on their own, various authors have called for studies that evaluate the transfer method itself in a context of information of differing quality and in an actual policy setting (Desvousges et al. 1998; Smith 1992).

2. PREVIOUS RESEARCH

Two strands of literature are of particular relevance to this chapter. One strand discusses the convergent validity of benefit transfer using a variety of regression models from a single study site or pooled estimates from meta-analyses of several studies. The other covers the use of Bayesian theory to update welfare and regression parameter estimates as new information on the policy site becomes available.

S. Navrud and R. Ready (eds.), Environmental Value Transfer: Issues and Methods, 261–282.

2.1. Benefit Transfer and Convergent Validity

Perhaps the largest group of studies in this field test convergent validity of transferring study site benefit estimates to new policy sites (Bergland et al. 1995; Brouwer and Spaninks 1997; Downing and Ozuna 1996; EC 1999; Kirchhoff et al. 1997; Loomis 1992; Parsons and Kealy 1994). By and large these studies reject the statistical transferability of benefit function coefficients between sites, and in most cases also the statistical equality of mean consumer welfare estimates, whether 'raw' or conditioned on study site coefficients and policy site characteristics. Studies dealing with convergent validity have found that the similarity of study populations and environmental good characteristics usually leads to lower transfer errors, but that this is by no means a sufficient condition for statistical transferability. Transfer errors range from only a few percent to several orders of magnitude. Some authors advocate the use of more complex benefit function transfers for improving convergence (Desvousges et al. 1998), while other find that transferring unconditional consumer surplus estimates performs as well, or better than, conditional approaches (Parsons and Kealy 1994). In anticipation of such results other authors emphasise evaluating the effect that transfer error has on policy decisions – 'importance tests' – rather than statistical significance tests in judging the applicability of benefit transfer (Smith 1992).

As valuation studies become more numerous, examples of meta-analyses on the contingent valuation, travel cost and hedonic pricing literature have appeared over the last decade (Brouwer et al. 1997; Desvousges et al. 1998; Kirchhoff 1998; Loomis and White 1996; Schwartz 1994; Smith and Huang 1995; Smith and Kaoru 1990; Sturtevant et al. 1995; Walsh et al. 1992). Transfer of a meta-benefit function or 'mother distribution' assumes that no single 'best' study is available, providing a different starting point for convergent validity tests. Common for both of these approaches in the literature is that they test convergence of particular transfers with original policy site estimates, but not the validity of the transfer method itself in a policy context (Desvousges et al. 1998). In a comparative study, Kirchhoff et al. (1998) find that single benefit function transfer outperforms meta-function transfer in most cases, but that single study transfers have mostly been 'experimental', designed to improve conditions of transferability under as similar site conditions as possible. In a rare case of policy application, Desvousges et al. (1998) use Monte Carlo simulation techniques to combine various econometric approaches to meta-analyses for the different health consequences of air pollution from a power utility, and then evaluate the policy consequences for power plant siting. The best single meta-benefit function transfers come within about 50% of the mean of this combined simulation method, although the authors believe this may be acceptable to decision-makers.

A review of all these transfer studies gives the feeling that the jury will be out for some time on the reliability of benefit transfer, perhaps awaiting a consensus on the reliability of non-market valuation methods in general in environmental policy-making. While we wait, an eclectic approach is called for. While most benefit transfer studies opt for one particular transfer method, a more realistic situation is that several approaches are used sequentially and complementarily as more policy-site information becomes available.

2.2. *Baysian Updating in Transfers*

Bayesian estimation provides one answer to modelling the use of information in decision-making processes. Various authors have proposed and applied Bayesian updating in modelling the response incentive or anchoring effects in contingent valuation surveys (Horowitz 1993; León and Vázquez Polo 1998; McLeod and Bergland 1995; Viscusi 1985). In this literature, respondents are modelled as Bayesian decision-makers updating their preferences using the additional cost information provided by follow-up valuation questions, such as in the double-bounded dichotomous choice (DBDC) or iterative bidding formats. An obvious question is whether the empirical findings on individual CV respondents as Bayesian decision-makers also apply to CV practitioners and their clients who use prior information on welfare estimates in benefit transfer.

A handful of studies have explored this question. Atkinson et al. (1992) use a meta-analysis of 15 hedonic pricing studies to obtain prior estimates of coefficients which are then updated using varying sample sizes at the policy site. The authors test the assumption of 'exchangeability' of coefficients, or what they term a 'grand model', across sites. They illustrate this Bayesian decision-problem through a loss-function where the benefit of decreased variance is weighed against the cost of additional observations at the policy site. The authors find that the gains in research efficiency from increasing sample size at the policy site generally dominate the gains from shrinking the variance of priors by updating using parameter estimates from meta-analysis.

Parsons and Kealy (1994) use Bayesian updating of predictions from a random utility model of recreation. Contrary to the recommendations to use more complex specifications in function transfer (Atherton and Ben-Akiva 1976; Desvousges et al. 1998), the authors find that simple unadjusted transfers sometimes predict site visitation as well as more complex behavioural models due to variable offsetting effects of the latter. Bayesian updating of study site models outperforms transfer of the study site model without updating, but only modestly so. Because their data comes from the same survey, the authors call for studies of transfers under more realistic conditions across data sets of varying sample size, from different regions, and using different variables and structure.

León et al. (1997) argue that the approaches described above require the availability of coefficients from study sites and matching information at the policy site which may not be available. They propose a Bayesian updating scheme using the more commonly available mean welfare estimates which are modelled using a 'very informative prior' beta-distribution. Using DBDC willingness-to-pay data from identical surveys at three national parks they also tested the effects of a low information 'maximum entropy' prior. Results show that the posterior mean is robust (consistent) to the choice of prior, *given* that the transfer is between parks that have very similar attributes. In these cases updating using a beta-distribution[1] reduces the variance of the posterior mean and can be a way of reducing sampling cost at the policy site. However, with reductions in sample size at the policy site the accuracy of the posterior mean becomes increasingly sensitive to assumptions about the prior distribution (León et al. 1997). The study demonstrates how sample size costs can be reduced to 10% using a highly informative beta-distribution as a prior, with small losses in efficiency.

In this study, we follow up several of the recommendations from Atkinson et al. (1992), Parsons and Kealy (1994) and León et al. (1997). We use a Bayesian estimator to update various priors of WTP, evaluating the different information sources' relative precision and bias, and their effects on policy decisions in a developing country context.

3. THEORY AND METHODS

3.1. Bayesian Updating

First we review the theory behind Bayesian updating in a contingent valuation setting and touch on some of the key problems behind using datasets arising from different random processes.[2] Our starting point is that we wish to estimate true individual WTP at the policy site *i*, which for the simple Bayesian updating we will assume is normally identically, and independently distributed at each study site:

$$\theta_i \sim N(\beta_\pi X_i, \sigma_\pi^2) \qquad (1)$$

where β_π and σ_π are unknown parameters and X_i is an observable data matrix. With a CV survey we can only hope to observe stated WTP (w_i) which includes measurement error (σ_m^2):

$$w_i \sim N(\theta_i, \sigma_m^2) \qquad (2)$$

Note that θ_i has probability density function conditional on site characteristics, $\Pi(\theta_i|X_i)$, while observed WTP is conditional on true WTP with the likelihood, $f(w_i|\theta_i)$. We can define the joint density of (w_i, θ_i) as:

$$h(w_i, \theta_i) = f(w_i|\theta_i)\Pi(\theta_i|X_i) \qquad (3)$$

The marginal density of w_i is then:

$$m(w_i) = \int h(w_i, \theta_i)d\theta \qquad (4)$$

In 'classical' CVM we obtain a welfare estimate where observed classical variance, σ_c^2, is the sum of true variation and measurement error:

$$w_i \sim N(\theta_i, \sigma_\pi^2 + \sigma_m^2) = N(\beta_\pi X_i, \sigma_c^2) \qquad (5)$$

However, if we have any prior knowledge about the true WTP distribution $\Pi(\theta_i|X_i)$, we can formulate our problem as one of updating this knowledge with the likelihood $f(w_i|\theta_i)$ of new observations. According to Bayes theorem the posterior distribution of WTP can be defined as:

$$(6) \qquad \Pi(\theta_i|X_i) = \frac{h(w_i, \theta_i)}{m(w_i)} = \frac{f(w_i|\theta_i)\Pi(\theta_i|X_i)}{\int h(w_i|\theta_i)d\theta}$$

Given the assumptions about the prior, the Bayesian posterior WTP estimate is also normally distributed $\mu_i \sim N(\mu w_i), \sigma_\mu{}^2)$ (Gourieroux and Monfort 1995), with a generalised maximum likelihood estimator consisting of the sum of prior and new information[3] on WTP weighted by their respective precisions (Desvousges et al. 1998):

$$(7) \qquad \mu(w_i) = w_i - \frac{\sigma_m^2}{\sigma_m^2 + \sigma_\pi^2}(w_i - \theta_i) = \frac{\frac{w_i}{\sigma_m^2} + \frac{\theta_i}{\sigma_\pi^2}}{\frac{1}{\sigma_m^2} + \frac{1}{\sigma_\pi^2}}$$

$$(8) \qquad \sigma_\mu{}^2 = \frac{1}{\frac{1}{\sigma_m^2} + \frac{1}{\sigma_\pi^2}}$$

In classical CVM the hyper-parameters $\sigma_\pi{}^2$, $\sigma_m{}^2$, and β_π are not known *a priori* (before a pilot survey is conducted), in which case the posterior parameters describing WTP depend entirely on the sample information from the main survey: $\mu(w_i) = w_i$ and $\sigma_\mu{}^2 = \sigma_c{}^2$. The hyper-parameters can also be interpreted as a second stage prior describing a mother-distribution of WTP:

$$(9) \qquad \mu_i \sim N(\theta_i|\sigma_m{}^2, \sigma_\pi{}^2, \beta_\pi)$$

$$(10) \qquad \text{where} \qquad \beta_\pi \sim N(\lambda_\beta, \tau_\beta{}^2)$$

$$(11) \qquad \text{and} \qquad \sigma_\pi{}^2 \sim N(\lambda_\sigma, \tau_\sigma{}^2)$$

constitute an 'informative prior' recovered from pooled datasets (León et al. 1997; Parsons et al. 1994), or a meta-analysis of point estimates from several studies (Atkinson et al. 1992).[4] Alternately, with a 'low information prior' we may assume a maximum entropy prior using only the range within which we believe welfare estimates lie (León et al. 1997).

Note that for simple Bayesian updating our posterior parameters μ and σ_μ in equations 7 and 8 become the subjective priors in the next updating step, replacing $(\theta_{i,} \sigma_\pi)$. The parameters (w_i, σ_m) then represent new information on WTP gleaned from primary survey data. If the assumptions of distribution normality, independence and identity do not hold, updating priors with new observations on mean and standard error using equations (7–8) will lead to biased posterior confidence intervals of WTP. If the random process that generates new observations (w_i), is *different* from the prior we have to solve for a joint marginal density[5] and double integrals to find the Bayesian posterior mean:

$$\text{(12)} \qquad \text{if } w_i = \xi(\theta_i), \text{ then } \mu(w_i) = E(w_i) = \iint \xi(\theta_i)\pi(\theta_i|\lambda)d\theta\, d\lambda$$

Combining different distributions thus greatly complicates the analysis with no closed form or analytical solution to the integrals.[6] Using equations (7–8) provides what is called the "simple" approach to Bayesian updating and is a first-cut approximation. A future research problem is how biased the confidence intervals of such a 'quick and dirty' approach are compared to the more flexible approach in equation (12), and under what conditions it would affect the policy decision differently.

3.2. Decision-Making Criteria

When non-market valuation methods inform environmental policy over time, a pluralistic approach is required which integrates previous and new expectations. Figure 1 compares the 'classical' contingent valuation approach to several starting points for "Bayesian" priors which are updated as new data becomes available. The philosophy of updating is that benefit transfer is a complement to primary studies, rather than a substitute. Ideally, at each step a decision would be made whether to continue data collection based on a welfare loss function. Atkinson et al. (1992) one can define this as the trade-off:

$$\text{(13)} \qquad \min_n W = p\sigma_\mu{}^2 + gC(n)$$

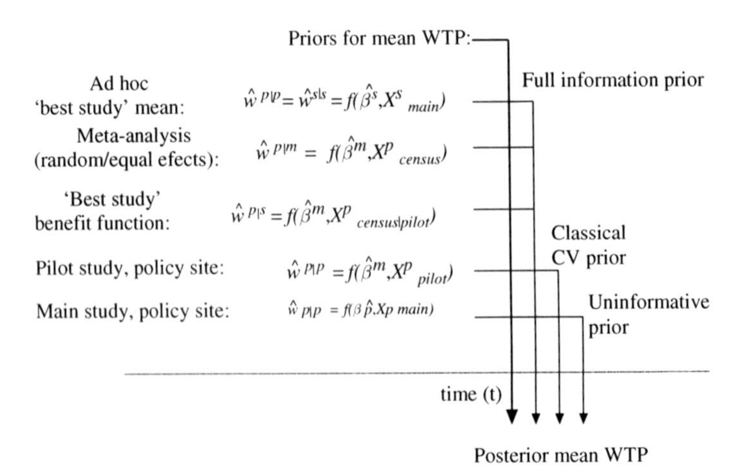

Figure 1. Strategies for Bayesian updating of contingent valuation estimates

where

σ_μ^2 = variance of the current subjective prior of WTP

p = implicit pecuniary weight (price) of an increase in the variance of net benefits

g = implicit pecuniary weight (price) of an increase in information effort

C(n) = total information effort spent on the nth additional benefit transfer or valuation study, including the opportunity costs of delaying decision on project implementation

Atkinson and colleagues show that the decision to use the current prior, $[\mu, \sigma_\mu^2]$, or to update it with any additional valuation information on $w_i \sim N(\theta_i, \sigma_c^2)$, depends on the relative uncertainty of the prior, and the additional information. As long as $\sigma_\mu^2 > \sigma_c^2$ they argue there will be an incentive to take one more step in Figure 1.

The value of reduced variance must also be balanced against additional information costs. However, equation (13) isn't especially intuitive, because few policy-makers think in terms of an implicit price (p) on the reduction of the variance of WTP. A more policy-relevant interpretation would be to minimise the absolute value of the deviation of estimated net benefits (w-c) from that of the true net benefits $(\theta - \kappa)$ of the policy. This is also known as minimising the expected opportunity costs of a wrong decision:

$$(14) \qquad \min_n E(W)_{w|\theta, c|\kappa} = \iint |(w-c)-(\theta-\kappa)| f(w|\theta) \cdot g(c|\kappa) dw\, dc + gC(n)$$

where $f(w|\theta)$ and $g(c|\kappa)$ are probability density functions for WTP and policy costs. If we decide to implement the project when net benefits are actually negative, we will incur *avoidable net costs*, while in the opposite situation we will have *foregone benefits*. In both cases the opportunity cost is the absolute value of net benefits, assuming that society is risk neutral.[7] Under uncertainty a representative social planner will seek to minimise the expected opportunity cost of making the wrong decision, balanced against the cost of reducing the probability that the wrong 'guesstimate' about net benefits will be made. We can rewrite the decision-criterion as:[8]

$$(15) \qquad \underset{n}{\text{Min}}\, E(W) = POP|\mu - c|\Phi\left(\frac{0-|\mu-c|}{\sigma_{\mu-c}}\right) + g\ R\ C(n)$$

where POP = number of households affected by the policy; $|\mu - c|$ = absolute value of updated net benefits per household per month; Φ = normal cumulative distribution returns the probability that updated household net benefits are different from 0, i.e. the probability of foregone net benefits/avoidable net costs with an incorrect decision (significance of a Type I error); $\sigma_{\mu-c}$ = standard error of updated net benefits; correct specification of study costs require using an amortising factor $R = r/(1-(1+r)^{-T})$, where r = discount rate per period and T = periods in project life. We assume that a new valuation study *n* provides no new information on project/policy cost. Using an *information efficiency* criteria we would continue updating benefit estimates as long as the *change in* expected opportunity cost outweighed the information cost of an additional valuation study:

$$(16) \qquad -\mathrm{POP}|\frac{d\mu}{dn} - c|\Phi\left(\frac{0-|\frac{d\mu}{dn}-c|}{\frac{d\sigma_{\mu-c}}{dn}}\right) > \mathrm{g}\ \mathrm{R}\frac{dC(n)}{dn}$$

given $\frac{d\sigma_{\mu}}{dn} < 0$ and $\frac{dC(n)}{dn} > 0$

The left hand side of equation (16) can be interpreted as a form of *quasi-option value* due to delaying the project implementation decision by one time step.[9] Quasi-option value is the positive value of information on delaying a decision with irreversible impacts, usually associated with conservation of unique species or habitats (Arrow 1974; Krutilla 1964).

Although building a wastewater treatment plant and sewerage network is not an irreversible decision in the original interpretation, it is economically speaking irreversible when a project has a 20 year lifetime and a social discount rate of for example 12%. There are opportunity benefits from delaying the decision on implementation, both through passive increase in the knowledge of net benefits and in actively seeking that knowledge (Freeman 1984). The right hand side of (16) shows that the active approach is not costless, and that we in fact are talking about *net quasi option value*. Another aspect of the information efficiency criteria is that once we have a prior, and as policy site populations become smaller, new valuation studies become harder to justify.

As an alternative, decision-makers with a feel for statistics will be more used to an *information sufficiency* criteria, which is slightly different. We should continue updating until we can decide *for* or *against* a project with a requisite level of confidence:

$$(17) \qquad |\mu - c| - Z_{CL\%}\cdot\sigma_{\mu-c} > 0 \vee |\mu - c| + Z_{CL\%}\cdot\sigma_{\mu-c} < 0$$

where $Z_{CL\%}$ = the inverse of the cumulative standard normal distribution at confidence level CL%. Note that the reduction in expected opportunity cost cannot be known until we know σ_m,and therefore $\sigma_{\mu-c}$, *ex post*. Prediction of $d\sigma_{\mu-c}$ with the next information step is difficult if new parameter estimates come from different valuation methods with different random processes.[10] Although we strive for optimal information, finding that optimum *ex post* is an academic exercise rather than a practical policy tool. *Ex ante*, the decision-maker can only determine whether the existing information is sufficient to make a decision on implementation, given her level of confidence in the available valuation estimates.

3.3. Choosing a Prior

The temptation for any consultant in a hurry is to *ad hoc* use the mean of the study she believes to be the most representative of the policy site and which uses the 'best' available methodology (Figure 1). A more laborious prior is based on prediction from a meta-analysis of several relevant studies. With only a limited number of valuation studies, a 'random effects' regression model can be used assuming we believe there is exchangeability or an underlying 'mother distribution' for WTP across study sites (Desvousges et al. 1998). The random effects model is used to predict WTP at the policy site, by assuming normality, independence and exchangeability:[11]

$$\theta_{ji} = \alpha + \beta_{\pi}'\mathbf{x}_{ji} + u_j + e_{ji} \quad (18)$$

where α as a constant term, and u_j is a random error term specific to each study j with zero mean and variance σ_u^2. Once systematic differences across studies have been controlled for $\beta_{\pi}'\mathbf{x}_{ji}$, remaining differences arise from draws on each u_j. This random effects model predicts mean WTP, $\hat{\theta}$, and standard error, $\hat{\sigma}_{\pi}$, underlying mother distribution from which the individual study estimates are drawn.[12]

Note that in order to make a prediction in the random effects model one must assume that the policy site belongs to a particular effect (e.g. a country already represented in the dataset). The available CV studies for that effect will strongly influence predicted WTP, although variance is drawn from the underlying mother distribution. If we make no assumption regarding a random effect which applies at the policy site (i.e. we have no "best" study), an equal effects model without the random component, u_j, can be used. A fixed effect model, with effects for example country, author or study requires many more degrees of freedom than we often have available. If the meta-analysis reveals a 'best study' containing regression information on the variance-covariance matrix, and the analyst also has representative policy site data for the significant variables (e.g. census data or a pilot study), a benefit function transfer would be an appropriate prior. A further assumption is that ad hoc 'best study' information is less costly to obtain per unit of reduced expected opportunity cost, than from meta-analysis data. In what we call 'classical' contingent valuation, open-ended (OE) willingness to pay is used as a prior for the distribution of bids in a dichotomous model, although not in the Bayesian updating sense discussed above. The cheapest prior is of course an uninformative one when we proceed straight to the main survey.[13]

4. DATA

The following information became available during the course of the study (Table 1). The costs *of this particular study* are summarised as they accrued over time for the steps outlined in Figure 1. A discussion of the validity and reliability of the pilot and main surveys at the 'best' study and policy sites is reported elsewhere (Barton 1999).

Some words on the studies in the meta-analysis are in order. With the Jaco CV study completed, a search was conducted for contingent valuation studies on sanitation and water quality, using colleagues, library resources and the Environmental Valuation Reference Inventory (EVRI™).[14] From this set we excluded those studies not conducted in coastal areas, as well as those focusing exclusively on household sanitation. We were left with 40 WTP observations from 16 countries (appendix 1).

Table 1. Benefit valuation data availability over time

Time	Data	Information cost	Comment
April 1998	Ad hoc 'best' study	negligible	Mean WTP (DB-DC) from CV study conducted in Jaco, Costa Rica (Barton 1998)
July	Meta-analysis, study sites	US$3077 (4 manweeks)	32 observations from 16 countries on mean WTP for water quality improvement in coastal areas (appendix 1).
Aug.	Official statistics, policy site	US$1538 (2 manweeks)	1992 socio-demographic household characteristics (means, no dispersion info)
	Benefit function 'best study' site		Full data set and variance-covariance matrix from the Jaco CV study[a]
Oct.	Focus groups and Pilot survey, policy site	US$6205 (5 manweeks, flight, surveys)	Open-ended (OE) CV survey, N = 82 responses WTP > 0
Nov.	Main survey, policy site	US$9935 (4 manweeks, surveys)	Double bounded dichotomous choice (DB-DC) CV survey, N = 691 responses WTP > 0

Notes: Local study costs converted using 1US$ = 268 colones (Nov. 1998) Note to Table 1: [a] This is the only departure from the true information situation in the study. Full data sets and regression coefficients were available for the Jaco, Costa Rica, study several months before. However, we wanted to simulate a typical situation where full datasets and unpublished regression results are obtained from authors *after* a meta-analysis has indicated the 'best study', rather than before as with an ad hoc approach.

Where a single study used multiple econometric models we selected the 'maximum information' estimates[15] from independent samples (e.g. resource users and non-users). This selection bias was unavoidable because we had too few studies to control for study specific methodological assumptions using a fixed effects model. The 32 remaining observations covered use and non-use CV values for improvements in coastal surface water quality. The aim of the benefit-cost analysis was to determine whether additional sewerage and wastewater treatment could be justified based on the water quality improvements perceived by households.[16]

Where available, we also collected information on mean socio-demographic site characteristics of the type we would also expect to find at any policy site covered by a national census. Unfortunately, very few studies reported descriptive statistics.[17] All values were first brought to mid-1997 prices using local consumer price indices and then converted to US$ using a PPP-adjusted exchange rate.[18] This left us with 21 comparable observations that could be used for regression analysis on one single variable, income.[19] Statistically speaking this is unimpressive, but seems to be a typical situation for most applications of contingent valuation, and serves our purpose of illustrating a typical prior.

Sewerage and treatment cost data were obtained using a prediction made by the SANEX® expert system for comparing sanitation options in developing countries (Loetscher 1999). SANEX input came from interviews with local sanitation engineers, as well as local building, land and water costs which were parameters required by the expert model. SANEX suggested settled sewerage with activated sludge treatment as a feasible option. Confidence bounds on the costs of this option were obtained by assuming that 95% of aggregate cost predictions will fall within 30% of the model's predicted value, which is a typical uncertainty level of transferring historical cost data and one achieved by SANEX in other contexts (Loetscher 1999). Aggregate project costs can be assumed approximately normally distributed when there is uncertainty about a number of individual cost components (Vaughan Jones 1991). We also had access to historic cost of a very similar sewerage and wastewater plant to the one evaluated here, which could be compared to expert system predictions.[20] Although cost uncertainty is not the focus of this paper, these assumptions illustrate the uncertainty typical of 'cost transfers' in a pre-feasibility study.

While a full census has not been conducted since 1984, census-type data was available from a household survey conducted in 1992 in Puntarenas by the Department of Statistics and Census (DGEC 1992). No dispersion measures for population characteristics were available in the census. All variation in a benefit transfer estimate based on population averages would therefore stem from uncertainty about parameters at the study site. A more refined picture would include variation of population characteristics as observed in a pilot study at the policy site. A pilot CV survey using open-ended WTP was conducted one month before the main survey. The main CV survey used the double-bounded dichotomous choice format, the details of which are reported elsewhere (Barton 1999). The bid distribution in the main survey in Puntarenas roughly represented quintiles in the distribution of open-ended WTP from the Jaco study site pilot, rather than the Puntarenas pilot study. This is not usually the case for classical CV, but was done in order to simulate a realistic stepwise use of information as it arises, rather than construct an ideal benefit transfer experiment.[21] It makes censored WTP responses in the Puntarenas main survey independent from the open-ended Puntarenas pilot data. Independence is a handy definition of what constitutes 'new' information when considering Bayesian updating.

5. RESULTS

The random effects regression was run with declared household income (INC) as the only explanatory variable, and study country as random effect. Each study in the meta-analysis was weighted by its inverse standard error (precision) obtaining the following regression model:

$$\theta_j = 83.3924 + 0.0045\ \mathrm{INC}_j,\ \hat{\mathbf{u}}'\hat{\mathbf{u}} = 21111.8431,\ \hat{\mathbf{e}}'\hat{\mathbf{e}} = 47.0140$$

$$(t = 1.73)(t = 2.44)$$

Household income is a significant (5%) predictor of WTP for coastal water quality. Log-transforming WTP and income in the same model yields an income elasticity of WTP across studies of $0.55 (t = 1.79)$. We also ran a small equal effects model replacing the random effects component with a dummy variable (GNPCAT) which was set to one for all countries with PPP-adjusted GNP/capita less than \$15,000/year, and zero otherwise. This definition roughly captures the developing countries in the sample.[22]

$$\theta_j = -129.7365 + 0.0068\ \mathrm{INC}_j + 178.6734\ \mathrm{GNPCAT}_J, \hat{\mathbf{e}}'\hat{\mathbf{e}} = 1444.4624$$
$$(t = -1.38)(t = 2.42)(t = 2.18)$$

Predicted WTP based on census figures for household income are given in Table 2 for the random and equal effects models. The assumptions that the Jaco study has the same underlying random effect as the policy site, as well as the small income differences between sites, lead to practically no divergence between predictions of the random effects model and the ad hoc 'best study' approach (Scenario A). When no such assumption is made the meta-analysis results in a much lower prior (Scenario B). Appendix 1 shows that the Jaco study, although considered 'best', is an upper-end estimate relative to previous studies and would not be a prudent choice of prior without detailed study information.

Table 2. Simple Bayesian updating of WTP

		Point estimates		Updated priors			
				Scenario A: full information		Scenario B: no best study information	
Valuation approach / information step	Policy site information	Mean WTP	St. error	Mean WTP	St. error	Mean WTP	St. error
0.Ad hoc 'best study'	None	3156.57	201.82	3156.57	201.82	–	–
1.1.Meta-analysis, random eff.	mean inc.*	3142.66	294.69	3152.13	166.51	–	–
1.2 Meta-analysis, equal eff.	mean inc.*	1139.22	235.80	–	–	1139.22	235.80
2. 'Best study' benefit function	socio-demo.*	2905.83	210.11	3057.11	130.50	–	–
3. 'Best study' benefit function	Pilot data	3001.00	181.75	3038.03	106.00	–	–
4. Pilot study mean (OE)	Pilot data	1759.20	171.79	2685.37	90.21	1544.24	138.85
5. Main study mean (DBDC-Normal)	Main data	2403.50	80.55	2528.54	60.08	2187.14	69.67

Note: All values are per month in November 1998 Colones. All regressions assume truncated normal distribution of errors (see appendix 2). Standard errors are bootstrapped (Bergland et al. 1990). *Income and socio-demographics statistics refer to official census type data with no dispersion information. Meta-analysis results converted using PPP-adj. exchange rate 1997: 96.1 Colones/\$ (World Development Indicators 1998); CPI ratio (July 97-November '98): 1.154 (Central bank of Costa Rica 1999).

Note that we have updated benefit estimates twice based on almost the same pilot study information. At Step 3 we use only the descriptive variables from the pilot, while at step 4 we use the OE WTP information as well. Through the Bayesian weighting of the DBDC benefit transfer estimate and the conditional OE pilot WTP estimate (Step 4) in Scenario A, we explicitly judge the relative validity of the two estimates by their precisions, and nothing else. However, because these two estimates are conditional on policy site characteristics we are violating the independence condition of the simple Bayesian updating approach – that all new information really is new information. This results in a small downward bias in standard error of the updated estimate, relative to true information content. This is another classical bias-variance trade-off we could avoid if we knew which of the WTP formats was the least biased relative to true unobserved WTP.

Table 3 illustrates the trade-off between reducing uncertainty about net benefits of the project versus additional study cost. Where the reduction in the confidence interval of net benefit is smaller than the study cost it would not be efficient to spend any more on obtaining WTP information for a further update. The results in Table 3 depend in part on our study cost accounting, which is amortised over the 20 years of the project lifetime at a 12% annual discount rate.[23] The probability of opportunity cost refers to the significance level with which project benefits exceed costs or vice versa.

Note that all costs of air travel, focus groups and pilot are attributed to step 3. Once we have done the pilot survey, updating benefit transfer estimates using the OE WTP response is assumed to be costless. However, the reduction in expected

Table 3. Application of the information efficiency criteria in benefit transfer

Valuation approach/ updating step	Probability of opportunity cost (%) $\Phi_* 100$	Reduction in aggregate expected opportunity costs per month $POP_* \lvert d\mu/dn - c\rvert_* d\Phi/dn$	Additional information cost amortised per month $g_* \frac{dC(n)}{dn}{}_* R$
Scenario A: Full information			
0. Ad hoc 'best study'	18.69	infinity	negligible
1.Meta-analysis (random effects)	17.14	47822	9080
2. Benefit function (official statistics)	13.11	65680	4539
3. Benefit function (pilot data)	10.78	81767	18310
4.Pilot study mean (OE)	3.65	219505	0
5.Main study mean (DBDC)	1.55	117220	29317
Scenario B: No 'best' study information			
1.Meta-analysis (equal effects)	0	infinity	9080
4.Pilot study mean (OE)	0	0	18310
5.Main study mean (DBDC)	0.2	−185304	29317

Note: all values in Nov.1998 colones. Population covered by wastewater treatment project = 6000 households. T = 20 years, r = 1% p.m.

opportunity costs of the benefit function transfer at step 3 still justifies the full cost of the pilot. Had we known *ex ante* by how much we could reduce uncertainty about opportunity costs, we would have continued updating up to and including the main survey (Step 5, Table 3). However, we would have sufficient information on net benefits to reject the project with 95% confidence level after having conducted the pilot study (see step 4, Figure 3).

A main survey would be necessary only if the policy-maker was so risk averse as to want less than a 5% chance of making the wrong decision. Note in Figure 2 that a simple inspection to see whether confidence intervals of willingness-to-pay and project costs overlap would lead us to conducting a full CV (Step 5) – and we would still believe we did not have requisite confidence in the results to make a decision, perhaps commissioning an increase in the sample size. However, the variance of net benefits is actually lower than the sum of WTP or cost variances due to the same properties as those of Bayesian updating in Equation (8).[24] Table 3 also shows that the population affected by the project needs to be less than about 25% of its current size (6000) for our main contingent valuation study not to increase expected welfare (Step 5 Scenario A).

In this particular study the opportunity costs of a wrong decision are relatively high compared to the study cost amortised over 20 years. Although we have assumed that no additional information on project costs is obtained during the study, it would probably have been more efficient to reduce cost uncertainty, which is high relative to that of WTP.

For completeness, Table 3 also includes expected reduction in opportunity cost when we assume no knowledge of a best study (Scenario B). The equal effects model predicts willingness-to-pay that is about 30% of costs and we would decide against the project based only on our equal effects meta-analysis. Collecting

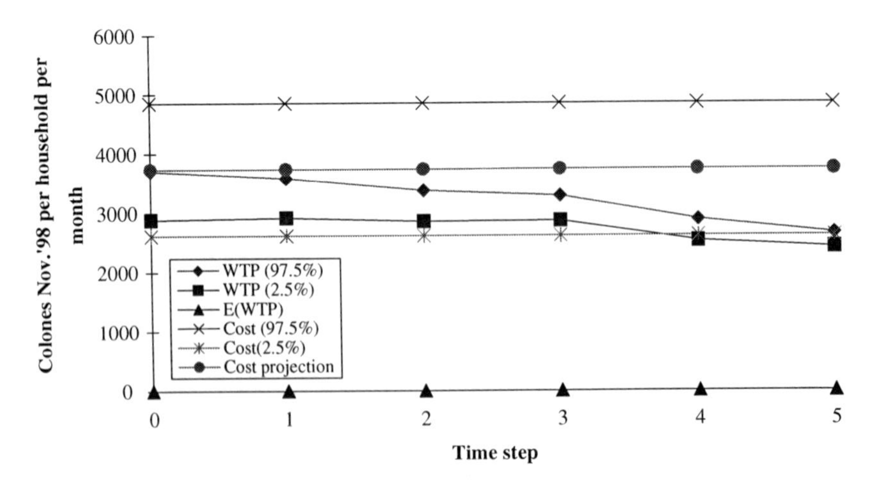

Figure 2. Predicted benefit and costs of sewerage and waste water treatment
Note: WTP calculated using values for Scenario A, Table 2. Sewerage and wastewater treatment cost projection ($c = 3731$ colones/hh month) using SANEX® and 95% confidence intervals at $+/-30\%$ of predicted mean. Historic costs of an identical project was Colones 2404 /hh month.

more information would not change that conclusion. Had we done a full CV survey anyway, we would actually have introduced a little uncertainty, posterior willingness-to-pay at the policy site being about 65% of project costs, with 0.2% probability of a mistaken decision.

If we had based ourselves on historic costs of a very similar sewerage and treatment system, instead of the expert system prediction, our policy conclusions would have been the opposite. We would have implemented under any type of Bayesian updating scenario, as can be seen by plotting historic cost of Colones 2404 per hh/month in Figure 2. Had we simply conducted a 'classical' CV study, using only the DBDC estimate ($E(w) = 2303.5$, $\sigma_c = 80.5$) we would have a project that could be expected to more or less 'break-even'. With symmetrical uncertainty about benefits, opportunity costs of a wrong decision here are approximately zero. Needless to say, this is a coincidence of this particular case study. However, it illustrates the general point that even our maximum *ex ante* information may turn out *ex post* to have lead us to a mistaken implementation decision and over/under-investment in information acquisition. The updating framework only makes such mistakes *less likely*, and helps to clarify ex post where we went wrong.

Assuming constant project cost uncertainty, Bayesian updating shows benefit transfer followed by original policy site studies to be attractive in this particular study. However, because our initial priors have such high accuracy, each new study reduces uncertainty by relatively little. The reduction of bias of policy site information has a much stronger effect on reducing expected opportunity cost of a wrong decision (Figure 3), relative to WTP predicted by the full CV at the policy site. The assumptions of either random or equal effects lead to over- or

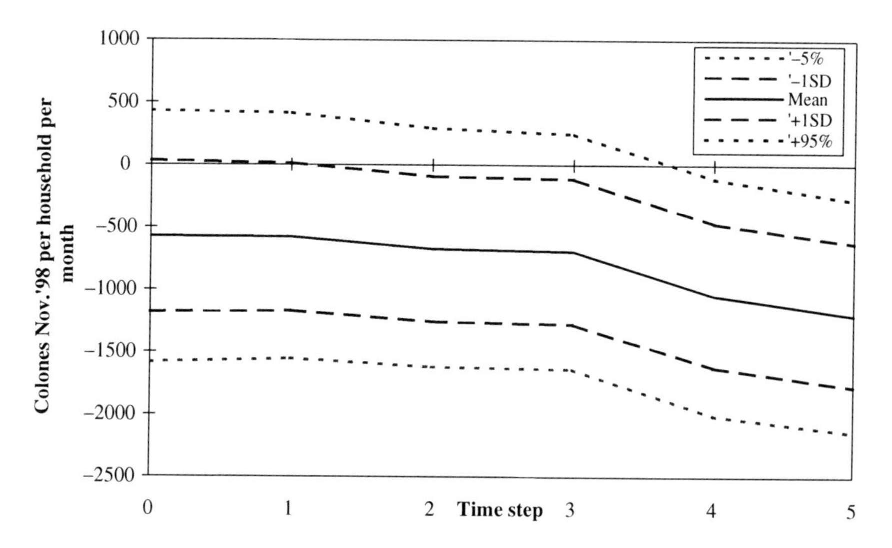

Figure 3. Applying the information sufficiency criteria to expected net benefits
Note: Mean net benefits with one standard deviation and 95% confidence interval. Simulation using Latin Hypercube sampling (Vose 1996) from parameters of normal WTP and cost distributions.

underestimation relative to the point estimate of the main CV survey at the policy site. This is a caution to conducting 'desk-top' valuation studies with no on-site information on WTP. Expected opportunity costs are very sensitive to the choice of prior, as León et al. (1997) have shown. By using a meta-analysis regression we impose a lot of structure and certainty that may not be there, for example that income elasticity of WTP between existing study sites applies to predictions at the policy site. 'Lower information' alternatives would be to estimate an unconditional normal mother distribution for WTP (Desvousges et al. 1998), or use a beta-distribution where we only know the upper and lower bounds of WTP (León et al. 1997). Both approaches would have wider confidence intervals and make the advantages of new information much clearer than shown in our example.

What could we have done differently in this particular case? A full policy-site CV study would have been justified from a local budget-constrained planners point of view only if she could amortise study costs over the lifetime of the project. However, a pilot survey would have been sufficient for the planner to be able to say with 95% certainty that the project would not be economically feasible. To do this, we would have to assume that open-ended responses are unbiased, while many authors believe dichotomous choice (DC) model to be more incentive compatible (Carson et al. 1999). As a compromise, open-ended responses could be updated with a smaller sample using single or double-bounded DC, rather than such a large sample study as we had here.

6. CONCLUSIONS

While contingent valuation is uncommon in development project feasibility studies, explicit consideration of information value is (almost) never undertaken. Unless surplus research funds can be cycled into a new project, the question of whether a study is "worth it" is an irrelevant when research funding is awarded as a lump-sum. For development agencies with in-house economists vying for research resources over time, the stepwise updating approach suggested here is more relevant.

A particular result of this study is that the local planner will find a full contingent valuation worthwhile, provided study costs can be spread over the project lifetime. In general, the smaller the affected population and the shorter the project horizon, the harder study costs will be to justify. With the accuracy and reliability of the willingness-to-pay data presented here, a planner in a town of less than about 1500 households could not justify a CV of the size we conducted. It would imply sampling over half the population.

This paper shows that CV studies are relatively inexpensive in a country such as Costa Rica for sewerage and wastewater projects. With higher study costs than ours, the following criteria should be fulfilled before conducting a full policy-relevant study: (1) the affected population should be 'large', or (2) the expected net benefits per household should be potentially large (positive or negative), but with uncertainty regarding household WTP so high that for example, sampling errors typical of a pilot study could lead to a mistaken decision. Such uncertainty may arise because of large variations in socio-demographic, institutional, or environmental conditions within the population affected by the policy. Uncertainty may also be high if the environmental good in question is unknown or the means of provision unclear.

The 'simple' Bayesian approach is satisfactory if we can make assumptions of approximate normality. If not, a Bayesian estimator which allows for flexible distributional assumptions is more appropriate. A future study should perform a significance test of the hypothesis that a more flexible Bayesian estimator changes the 'sufficient' and 'efficient' information levels.

A Bayesian approach to valuation assumes that the CV practitioner works hand in hand with the policy-maker, rather than being commissioned and coming back with 'the answer' several months (or years) later. It is a structured way of making as many subjective valuation assumptions as possible explicit for decision-makers. It exposes inconsistencies in how we use the information of different non-market valuation methods in environmental policy-making over time. CV practitioners are required to think through the relative merits and dangers of benefit transfer versus full blown studies at the policy site.

David Barton is research scientist, Norwegian Institute for Water Research. The author thanks Olvar Bergland and Ståle Navrud of the Norwegian University of Life Sciences for their inputs into this research. This study was financed by the Norwegian Research Council.

7. NOTES

[1] Parameters are determined from the minimum, maximum, mode and mean values of WTP values from previous studies.

[2] I am indebted to Olvar Bergland for several discussions on Bayesian analysis.

[3] What information is contained in the 'subjective prior', the likelihood, and posterior distribution depends on where we are in the information gathering process for the valuation study, i.e. at which step in the iterative updating of welfare estimates we find ourselves.

[4] Measurement error, σ_m^2, can be assumed to have mean zero for a large number of independently and randomly generated studies in the meta-analysis.

[5] Formally, if the prior distribution is not a conjugate form of the likelihood function the posterior probability distribution cannot be found in closed form (León et al. 1998).

[6] León et al. (1998) solve the problem of no closed form of $\mu(w_i)$ by using a numerical technique called Integration Method by Monte Carlo. Intuitively, the problem is one of how to weight the new information on parameters (e.g. from a survey) against the prior parameters (e.g. from the meta-analysis).

[7] Another interesting research question is an empirical evaluation of environmental policy-makers risk aversity, set in terms of the decision-criteria outlined here.

[8] A reviewer has observed that, more generally, the probability of $(\mu - c)$ leading to a wrong decision about the magnitude of net benefits is, $\Phi\left(\frac{-[|\theta-\kappa|+|\mu-c|]}{\sigma_{\mu-c}}\right)$, giving the loss function the following form,

$$\operatorname*{Min}_{n} E(W) = POP \int |\theta - \kappa| \theta \left(\frac{-[|\theta-\kappa|+|\mu-c|]}{\sigma_{\mu-c}}\right) d(\theta - \kappa) g\, R\, C(n),$$

with limits to integration being $\int_0^\infty$ if $(\mu - c) <$, and $\int_{-\infty}^0$ if $(\mu - c) > 0$.

The reviewer argues that loss function (15) underestimates the probability of a wrong decision (personal communication, Jose Lima Santos, 2000). The implications of this formulation are being explored.

[9] It is multiplied by -1 because a decrease in the probability of opportunity costs is seen as having positive value.

[10] When we do Bayesian updating using only formulae 7-8 we are making the assumption that all new benefit information comes from the same normal mother distribution.

[11] Assumes the studies share the same estimation protocol and have been conducted independently.

[12] The predicted mean and standard error of WTP are drawn in two seperate processes in the random effects model: $\hat{\theta}$ and $\hat{\sigma}_{\pi}$ are determinedby the true mean, θ_j, and variance, σ_j^2, of the distribution of WTP at each study site from which they are drawn. It further assumes that the true mean and variance at each study site are drawn from the underlying hyper-parameters of a mother distribution (Desvousges et al., 1998). True study site parameters are corrected for any systematic differences between studies, β_{π} such as in income levels. If there is no variation in **X**, all $\beta_{\pi} = 0$, and the mother distribution of WTP is independent of site characteristics such as the income range of the average household across studies.

[13] In general, a first prior will always be worth the information cost if we are completely ignorant to start off with.

[14] Administrated by Environment Canada, the EVRI is a database containing summaries of valuation case studies with the aim of providing a low-cost way of putting together meta-analyses and identifying 'best studies' for benefit transfer.

[15] i.e. where there was a choice between open-ended and DC we selected open-ended; between single and double bounded we selected double bounded, etc.

[16] Almost all households already have in-house sanitation.

[17] To Desvousges et al. (1998) plea for publishing the variance-covariance matrix, we add a call for census-standard socio-demographic descriptive statistics in all valuation studies.

[18] Large currency devaluations and/or periods of hyperinflation between the study date and 1997 made it difficult in some cases to find reliable consumer price indices and exchange rates. Where in doubt, the observation was excluded.

[19] We excluded observations that reported household expenditures.

[20] In El Roble, Barranca de Puntarenas. Source: Licitacion Publica Internacional No-.86-1219 (AyA, Division de Obras por Contrato) and personal communications with Engineer Leonardo Moya.

[21] If censored WTP is correlated with bid distribution, the Puntarenas main survey contains prior information from the Jaco pilot. An 'ideal', rather than 'realistic', benefit transfer experiment would compare WTP censored by bids obtained from separate distributions from pilot studies at the 'study' and 'policy' sites.

[22] Household income and the GNP/cap dummy are strongly correlated. However, we keep both variables for the prediction because without the fixed effect for developing countries, household income is not a significant predictor of WTP.

[23] The choice of discount rate is arbitrary, although representative of the rate used in some development project feasibility studies.

[24] It can be shown that the simultaneous distribution of σ_{π}^2, σ_m^2 is identical to the Bayesian conjugate prior σ_{μ}^2, composed of these two distributions (Judge et al. 1988).

8. REFERENCES

Arrow KJ, Fisher AC (1974) Environmental preservation, uncertainty, and irreversibility, Quarterly Journal of Economics (88):312–19

Atherton TJ, Ben-Akiva ME (1976) Transferability and updating of disaggregate travel demand models, Transportation Research Record (610):12–18

Atkinson SE, Crocker TD, Shogren JF (1992) Bayesian exchangeability, benefit transfer, and research efficiency, Water Resources Research, 28(3):715–722

Barton DN (1999) The Quick, the Cheap and the Dirty Benefit Transfer Approaches to the Non-market Valuation of Coastal Water Quality in Costa Rica Doctor Scientiarum Thesis 1999:34, Agricultural University of Norway

Bergland O, Magnussen K, Navrud S (1995) Benefit Transfer: Testing for Accuracy and Reliability. Discussion Paper #D-03/1995. Norway: Department of Economics and Social Sciences, Agricultural

University of Norway, 21p. Published as Chapter 7 (pp. 117–132) in Florax, R.J.G.M., P. Nijkamp and K. Willis (eds.) 2002: Comparative Environmental Economic Assessment. Edward Elgar Publishing, Cheltenham, UK

Bergland O, Romstad, E, Kim, S-W, McLeod, D (1990) The Use of bootstrapping in contingent valuation studies, Department of Agriculture and Resource Economics, Oregon State University

Brouwer R, Langford, I H, Bateman, I J, Crowards, T C, Turner, R K (1997) A Meta-analysis of wetland contingent valution studies Working Paper GEC 97–20, CSERGE, London

Brouwer, R, Spaninks, F (1997) The Validity of transferring environmental benefits: further empirical testing Working Paper GEC 97-07, CSERGE, London

Carson, RT, Groves T, Machina MJ (1999) Incentive and informational properties of preference questions Plenary address European Association of Resource and Environmental Economists, Oslo, Norway, June 1999

Desvousges WH, Johnson FR, Banzhaf HS (1998) Environmental policy analysis with limited information Principles and applications of the transfer method, Edward Elgar, Cheltenham, UK

DGEC (1992) Encuesta de hogares de propositos multiples del Gran Puntarenas, Direccion General de Estadistica y Censos, Costa Rica

Downing M, Ozuna TJ (1996) Testing the reliability of the benefit function transfer approach Journal of Environmental Economics and Management, 30(3):316–22

EC (1999) Benefit transfer and economic valuation of environmental damage in the European Union: with special reference to health Final Report to the DG-XII, European Comission contract ENV4-CT96-0227

Freeman AM (1984) The Quasi-option value of irreversible development Journal of Environmental Economics and Management, 11:292–295

Gourieroux C, Monfort A (1995) Statistics and econometric models Volume 1, Cambridge University Press, Cambridge

Horowitz J (1993) A New model of contingent valuation American Journal of Agricultural Economics, 75(5):1268–72

Judge GG, Hill RC, Griffiths WE, Lütkepohl H, Lee T-C (1988) Introduction to the theory and practice of econometrics, John Wiley and Sons, New York

Kirchhoff S (1998) Benefit function transfer vs meta-analysis as policy-making tools: a comparison Workshop on Meta-analysis ad benefit transfer: state of the art and prospects, Tinbergen Institute, Amsterdam

Kirchhoff S, Colby BG, LaFrance JT (1997) Evaluating the performance of benefit transfer: an empirical inquiry Water Resources Research(33):75–93

Krutilla JV (1964) Conservation reconsidered American Economic Review, 57:777–786

Leamer EE (1982) Sets of posterior means with bounded priors Econometrica(50):725–736

León CJ, Vázquez-Polo FJ, Riera P, Guerra N (1997) Testing benefit transfer with prior information Discussion paper, Department of Applied Economics, University of Las Palmeras de Gran Canaria, Spain

León J, Vázquez Polo FJ (1998) A Bayesian approach to double bounded contingent valuation (Modelización del aprendizaje en valoración contingente) Discussion paper, Department of Applied Economics, University of Las Palmeras de Gran Canaria, Spain

Loetscher T (1999) Appropriate sanitation in developing countries: the development of a computerised decision aid, PhD thesis, University of Queensland, Brisbane

Loomis JB (1992) The Evolution of a more rigorous approach to benefit transfer: benefit function transfer Water Resources Research, 28(3):701–705

Loomis JB, White DS (1996) Economic benefits of rare and endangered species: summary and meta-analysis Ecological Economics 18(3):197–206

McLeod DM, Bergland O (1995) Willingness-to-pay estimates using the double-bounded dichotomous-choice contingent valuation format: a test for validity and precision in a Bayesian framework Land Economics, 75(1):115–125

Parsons, G R, Kealy, M J (1994) Benefits transfer in a random utility model of recreation, Water Resources Research, 30(8):2477–2484

Schwartz, J (1994) Air pollution and daily mortality: A review and meta-analysis Envrionmental Resources (64), 36–52

Smith, V K (1992) On separating defensible benefit transfers from smoke and mirrors Water Resources Research, 28(3), 685–94

Smith, V K, Huang, J (1995) Can markets value air quality? A Meta-analysis of hedonic property value models Journal of Political Economy, 103(1), 209–27

Smith, V K, Kaoru, Y (1990) What have we learned since Hotelling's letter? A meta-analysis Economics Letters(32), 267–272

Sturtevant, L A, Johnson, F R, Desvousges, W H (1995) A meta-analysis of recreational fishing, Triangle Economic Research, Durham, North Carolina

van de Walle, D, Gunewardena, D (1998) How dirty are quick and dirty methods of project appraisal? Policy Research Working Paper 1908, Development Research Group, World Bank

Vaughan Jones, C (1991) Financial risk analysis of infrastructure debt The Case of water and power investments, Quorum Books, New York

Viscusi, W (1985) Are individuals Bayesian decision-makers? American Economic Review, 75, 381–85

Vose, D (1996) Quantitative risk analysis A Guide to Monte Carlo simulation modelling, John Wiley & Sons, Chichester

Walsh, R G, Johnson, D M, McKean, J R (1992) Benefit Transfer of Outdoor Recreation Demand Studies 1968–1988 Water Resources Research, 28(3), 707–713

Appendix 1. A meta-analysis of WTP for waste water treatment in coastal areas

Author (year)	Country	Notes on Bid Models, Sample, Scenario	Sample size (N)	Mean WTP (p.a. PPP-$1997)	St. dev. of WTP_i (p.a. PPP-$1997)	**CV**	WTP % of Household income	WTP of PPP-adj. GNP/cap.
Breivik and Hem (1986)	Norway (Kristiansand)	IB,A,W	300	48.35	80.96[#]	1.67	n.a.	0.20
Breivik and Hem (1986)	Norway (Kristiansand)	IB,U,W	143	46.45	118.90[#]	2.56	n.a.	0.19
Dalgard (1989)	Norway (Drammen)	PC,U,W	156	78.17	127.67	1.63	0.23	0.32
Dalgard (1989)	Norway (Drammen)	PC,N,W	127	47.94	73.71	1.54	0.14	0.20
Aarskog (1987)	Norway (Oslo)	PC,N,W	43	72.04	92.47	1.28	0.16	0.30
Aarskog (1987)	Norway (Oslo)	PC,U,W	150	109.13	139.24	1.28	0.25	0.45
Bergland et al. (1995)	Norway (Vannsjø-Hobøl)	DB,A,W	258	223.39	345.64[#]	1.55	n.a.	0.92
Bergland et al. (1995)	Norway (Orre)	DB,A,W	236	323.52	419.52[#]	1.30	n.a.	1.33
Lindsey (1994)	US (Baltimore Country)	PC,A,W	824	54.73	97.25[#]	1.78	n.a.	0.19
Bockstael et al. (1989)	US (Cheasapeake Bay)	SB,U,W	412	188.85	1156.28[#]	6.12	n.a.	0.65
Bockstael et al. (1989)	US (Cheasapeake Bay)	SB,N,W	547	59.31	304.79[#]	5.14	n.a.	0.20
Wood et al. (1996)	Canada (Nova Scotia)	PC,A,W	267	300.39	481.68	*n.a.*	0.88	1.38
LeGoffe (1995)	France (Brest)	PC,A,W	607	37.72	60.48	*n.a.*	0.17	0.17
Georgiou et al. (1996)	UK (Great Yarmouth)	OE,U,W	197	23.04	47.59[#]	2.07	0.10	0.11
Georgiou et al. (1996)	UK (Lowestoft)	OE,U,W	203	26.10	41.56[#]	1.59	0.11	0.13
Vasquez (1998)	Chile (Dichato)	SB,U,W	370	290.82	440.85[#]	1.52	0.98	2.38
Niklitschek & Leon (1996)	Chile (coastal city)	SB,A,W	1247	241.27	386.88	*n.a.*	1.22	1.97
Darling et al. (1992)	Barbados	SB,A,WS	433	235.25	377.23	*n.a.*	n.a.	2.24
Darling et al. (1992)	Barbados	SB,A,W	277	14.54	23.31	*n.a.*	n.a.	0.14
McConnel & Ducci (1989)	Uruguay (coastal city)	SB,A,W	1500	19.48	31.23	*n.a.*	0.44	0.21
Whittington et al. (1993)	Ghana (Kumasi)	IB-OE,U,S	1224	76.67	62.04	0.81	1.89	4.76

(Continued)

Appendix 1. *(Continued)*

Author (year)	Country	Notes on Bid Models, Sample, Scenario	Sample size (N)	Mean WTP (p.a. PPP-$1997)	St. dev. of WTP_i (p.a. PPP-$1997)	CV	WTP % of Household income	WTP of PPP-adj. GNP/cap.
Hoehn & Krieger (1998)	Egypt (Cairo)	SB,A,S	903	118.70	96.14#	0.81	n.a.	3.85
Hoehn & Krieger (1998)	Egypt (Cairo)	SB,A,W	1008	34.78	35.33#	1.02	n.a.	1.13
Choe (1998)	Georgia (Tblisi)	IB-OE,A,W	342	20.40	32.71	n.a.	0.54	1.03
Choe (1998)	Georgia (Kutasi)	IB-OE,A,W	119	31.00	49.71	n.a.	1.12	1.57
Choe et al. (1996)	Philippines (Davao)	IB-OE,U,W	174	74.17	82.19	1.11	0.81	2.02
Choe et al. (1996)	Philippines (Davao)	IB-OE,N,W	389	52.12	70.16	1.35	0.57	1.42
Choe et al. (1996)	Philippines (Davao)	SB,U,W	174	60.14	72.17	1.20	0.66	1.64
Choe et al. (1996)	Philippines (Davao)	SB,N,W	389	2.00	62.14	31.00	0.02	0.05
Choe et al. (1996)	Philippines (Davao)	DB,U,W	174	102.24	60.14	0.59	1.12	2.79
Choe et al. (1996)	Philippines (Davao)	DB,N,W	389	70.16	32.07	0.46	0.77	1.91
Lauria et al. (1999)	Philippines (Calamba)	IB-OE,A,S	374	72.17	98.23	1.36	0.58	1.97
Lauria et al. (1999)	Philippines (Calamba)	IB-OE,A,WS	354	110.26	106.25	0.96	0.89	3.00
Lauria et al. (1999)	Philippines (Calamba)	SB,A,S	374	90.21	68.16	0.76	0.73	2.46
Lauria et al. (1999)	Philippines (Calamba)	SB,A,WS	354	134.31	70.16	0.52	1.08	3.66
Altaf and Hughes (1994)	B. Faso (Ouagadougou)	SB,A,S	605	238.50	382.43	n.a.	3.84*	23.85**
Altaf (1994)	Pakistan (Gujaranwala)	IB,A,S	986	54.32	87.10	n.a.	0.78	3.44
World Bank (1992)	Brasil (Sao Paolo)	SB,A,W	550	409.13	656.04	n.a.	6.34	6.44
Barton (1998)	Costa Rica (Jaco)	DB,A,WS	271	353.51	372.07#	1.05	3.04	5.43

Abreviations: PPP=purchasing power parity, U=surface water resource users, N=non-users, A=no distinction/all users and non-users, OE=open-ended, SB/DB=single/double bounded dichotomous choice, IB=iterative bidding, P=payment card, IB-OE=iterative bidding with open follow-up; S=WTP for sewerage connection *only*. W=wastewater treatment *only*, SW=sewerage connection *and* waste water treatment. CV=coefficient of variation=St.dev.WTP/Mean WTP. **Notes:** *as % of household expenditure. **Due to currency overvaluation and problems in finding a reliable consumer price index, 1997 PPP-adjusted values are unreliable.; # calculated from reported standard error of mean WTP as St.dev. individual $WTP_i = \sqrt{N}$* St.error mean WTP

S. NAVRUD AND R. READY

LESSONS LEARNED FOR ENVIRONMENTAL VALUE TRANSFER

1. POLICY USE OF VALUE TRANSFER METHODS

Environmental valuation studies have four main types of use (Navrud and Pruckner 1997):

i) *Cost-benefit analysis (CBA)* of investment projects and policies,
ii) *Environmental costing* in order to map the marginal environmental and health damages of e.g. air, water and soil pollution from energy production, waste treatment and other production and consumption activities. These marginal external cost can be used in public investment decisions and policy (e.g. as the basis for "green taxes")
iii) *Environmental accounting* at the national level (green national accounts) and firm level (environmental reporting and accounting)
iv) *Natural Resource Damage Assessment (NRDA)/Liability for environmental damages;* i.e. compensation payments for natural resource injuries from e.g. pollution accidents

Environmental valuation techniques have mostly been used in CBAs, but are also used to scale resource compensation in NRDAs in the US; environmental costing of electricity production from different energy sources in both the US and Europe (see e.g. Rowe et al 1995; Desvousges et al 1998; and European Commissions – DG XII, 1995, 1999); and green national accounting exercises, e.g. the Green Accounting Research Project (GARP) of the European Commission (Markandya and Tamborra 2005). The accuracy needed increases, and thus the applicability of value transfer techniques decreases, as we move down the list of potential policy uses of valuation studies (Navrud and Pruckner 1997).

CBA has a long tradition in the US as a project evaluation tool, and has also been used extensively as an input in decision making ever since President Reagan issued Executive Order 12291 in 1981, necessitating a formal analysis of costs and benefits for federal environmental regulations that impose significant costs or economic impacts (i.e. Regulatory Impact Analysis). In Europe, CBA has a long tradition in evaluation of transportation investment projects in many countries, but environmental valuation techniques were in most cases not applied. There seems to be no legal basis for CBA in any European country, but the UK Environment Act requires a comparison of costs and benefits. Some countries have administrative CBA guidelines for project and policy evaluation, and in a few cases these includes a section on environmental valuation techniques.

Paragraph 130r of the Maastricht Treaty, which focuses on EU's environmental goals, environmental protection measures and international cooperation in general,

S. Navrud and R. Ready (eds.), Environmental Value Transfer: Issues and Methods, 283–290.

says that the EU will consider the burden and advantage of environmental action or non-action. Furthermore, the "Fifth Activity Programme for Environmental Protection Towards Sustainability" (1993–2000) says:

> In accordance with the Treaty, an analysis of the potential costs and benefits of action and non-action will be undertaken in developing specific formal proposals within the Commission. In developing such proposals every care will be taken as far as possible to avoid the imposition of disproportionate costs and to ensure that the benefits will outweigh the costs over time (European Community 1993, p. 142).

The 1994 Communication from the Commission to the Council of the European Parliament, entitled:"Directions for the EU on Environmental Indicators and Green National accounting – The Integration of Environmental and Economic Information Systems" (COM (94)670, final 21.12.94) states a specific action for *improving the methodology and enlarging the scope for monetary valuation of environmental damage*. More recently, the European Commission (EC)'s Green Paper, entitled: "For a European Union Energy Policy", states that *internalisation of external costs is central to energy and environmental policy*. EC DG Environment has prepared guidelines on benefit assessments for all DG Environment policy and project assessments. Their recent cost-benefit analysis of the Clean Air for Europe (CAFE) programme includes extensive transfer of morbidity and mortality values (Holland et al 2005). CAFE is a programme of technical analysis and policy development that underpinned the development of the Thematic Strategy on Air Pollution under the EU Sixth Environmental Action Programme (2001–2010). In April 2004 the Environmental Liability directive, that covers environmental damage, came into force. Thus, there seems to be an increased interest within the European Commissions in using environmental valuation and value transfer for all four potential policy uses: CBA, environmental costing, environmental accounting and to assess natural resource damages for environmental liability.

International organisations like the OECD, the World Bank and regional development banks and the United Nations Environment Program (UNEP) have produced guidelines on environmental valuation techniques; e.g. OECD (1989, 1994, 1995); Asian Development Bank (1996), and UNEP (1995, Chapter 12). In many cases they have used valuation techniques as an integral part of CBA of investment projects, e.g. the World Banks evaluation of water and sanitation projects (Whittington 1998). UN's statistical division UNSTAT has also actively supported the development of resource accounting systems (e.g. the Handbook on Integrated Environmental Economic Accounts). Even though there have been numerous environmental valuation studies of biodiversity and ecosystem functions in the US and in Europe (see Navrud 1992,1999 for an overview of European valuation studies), the policy use of valuation studies have historically concentrated on air and water pollution impacts and policies (Navrud & Pruckner 1997), and these are still the policy areas where value transfer is used the most in Europe while in the US they are also used to assess recreational benefits.

2. CHALLENGES IN VALUE TRANSFER

Value transfer is less than ideal, but so are most valuation efforts in the sense that better estimates could be obtained if more time and money were available. Analysts must constantly judge how to provide policy advice in a timely manner, subject to the resource constraints they face. Analysts should compare the cost of doing a new, original valuation study with the potential loss of making the wrong decision when using the transferred estimate. Decision Theory and Bayesian Analysis could be used to assess the need for further information about both monetary values and other inputs into the analysis.

Value transfer methods may be particularly useful in policy contexts where rough or crude economic value estimates are sufficient to make a judgment regarding the advisability of a policy or project. Thus, value transfer could be used in cost-benefit analyses of projects and policies, but one should be more careful in using transferred values in environmental costing and accounting exercises at the national and firm levels, and particularly in calculating compensation payments for natural resource injuries.

We see five main difficulties or challenges in value transfer:

1. *Availability and quality of existing studies*
2. *Mismatch between the good as it is valued in existing studies, and the good as it will change as a result of the policy*. For example,
   - The expected change resulting from a policy is *outside the range* of previous experience
   - Previous studies value a *discrete* change in environmental quality; how can that be converted into *marginal* values to value the new policy?
   - Previous studies value a *gain* in environmental quality; how can that be converted to value *losses* in environmental quality?
3. *Differences in the study site(s) and policy site* that are not accounted for in the specification of the value function or in the procedure used to adjust the unit value.
4. Determination of the *extent of the market*. To calculate aggregated benefits the mean value estimate has to be multiplied by the total number of affected households. There is a need for guidelines on how to determine the size of the affected population, since this can have a greater affect on aggregate value than uncertainty in the mean value estimate.
5. While original valuation studies can be constructed to value many benefit (or cost) components simultaneously, value transfer studies would often involve transfer and then *aggregation of individual components*. Simply adding them assumes independence in value between the components. If components are substitutes or complements, this simple adding-up procedure would over- and under-estimate the total benefits (or costs), respectively. Thus, correction factors to take these interdependencies into account have to be applied. It remains to be seen whether it is possible to construct general sets of correction factors for groups of environmental goods.

3. LESSONS LEARNED

It has been more than 30 years since the first formal value transfer exercise (Krutilla and Fisher, 1975), and nearly 15 years since attention was focused on value transfer techniques in the special issue of Water Resource Research (Brookshire and Neill, 1992) and at the 1992 Association of Environmental and Resource Economists Workshop at Snowbird, Utah. What have we learned? There are certain stylised facts that have become conventional wisdom about value transfer. Like all conventional wisdom, they contain some truth, but must not be completely accepted without reservation. The contributions in this book help give us a more nuanced view.

1. *The validity of a value transfer will be higher if the good that was valued in the source study is similar to the good that will be changed at policy site, in terms of the definition of the good itself, the degree to which it will change, and the population affected.* Rosenberger and Phipps, in their review of the literature, make a convincing case that this is true. Other case studies in this volume also demonstrate that transfer tends to be more valid when the policy site and the study site are more similar. However, many of the transfer tests included here demonstrate that the valuation method used in the source study may be even more important than the precise definition of the good or the characteristics of the population affected. When choosing a source study, differences between stated preference and revealed preference studies, or among stated preference studies conducted using different elicitation methods, may have a greater impact on the resulting value transfer than differences in how the good was defined, or the characteristics of the affected population.
2. *The more information we use in the value transfer, the better it will predict value at the policy site.* This intuition was the motivation behind advocating value function transfer as opposed to unit value transfer, and the motivation behind advocating value functions based on meta-analysis of multiple studies, as opposed to value functions based on a data collected from one study only. However, empirical evidence has been mixed. Ready and Navrud found that value function transfer did not outperform simple unit value transfer. It is a strong assumption that value will vary between the study site and the policy site in the same way that it varies within the study site. Santos found that value function transfer based on a meta-analysis of many studies did not outperform a unit value transfer from a single study. When you have access to a study that matches well the characteristics of the policy site, it is not necessarily an advantage to add information from studies that do not match the policy site.
3. *Value transfer will be more valid for use values than for non-use values.* We notice that value transfer is most common in situations where "use" can be measured in clearly identified units, such as a fishing day, or a coughing day. It is in those situations where unit value transfer is most common in applied policy analyses, such as the U.S. Forest Service's Resource Planning Assessment procedures, and the U.S. Environmental Protection Agency's procedures for estimating the benefits from health impacts of environmental regulations. The

presumption is that as the good becomes more qualitative (for example changes in resource quality, rather than quantity) or as the non-use component of value becomes more important, values become more context specific, making value transfer more problematic. Here too, however, the empirical evidence is mixed. Kristófersson and Navrud found lower transfer errors for non-use values than for use values. Transfer of non-use values is a topic that deserves more attention.

4. *It is better to conduct a new study at the policy site than to conduct a value transfer from existing studies.* Here, several caveats must be raised regarding this statement. First, as Barton points out, the question should not be "should I do value transfer, or should I do a new study?" Rather, the appropriate question is, given the information I already have about value at the policy site, is it worth the extra cost to collect new information, by conducting a new valuation study? Rather than throw away information contained in previous studies, the Bayesian frameworks demonstrated by Barton and by León, León and Vázquez-Polo allow new sample information to be efficiently combined with existing knowledge about value. Second, as Brookshire and Chermak point out, and Brookshire, Chermak and DiSimone demonstrate, uncertainty over value at the policy site may not be the most important source of uncertainty in the analysis. It may well be that the payoff, in a value of information sense, is greatest from reducing uncertainty over physical or ecological relationships.
5. *Value transfer too often fails convergent validity tests.* It is certainly true that all of the value transfer approaches discussed in this book sometimes fail convergent validity tests, in that transferred values can be statistically different from values estimated at the policy site. However, as many of the authors in this volume point out, statistical tests of equivalence may not be the most important criterion for deciding the usefulness of value transfer. Most authors now perform "importance tests," where the relative size of transfer error is the focus, not whether the error is statistically different from zero. Kristófersson and Navrud demonstrate a formal test of transfer validity that focuses on the policy relevance of the transfer error. While many of these importance tests find that transfer error can be frighteningly large (up to 1500% in one international transfer), several of the authors in this study find that errors can be small (on the order of 20–40%) when goods and populations are similar. Even where statistical validity tests reject equivalence, the relative size of the transfer error can be rather small. We believe that for most uses to which environmental values will be put, errors on the order of 20–40% are acceptable. They are certainly smaller than the differences that can arise from using different valuation methods, and are dwarfed by some of the uncertainties over biological, epidemiological, and physical relationships that are used to project the impacts being valued.

It is commonly stated that the quality of a value transfer is only as good as the quality of the source study. While this is certainly true, the concept also applies to tests of validity of value transfer. Poor performance of a value transfer can be due to poor quality of the source study, inappropriate transfer methods or assumptions, or to error in the value estimate that is taken as the "true" value at the policy site. If a transferred value differs meaningfully from a value generated in a study conducted at the policy site, it is not necessarily the transfer approach that is at fault. If we want to test the validity of value transfer, in isolation from concerns over methodological issues, then values should be generated using the same methods, and, ideally, the same survey instrument. Even then, we need to judge the performance of the transfer relative to the inherent variability of each of the two value estimates.

4. THE WAY FORWARD

In many ways, the discussion about value transfer over the last 10–15 years has followed the same course as discussions about non-market valuation more generally. The question, "is the method always valid?" has been replaced by the question, "under what conditions is the method valid?" The question, "does it work?" has been replaced by the question, "how can we make it work better?"

The purpose of this volume has been to take a snapshot of the research that is ongoing in the area of value transfer. The potential and need for methodological and empirical research on value transfer is by no means "played out." On the methodological side, we have seen a continuous advance in model sophistication, from unit value transfer to value function transfer to value functions based on meta-analysis, and now methods based on utility-theoretic value functions and on Bayesian approaches. That advance should continue. At the same time, empirical tests of the performance of these approaches should be conducted, to verify that they really do represent steps forward. Simple approaches should not be cast aside until we are confident that more complex approaches do perform better.

Value transfer is not always appropriate. As methods improve, we will also want to continuously re-examine the range of policy issues where value transfer is appropriate. Here, there is a need for more research identifying the boundaries for application of value transfer. Is transfer between countries valid? How similar must the countries be, and how do we define "similar?" How similar must the good and the population be?

Of course, the single most important factor for improving the performance and validity of value transfer is improving the range and quality of original valuation studies. Not only is it important that primary studies be conducted for a wider range of policy problems and situations, but that those studies will be available. Here, an important recent advance is the web-based database Environmental Valuation Reference Inventory (EVRI, www.evri.ca), which now contains more than 1800 valuation studies. The majority of these studies are from North America, but the proportion from other continents has been growing over time. There is, however, a need to increase the number of existing valuation studies captured in this database. Since many valuation studies are old and use outdated methodology and there are few studies for many environmental goods, there is also a great need for new, primary valuation studies using state-of-the-art methodology. These studies should be designed with value transfer in mind, and reported with enough detail to support development of more comprehensive value functions.

Finally, research on value transfer cannot occur in isolation from research on valuation methodology. As several of the contributions in this volume demonstrate, basic methodological issues in the source studies are at least as important as the specific methods used to transfer the resulting values. Continued research on basic valuation methodology will only benefit the quality of value transfer.

Ståle Navrud is associate professor, Department of Economics and Resource Management, Norwegian University of Life Sciences, Ås, Norway. Richard Ready is associate professor, Department of Agricultural Economics and Rural Sociology, The Pennsylvania State University, State College, PA, USA.

5. REFERENCES

Asian Development Bank (1996) Economic Evaluation of Environmental Impacts A Workbook Parts I and II Environment Division, Manilla, Philippines March 1996

Brookshire David S, Helen R Niell (1992) Benefit Transfers: Conceptual and Engineered Issues, Water Resources Research 28(March):651–655

Desvousges WH, Johnson FR, Banzhaf HS (1998) Environmental Policy Analysis with Limited InformationPrinciples and Applications of the Transfer Method New Horizons in Environmental Economics Edward Elgar, Cheltenham, UK and Northampton, MA, USA

European Commissions – DG XII (1995) ExternE – Externalities of Energy Vol 2 Methodology European Commission Directorate General-XII Science Research and Development, Report EUR 16521 EN, Brussels

European Commission – DG XII (1999) ExternE – Externalities of Energy Vol 7: Methodology 1998 Update European Commission (EC) – Directorate General (DG) XII Report EUR 19083, Brussels

European Community (1993) Towards sustainability: A European Community programme of policy and action in relation to the environment and sustainable development Office for Official Publications of the European Communities, Brussels

Krutilla JV, Fisher AC (1975) The Economics of Natural Environments: Studies in the Valuation of Commodity and Amenity Resources, Baltimore: Johns Hopkins Press

Markandya A, Tamborra M (2005) Green Accounting in Europe: A Comparative Study, Cheltenham: Edward Elgar

Mike Holland, Alistair Hunt, Fintan Hurley, Stale Navrud and Paul Watkiss 2005: Methodologies for the cost-benefit analysis for CAFÉ. Volume 1: Overview of Methodology. Report to DG Environment, European Commission, February 2005, AEA Technology Environment, Oxon, UK. http://ec.europa.eu/environment/air/cafe/pdf/cba_methodology_vol1.pdf

Navrud S (1992) Pricing the European Environment, Scandinavian University Press/Oxford University Press, Oslo, Oxford, New York

Navrud S (1999) Pilot Project to assess Environmental Valuation Reference Inventory (EVRI) and to Expands Its Coverage to the EU Report to the European Commission, DG XI – Environment Report July 1, 1999

Navrud S, Pruckner GJ (1997) Environmental Valuation – To Use or Not to Use?, A Comparative Study of the United States and Europe Environmental and Resource Economics, 10:1–26

OECD (1989) Environmental Policy Benefits: Monetary Valuation Organisation for Economic Co-operation and Development (OECD), Paris, ISBN 92-64-13182-5

OECD (1994) Project and Policy Appraisal: Integrating Economics and Environment Organisation for Economic Co-operation and Development (OECD), Paris, ISBN 92-64-14107-3

OECD (1995) The Economic Appraisal of Environmental Projects and Policies A Practical Guide Organisation for Economic Co-Operation and Development (OECD), Paris, ISBN 92-64-14583-4

Rowe, RD, Lang CM, Chestnut LG, Latimer DA, Rae DA, Bernow SM and White DE (1995) The New York Electricity Externality Study Volume I and II Hagler Bailley Consulting, Inc, Oceana Publications Inc

UNEP (1995) Global Biodiversity Assessment (Chapter 12: Economic values of Biodiversity) United Nations Environment Program (UNEP), Cambridge University Press ISBN 0-521-56481-6

Whittington D (1998) Administering Contingent Valuation Surveys in Developing Countries, World Development, 26(1):21–30

The Economics of Non-Market Goods and Resources

1. John Loomis and Gloria Helfand: *Environmental Policy Analysis for Decision Making*. 2001 ISBN 0-7923-6500-3
2. Linda Fernandez and Richard T. Carson (eds.): *Both Sides of the Border*. 2002 ISBN 0-4020-7126-4
3. Patricia A. Champ, Kevin J. Boyle and Thomas C. Brown (eds.): *A Primer on Nonmarket Valuation*. 2003 ISBN 0-7923-6498-8; Pb: 1-4020-1445-7
4. Partha Dasgupta and Karl-Göran Mäler (eds.): *The Economics of Non-Convex Eco-systems*. 2004 ISBN 1-4020-1945-9; Pb: 1-4020-1864-9
5. R.T. Carson, M.B. Conaway, W.M. Hanemann, J.A. Krosnick, R.C. Mitchell, and S . Presser: *Valuing Oil Spill Prevention: A Case Study of California's Central Coast*. 2004 ISBN: 0-7923-6497-X
6. Riccardo Scarpa and Anna A. Alberini (eds.):*Applications of Simulation Methods in Environmental and Resource Economics*. 2005 ISBN: 1-4020-3683-3
7. Nancy E. Bockstael and Kenneth E. McConnell: *Environmental and Resource Valuation with Revealed Preferences*. 2007 ISBN 0-7923-6501-1
8. Barbara J. Kanninen (ed.): *Valuing Environmental Amenities Using Choice Experiments*. 2007 ISBN 1-4020-4064-4
9. Ståle Navrud and Richard Ready (eds.): *Environmental Value Transfer*. 2007 ISBN 1-4020-4081-4

LaVergne, TN USA
16 April 2010
179566LV00002B/11/P

9 789048 105861